THE LONG SHADOWS

The Long Shadows

A GLOBAL ENVIRONMENTAL HISTORY OF THE SECOND WORLD WAR

Edited by
Simo Laakkonen
Richard P. Tucker
Timo Vuorisalo

Oregon State University Press Corvallis

Library of Congress Cataloging-in-Publication Data

Names: Laakkonen, Simo, editor, author. | Tucker, Richard P., 1938– editor, author. | Vuorisalo, Timo Olavi, editor, author.
Title: The long shadows : a global environmental history of the Second World War / edited by Simo Laakkonen, Richard P. Tucker, and Timo Vuorisalo.
Description: Corvallis : Oregon State University Press, 2017. | Includes bibliographical references and index.
Identifiers: LCCN 2016056186 | ISBN 9780870718793 (alk. paper)
Subjects: LCSH: World War, 1939–1945—Environmental aspects. | World War, 1939–1945—Influence.
Classification: LCC D744 .L58 2017 | DDC 940.53/1—dc23
LC record available at https://lccn.loc.gov/2016056186

∞This paper meets the requirements of ANSI/NISO Z39.48-1992 (Permanence of Paper).

First published in 2017 by Oregon State University Press
Printed in the United States of America

OSU
Oregon State
UNIVERSITY
OSU Press

Oregon State University Press
121 The Valley Library
Corvallis OR 97331-4501
541-737-3166 • fax 541-737-3170
www.osupress.oregonstate.edu

Contents

Foreword

It is commonplace to say that World War II changed the world. Film, fiction, and history books bolster living memory to keep alive an understanding of the devastation, the human horror, the personal tragedies, and the heroisms of the war. Every educated person carries some sense of how the war shifted the tectonic plates of high politics, crushing empires and raising up new superpowers. But how did World War II change the environment? Like no other book in existence, *The Long Shadows* reveals just how, and just how extensively, it did.

World War II scorched broad regions of Europe, Asia, and the Pacific. It touched, directly or indirectly, every continent and almost every country. It dislodged tens of millions of refugees and killed tens of millions of people. It was the most global war ever fought. The technologies of World War II, particularly in artillery and aerial bombing, made it especially destructive, of lives, buildings, and infrastructure.

Those technologies also made it consequential from an ecological point of view. First, combat annihilated local landscapes, as at Stalingrad and Iwo Jima. Aerial bombardment obliterated large swatches of cities, revamping urban ecosystems and reconfiguring urban-rural relationships. Even the weapons of the weak, of partisans in Yugoslavia or Chinese guerillas, could result in burned forests and depleted wildlife. Second, the mobilization of resources for war efforts placed heavy strains upon the systems responsible for the production of raw materials, whether mineral, timber, food, fiber, or fuel. The war economies of the major combatants required maximal production as fast as possible. That demand trickled down into the resource sectors of neutral nations such as Sweden and Turkey and Venezuela.[1] Third, the urgency with which the metallurgical and petroleum industries operated led to enormous surges in pollution in and around the several military-industrial complexes that were churning out tanks, ships, aircraft, and munitions throughout the course of the war. Almost no one gave a thought to air or water quality, given the exigencies of those times. So while World War II changed the world, it also changed the planet.

The Long Shadows takes readers from Arctic latitudes to the South Pacific. It explores the environmental ramifications of World War II within the major Axis and Allied powers, but not only along the front lines. It looks far beyond the campaign theaters, to the farms of Tanganyika and the forests of Mexico. It reveals different sets of linkages that the war created—political and economic, of course, but ecological as well.

The war's long shadows lingered past 1945. Millions of refugees remained in limbo for years. Wartime reconstruction took many years, leaving its own imprint in ways large and small. Berlin's tallest hill, the Teufelsberg, is a heap of rubble first created by Allied bombing, and many German cities have smaller versions. The remarkable advances in public health advanced by the war, such as routine antibiotic use and vaccination programs, led to the extraordinary surge in human population that gathered steam in the late 1940s and is slowly winding down today. While the war's violence killed over fifty million, its health infrastructure and technologies ultimately permitted far more to survive in the postwar decades.

The persistent military-industrial complexes that survived after 1945, and grew during the Cold War, constitute another long shadow. Thanks to the pressures of the Cold War, military industry in the major powers was shielded from almost all environmental regulation. What the United States and the Soviet Union built in haste after 1941—the arsenals of democracy and of Communism—they maintained and expanded after 1945. Perhaps the longest shadow of all will prove to be the nuclear weapons programs of the postwar world, which descended in some sense from the American one that got under way in 1942 with the first self-sustaining nuclear reaction. Some of the radioactive legacy of nuclear testing will linger on earth for more than a hundred thousand years, long after all memory, and perhaps all knowledge, of World War II disappears.

John R. McNeill, Georgetown University

Notes

1 Venezuela and Turkey were officially neutral until February 1945.

Acknowledgments

This book is largely based on the workshop "The Long Shadows: An Environmental History of the Second World War," held in Helsinki, Finland, August 7–11, 2012. The Finnish capital is surrounded on three sides by the Baltic Sea, and the location for the workshop was the tiny island of Harakka, also known as Magpie Island, one of the three hundred islands or islets off the city's shoreline. But behind the glittering beauty of the northern coast lurk the shadows. Because of the proximity of the Russian border, most of the islands belong to the Finnish Defence Forces, and many are still today used as military bases.

The island of Harakka turned out to be not only practically but also symbolically suitable for our workshop. This island used to belong to Suomenlinna, one of the world's largest maritime fortresses, and was later the site of a chemical test laboratory of the Finnish Army. The small island served also as a base for operations during the Second World War and was hit by Soviet aerial bombings at the end of the war. Like many other militarized islands, Harakka remained closed from the public for centuries. Today, however, the former laboratory facility hosts artists, and the island itself serves as a municipal nature preserve and is enjoyed as a popular maritime recreational area among the inhabitants of Helsinki.

The original idea for compiling a book dedicated to the environmental history of the Second World War emerged in Finland, because of the great importance of that war in Finnish history and the increasing number of studies and publications on this theme in the country. The idea of making an international publication was first discussed at the sixth European Society for Environmental History (ESEH) conference in Turku, Finland, in the summer of 2011. Behind these discussions was long-term research on the subject, both at the University of Turku in Finland and at the University of Michigan in the United States. Cooperation between these universities enabled us to join forces and gather a remarkable group of scholars to discuss the Second World War and the environment at the Harakka Island workshop. Over twenty participants from several countries worked together to survey this emerging subject. The intensive discussions during and after the workshop helped the authors revise their ideas for this volume, integrate their work more effectively into the whole, and provide a global survey.

The Helsinki workshop was financially supported by the Foundation for Baltic and East European Studies (Östersjöstiftelsen). The Kone Foundation, Finland, supported preparation of some manuscripts. Editing work was partly supported by the Department of Contemporary History, Södertörn University, Sweden; Helsinki University's Centre for Environment (HENVI); and the University of Turku. The editors wish to express their warm gratitude to these sponsors for their crucial support of the research and editorial work.

We would also like to thank Lasse, Mengyan, Otto, and Otso for their invaluable help at the Helsinki workshop; Kristian London for language revision; and geographer Salla Jokela for drawing the figures and maps for this book. Our heartfelt thanks go to Mary Elizabeth Braun at Oregon State University Press. Mary Elizabeth has supported and clarified our work over the years in ways too varied to identify in one or two sentences. We are also grateful for the thoughtful comments of Professor Kurk Dorsey and the anonymous referees. We hope that readers will find this book as inspiring to read as we have found making it to be.

Simo Laakkonen, Richard P. Tucker, and Timo Vuorisalo,
in Helsinki, Ann Arbor, and Turku, respectively,
on Midsummer Night 2016

Illustrations

FIGURES

TABLES

Part I

INTRODUCTION

Chapter 1
The Long Shadows

SIMO LAAKKONEN, RICHARD P. TUCKER,
AND TIMO VUORISALO

> The site of one of the bloodiest, hardest battles of the Winter War . . . Summa. The battlefield of Summa is now spread out in front of my eyes. The whole landscape has been turned upside down, its forests devastated. Iron now covers the famed fields of Summa. Every step taken disturbs fragments of shells on the ground. As far as one can see there is one crater located next to another one. The ground has been churned up and is completely devastated. The terrible power of the modern machinery of war is demonstrated in the lingering traces of grenades, bombs, tanks and guns. In the midst of all this, the lines of the Finnish troops have been situated and have held. Shelling has wiped the village of Summa completely out from the face of the Earth, not even ruins remained. Now Summa offers only sad, disconsolate scenery. However, nature—the great nature—will heal all wounds, even here.[1]

Imperial Japan, fascist Italy, and Nazi Germany were not the only aggressors that started the Second World War. The Soviet Union was also one of the invaders, as people living in the Baltic States of Estonia, Latvia, Lithuania, and the eastern parts of Poland and Romania bitterly noted in the fall of 1939. After invading these countries, the Soviet Union attacked Finland on November 30, 1939, and started the Winter War, which took place during one of the coldest winters of the century. This time, however, the invasion did not go as Stalin had planned. The passage above, from a Finnish radio reporter, describes the battlefield after the Red Army attacked Finnish main defense lines in the Karelian Isthmus. These snowy fields of the small village of Summa experienced the hardest artillery fire, the most massive assaults, and the highest losses that the world had seen since World War I. Despite mobilizing what was finally an army of nearly one million men to pass the subarctic Thermopylae, the Soviet Union was not able to subdue Finland, and the war ended with a peace treaty on March 13, 1940.[2]

But unlike what the radio reporter predicted, nature did not heal the wounds. It did not heal all the wounds of the half million Finnish and Soviet men who were killed or wounded during the extremely bloody three months of the Winter War.[3] Nature did not heal all the wounds in the landscape, either. Despite the growth of forests since that time, bunkers and trenches, traces of bombardments, and remnants of both armies can still be found in these battlefields. Visitors from both Finnish and Russian sides come to the site every year to honor the fallen. Even though the Second World War slowly withers away in the sediments of time, wounds are still there—in the bodies, hearts, and minds of people, as well as in societies and in the landscape.

Similar battlefields, which can be found almost all over the world, were in many ways a miniature version of World War II, because a major characteristic of the war was its fragmented nature. In Finland and all other countries, memories are basically dependent on national, regional, local, or individual accounts of war. World War II was a macroscale war, yet its history is made of microhistories. This war began and ended in different times and ways and in different places. It was truly a global war, with sixty-one countries becoming engulfed in it. War between nations intertwined in many places with civil wars or local armed conflicts or resistance that continued after the war formally ended in 1945. It became a long war in Asia and to some extent also in Europe, but it lasted fewer than four years for other countries, such as the United States. It started in Asia in 1937 when Japan attacked China, and came to de facto end in 1949 when the People's Republic of China was founded. Yet it can be argued that World War II continued in Asia in the 1950s because the Korean War of the early 1950s and the Indochinese wars from the late 1940s to the early 1970s were direct consequences of World War II. In Europe, even the Balkan wars of the 1990s were directly related to the aftermath of World War II. As long as conflict lasted, each location suffered environmental consequences.

Warfare has a history as long as humans have lived in organized societies. In Europe alone, by one author's calculation, 5,800 wars were fought prior to

World War II.[4] Although World War II was exceptional owing to its scope and destructiveness, it is part of the historical phase in the development of modern warfare. In modern industrial warfare, communication, transportation, armament and other stockpiling and supply, and military operations themselves rely on technological-scientific product development, industrially manufactured products, and societies' economic production capacity. In industrial wars, new types of weapons—rapid-load firearms and explosives—have led to massive casualties in open battlefields. These unsustainable casualties, for their part, have forced the troops to break up, dig into the surrounding terrain, and build defensive devices. Because of the resulting cycle of armament, the victory in encounters between mass armies was to a large extent decided by their comparative industrial-economic production capacities.

Industrial warfare on a devastating scale began during the American Civil War (1861–65), the first war whose environmental history has been studied intensively.[5] Industrial warfare reached Europe during the same years, in the Crimean, Austro-Prussian, and above all the Franco-Prussian War (1870–71), in which Prussia's success was achieved by total warfare, the mobilization of all the country's material and immaterial resources to the service of war.[6] This process began to spread worldwide in the last decades of the century, as the industrial powers and local elites finally crushed native resistance—for example, in southern Chile (the Mapuche), the Great Plains (the Lakota and other tribes), British South Africa (the Matabele), and New Zealand (the Maori)—and competed with each other for control of African and Southeast Asian colonies and their natural resources. Yet there are few studies of the environmental history of warfare in Europe and colonialism prior to World War I.[7] The complex environmental legacy of World War I is only now being addressed in any focused way.[8] This is the case also with the Spanish Civil War and Japan's attack on the Chinese mainland in the 1930s.

Research interest in the environmental impacts of wars and even peacetime military operations was aroused by the radioactive fallout from the nuclear testing of the 1950s and 1960s, in the wake of the atomic bombs dropped on Hiroshima and Nagasaki.[9] In the 1960s and early 1970s, the world's first televised war brought to screens the deforestation that the United States caused in Vietnam with bulldozers and herbicides.[10] Concern for the environmental changes resulting from war arose again during the Gulf War (1990–91), when images of hundreds of oil wells ignited by the Iraqi Armed Forces were broadcast around the globe. In the late 1990s, all of Europe watched, hearts chilled by both the environmental and human toll, as the Balkans slid into civil war. The US-led allied assault on Iraq in 2003 was the first war preceded by discussions of its probable environmental consequences, even if studies conducted during the war itself could only be fragmentary. Wars fought in the developing countries of Africa, Latin America, and Asia have also shined a media spotlight on the ravages of oil drilling, deforestation, mining, and poaching of endangered animals such as the mountain gorilla.[11] Many were "resource wars," fought over control of valuable natural resources and the riches that came from extracting and exporting them. There has been much discussion on the relationship between natural resources and wars, but the historical perspective has been insufficient.[12]

Because of the dearth of research, in practice, all wars fought throughout history continue to be "unknown wars" from the environmental perspective, and the people who have worked on behalf of the environment during wartime conditions remain "unknown soldiers." Yet throughout the industrial era, many participants who were caught in conflicts paid written attention to the environmental effects of war. If scholars have neglected the relationship between war and its imprint on nature, soldiers and civilians have not. Diary entries, newspaper articles, reports, photographs, newsreels, films, and works of art demonstrate that both soldiers and civilians consciously observed wartime destruction in their immediate vicinities as well as in the natural world. For instance, soldiers monitored birds, installed nest boxes, reared fledglings, and even reported on birds' singing activity during military operations.[13] Apparently war did not prevent people from observing, enjoying, and loving nature; rather the contrary.

How has contemporary scholarship answered this challenge of environmentally "unknown" wars? World War II encompassed economic, social, and cultural dimensions. There is a vast literature on those dimensions of the conflagration.[14] The environmental consequences and legacies of modern wars are only beginning to be studied systematically.[15] As research on environmental history has focused on peacetime developments, until recently neither war nor wartime has been considered a prominent subject of environmentally themed study.[16] The environmental history of wars has thus ended up in an academic no-man's-land between two traditions of historical research: the history of war and environmental history.

In recent years, approaches to research on the history of war have expanded appreciably, but even so, the environmental consequences of mass violence have remained largely unexamined.[17] Military historiography and geography have long and carefully analyzed the environmental settings of conflict in the course of strategic planning. The environmental impact dimension has been studied only obliquely or fragmentarily, but nature conservationists have paid some attention to the importance of warfare in transport of potentially harmful species or substances between regions.[18]

Broad examinations of the interplay between war and the environment through history have been conducted primarily in the United States,[19] Germany,[20] and Finland.[21] The full range of industrial warfare has been addressed in two pioneering collections.[22] There are not enough relevant studies on wars, even recent ones, to form a comprehensive picture of their environmental effects.[23] And the deeper we travel into history, the less we know about these effects. This also holds for World War II; a few articles have been published about its environmental history, and some books have touched on the period.[24] There is no global environmental history of World War II, however, though the political and military history of World War II has been massively chronicled. This book is the first work that strives to open global perspectives specifically on the environmental history of that global war.

The magnitude and long geopolitical shadow of World War II are obvious reasons to study its environmental history. The industrialization of warfare came to a climax in World War II, which has a unique status among the wars of the world. It engulfed 1.7 billion people, three-quarters of the world's population. The war was waged on almost all continents and in the polar, temperate, and tropical climate zones. It was the world's most destructive war, claiming fifty to seventy million lives. In addition, the war injured hundreds of millions of people and innumerable other living creatures. The worldwide cost of World War II is estimated at about $1 trillion, making it in its day the most expensive war by far.[25] World War II changed the political map of the globe. It also propelled the human race into the Atomic Age.

THE LONG SHADOWS OF WORLD WAR II

The concept of long shadows formulates our overall approach to the environmental history of World War II. Because of a lack of previous research, our book is not a systematic study of the theme but rather an orientation into this new and vast field. Therefore the concept of long shadows is at the same time a question, an argument, and an ethical principle. First, it asks a demanding question that we attempt to answer in this volume: What were the main environmental impacts of World War II on the global level? Second, when seeking answers, the concept addresses the need to explore various long-term socio-ecological developments in prewar, wartime, and postwar eras and at different scales.[26] Each long shadow had its own particular chain of effects. Third, exploring the environmental history of warfare requires both knowledge and courage, because environmental problems caused by past and present wars are a politically sensitive topic in several countries. Nations do not, generally speaking, want to become responsible for the environmental destruction they cause, because rectifying it is embarrassing, demanding, and expensive.[27] Despite the tragic outcomes of every war, it is necessary to explore on equal basis the negative, contradictory, and possible positive impacts of war. We are only starting to grasp the impacts of the long and dark shadows that wars have imposed on societies and the environment. In the concluding chapter, we present some hypotheses of the global long-term environmental and societal impacts of World War II.

Even though our book for practical reasons focuses on only World War II, the notion of the long shadows takes into account the cumulative impacts of successive wars in order to understand a certain war and its impacts. World War II was the most massive war the world had seen, but it was in practice a relatively short outcome of long-term societal and military processes that had developed over time. The essential elements of World War II were formed by the wars of postrevolutionary France, wars of major European powers and the United States in the mid-nineteenth century, the First World War, the Spanish Civil War, the Japanese aggression in China, and related changes in respective societies. Consequently, World War II was merely one step, if a decisively important one, in the long-term development of warfare, which continues today and casts its long shadows on future wars, societies, and environment.

A key element of the long shadows of World War II is militarization of landscapes, or physical and cultural changes in environments and societies that serve military purposes.[28] Even though the essays in this book do not explicitly explore militarized landscapes, they are a highly useful frame for study. World War II and its continuation, the Cold War, brought about global militarization of various landscapes, affecting directly or indirectly all continents and leaving remnants and scars that are still visible all over the world, especially in ecologically sensitive areas.[29] Warfare and armed forces affected flora and fauna, air, water, and soil.[30] Soldiers from every continent fought in militarized landscapes, and after the war the refugees forced out of their homes ended up on every continent. Even though battles were not fought everywhere, people, other organisms, and various natural elements—water, earth, and air—carried the socioecological consequences of the Second World War to every corner of the globe. The global search for strategic energy and minerals resources led to militarization of large regions, trade routes, facilities, and ports.[31] Front lines, military bases, and bunkers all over the world as well as civil infrastructure in the countryside, towns, and cities were to a various extent built or destroyed according to military needs in the Northern and Southern Hemispheres. The first step toward militarization of space was also taken during the Second World War, when the first man-made object, the V2 ballistic missile, entered outer space.

Not only physical landscapes and infrastructure but also states and societies were militarized. War made science and technology humble servants of death and destruction. Professional armies, conscription, and various paramilitary organizations militarized millions of men annually after the war. School education was to legitimate and even glorify military action. Media, books, and films addressed the need of individual sacrifice. The memory of World War II was cherished in innumerable parades, cemeteries, mausoleums, museums, monuments, memorials, and public and private memories during and after the war.[32] Consequently, nearly the whole of the earth—both societies and nature—was to some extent militarized during and after World War II. The global militarization of physical and above all psychological landscapes naturally served the interests of the armed forces and military-industrial complex, strengthening their position and impact around the world. We live today in a world that is to a great extent made by World War II.

The outcome of World War II varied from country to country. Some old states disappeared as a result of the war, while new ones were established. The war brought democracy, work, and prosperity to some countries, whereas others were overrun by destruction, dictatorship, repression, and poverty. On the one hand, war gave rise to Communist dictatorships in Eastern Europe and Asia, violent neocolonialism in Africa and Southeast Asia, and the Cold War, which became a threat to the continuity of human institutions and beyond that to the biosphere as a whole. On the other hand, World War II weakened traditional colonial powers; ended Nazi Germany, imperial Japan, and fascist Italy; and started an unprecedented era of economic growth, strong public policy, social welfare, and political democracy in the Western world. In essence, World War II was a political revolution.

These deep structural changes brought about by World War II naturally caused severe environmental problems, but these changes also made industrial societies more receptive to environmental ideas and activities, and enabled public power to carry out necessary reforms. Thus World War II explains to a great extent the emergence of institutional changes that were prerequisites for the advance of environmental awakening in the Cold War period. This topic is further elaborated upon in the concluding chapter.

CONTENTS OF THE BOOK

This book consists of seventeen chapters. It is directed to readers interested in environmental, military, and global history, as well as students of and experts from various fields of study. Concentrating on a relatively brief period brings a unity to the book, whereas the breadth of perspectives creates diversity and even disagreement. That the essays vary widely in structure can be considered an advantage of the collection, as together they illustrate the wide range of issues and treatments that the war demands. Providing a broad research perspective on both a topical and geographic basis (all continents except Antarctica are covered) has been a conscious choice, so that this volume's contributions offer an innovative approach to further research on the relatively unknown environmental history of this war. Together these essays provide a far greater range of dimensions and background bibliography than anything that has been done before.

After this introduction, chapters 2 and 3 present global perspectives. In "Polemosphere: The War, Society, and the Environment," Simo Laakkonen frames subsequent chapters with a wide survey of the interactions among societies, warfare, and natural environments over time. Evan Mawdsley's survey in "World War II: A Global Perspective" describes the global reach and integration of military operations and their impacts on societies in the war years.

The following five chapters explore the war's systemic impacts on the basis of regional narratives of the war from the perspective of environmental costs. The long shadow that the prewar Depression years cast, eventually leading to the outbreak of hostilities, is explored in Simo Laakkonen's "Environmental Policies of the Third Reich." Paul Josephson writes about the impacts of war in the Soviet Union, the world's largest state with seemingly inexhaustible resources. Micah Muscolino studies the energetics of militarized landscapes in China. Henan Province, along with many other Chinese regions, suffered from war-induced floods and a resulting famine. Richard Tucker's chapter on the war's environmental legacy in northeastern India and Burma introduces us to the long-term ecological impacts of road, railroad, and airstrip construction in remote but strategically important areas. Carol MacLennan's essay on Pearl Harbor's aftermath discusses the impact of war in one of the main settings of war, the Pacific.

The next two chapters explore invisible elements of warfare, which were of fundamental importance in terms of life and death. Helene Laurent explores in her essay "The Great Louse War: Control of Typhus Fever" how differently Axis and Allied states reacted to this disease, which has traditionally been one of the main killers during military campaigns. Chapter 10 introduces a truly pioneering approach to warfare, acoustic ecology. In "Perspectives on the Acoustic Ecology of War," Outi Ampuja examines the different meanings and impacts of soundscapes for soldiers.

Five chapters discuss the question of resources—their availability, use, and depletion—during the conflict and in postwar years. The topics include the aluminum economy (by Matthew Evenden), Mexican forests (by Chris Boyer), wartime lack of food security in Tanganyika (by Gregory Maddox), the great opening of the circumpolar region to widespread resource utilization during and after the war (by Ilmo Massa and Alla Bolotova), and fisheries and whaling by Japanese vessels (by William Tsutsui and Timo Vuorisalo). One common theme discussed in all these chapters is the intensity of efforts to mobilize natural resources for warfare and peacetime defense. There is a wide literature on the economics, industrialization, and social mobilization that regimes pursued for the war.[33] But until now, this central dimension of a global resource-consuming war has not been considered from the perspective of the depletion of major natural resources or the environmental scars that inevitably accompany the extraction of minerals, fossil fuels, and food.

Finally, World War II, like its predecessor, had huge and complex impacts on natural ecosystems, wildlife, and species diversity. Moreover, the global wildlife conservation movement, which had emerged in the late 1800s, was to some extent paralyzed during both world wars, but emerged with new determination after each of the two wars. Anna-Katharina Wöbse presents a unique overview of the struggles of the movement in "International Conservation after the Two World Wars."

In the concluding essay, the editors address two themes: the long environmental shadows of World War II and following the Cold War as well as themes that were only briefly considered in the contributions to this volume, demanding further research as we move toward a fully satisfactory understanding of the environmental legacy of the Second World War.

Notes

1 "Kuvaus Summan taistelumaastosta," YouTube video, 4:55, posted by Suomi Sodassa, January 15, 2015, https://www.youtube.com/watch?v=i8S_lJnaHRE.

2 William R. Trotter, *A Frozen Hell: The Russo-Finnish Winter War of 1939–1940* (Chapel Hill, NC: Algonquin Books, 2000); for a military analysis, see Gordon F. Sander, *The Hundred Day Winter War: Finland's Gallant Stand against the Soviet Army* (Lawrence: University Press of Kansas, 2013).

3 Of the Red Army, about 167,000 men perished and 260,000 were wounded, while of the Finnish Army, over 26,000 perished and 45,000 were wounded. Pavel Petrov, Antero Uitto, Carl-Fredrik Geust, and Pauli Kruhse, *Venäläinen talvisotakirjallisuus: Bibliografia 1939–1945* (Jyväskylä: Docendo, 2013); Riitta Lentilä and Antti Juutilainen, "Talvisodan uhrit," in *Talvisodan pikkujättiläinen*, ed. Jari Leskinen and Antti Juutilainen (Porvoo, Helsinki: WSOY, 1999), 816–28.

4 Douglas Holdstock, "Morbidity and Mortality among Soldiers and Civilians," in *War or Health? A Reader*, ed. Ilkka Taipale et al. (Helsinki: Physicians for Social Responsibility, 2002), 184.

5 Lisa Brady, *War upon the Land: Military Strategy and the Transformation of Southern Landscapes during the American Civil War* (Athens: University of Georgia Press, 2012); Mark Fiege, "Gettysburg and the Organic Nature of the American Civil War," in *Natural Enemy, Natural Ally: Toward an Environmental History of War*, ed. Richard P. Tucker and Edmund Russell (Corvallis: Oregon State University Press, 2004), 93–109; Kathryn Shively Meier, *Nature's Civil War: Common Soldiers and the Environment in 1862 Virginia* (Chapel Hill: University of North Carolina Press, 2013); Brian Allen Drake, ed., *The Blue, the Gray, and the Green: Toward an Environmental History of the Civil War* (Athens: University of Georgia Press, 2015).

6 An important exception is Chris Pearson, *Mobilizing Nature: The Environmental History of War and Militarization in Modern France* (Manchester: Manchester University Press, 2012).

7 For example, see Thaddeus Sunseri, "Reinterpreting a Colonial Rebellion: Forestry and Social Control in German East Africa, 1874–1915," *Environmental History* 8, no. 3 (July 2003): 430–51; Daniel R. Headrick, *Power over People: Technology, Environments, and Western Imperialism, 1400 to the Present* (Princeton, NJ: Princeton University Press, 2010).

8 See Tait Keller, J. R. McNeill, Martin Schmid, and Richard Tucker, eds., *Environmental Histories of World War I* (forthcoming).

9 Mark D. Merlin and Ricardo M. Gonzalez, "Environmental Impacts of Nuclear Testing in Remote Oceania, 1946–1996," in *Environmental Histories of the Cold War*, ed. J. R. McNeill and Corinna R. Unger (Cambridge: Cambridge University Press, 2010), 167–202; Ryan H. Edgington, *Range Wars: The Environmental Contest for White Sands Missile Range* (Lincoln: University of Nebraska Press, 2014); Gretchen Heefner, *The Missile Next Door: The Minuteman in the American Heartland* (Cambridge, MA: Harvard University Press, 2012).

10 Recent studies include David A. Biggs, *Quagmire: Nation-Building and Nature in the Mekong Delta* (Seattle: University of Washington Press, 2010), and Edwin A. Martini, *Agent Orange: History, Science, and the Politics of Uncertainty* (Amherst: University of Massachusetts, 2012).

11 Mark Jenkins, "Who Murdered the Virunga Gorillas?" *National Geographic* (July 2008): 34–65.

12 T. F. Homer-Dixon, *Environment, Scarcity, and Violence* (Princeton, NJ: Princeton University Press, 1994); special issue on environmental conflict, *Journal of Peace Research* 35, no. 3 (1998); T. H. Hastings, *Ecology of War and Peace: Counting Costs of Conflict* (Lanham, MD: University Press of America, 2000); Jay E. Austin and Carl E. Bruch, eds., *The Environmental Consequences of War: Legal, Economic and Scientific Perspectives* (Cambridge: Cambridge University Press, 2000); Philippe Le Billon, "The Political Ecology of War: Natural Resources and Armed Conflict," *Political Geography* 20, no. 5 (June 2001): 561–84; see the series of works edited by Günther Bächler et al., *Kriegursache Umweltzerstörung*, vol. 1–3 (Zürich: Rüegger, 1996); on riding on wartime propaganda, see David Day, *The Ecowars: A Layman's Guide to the Ecology Movement* (London: Paladin, 1989).

13 Simo Laakkonen and Timo Vuorisalo, eds., *Sodan ekologia: Nykyaikaisen sodankäynnin ympäristöhistoriaa* [Ecology of war: environment history of modern warfare] (Helsinki: Suomalaisen Kirjallisuuden Seura / Finnish Literature Society, 2007).

14 Already in 1992 the Library of Congress computer catalog included 50,000 titles on World War II. "World War II: A Selected List of References," Library of Congress, accessed October 15, 2012, http://www.loc.gov/rr/main/gopher/WWII_bib.html.

15 Military historians and geographers have long studied the impact of nature on the landscapes and outcomes of battles, as well as on the logistics of supply. But the consequences of military operations for the natural world have generally been peripheral to their purposes.

16 Some exceptions to the rule exist. A. Corvol and J. P. Amat, eds., *Forêt et guerre* (Paris: Groupe d'histoire des forêts françaises, L'Harmattan, 1994).

17 For example, see Manus I. Midlarsky, ed., *Handbook of War Studies* (Boston: Unwin Hyman, 1989).

18 Shipping related to regional or world wars has led to the introduction and invasion of mollusk species to new areas. James T. Carlton, "Introduced Marine and Estuarine Mollusks of North America: An End-of-the-20th-Century Perspective," *Journal of Shellfish Research* 11, no. 2 (1992): 489–505. The best-known example may be the brown tree snake (*Boiga irregularis*), which was introduced to Guam after World War II with military equipment. David Pimentel, Rodolfo Zuniga, and Doug Morrison, "Update on the Environmental and Economic Costs Associated with Alien-Invasive Species in the United States," *Ecological Economics* 52, no. 3 (February 2005): 273–88; Roy MacLeod, "Strictly for the Birds: Science, the Military and the Smithsonian's Pacific Ocean Biological Program," *Journal of the History of Biology* 34, no. 2 (June 2001): 315–52.

19 Susan D. Lanier-Graham, *The Ecology of War: Environmental Impacts of Weaponry and Warfare* (New York: Walker, 1993), offers a historical overview, primarily addressing American wars. More recent overviews include Tucker and Russell, *Natural Enemy*; Richard P. Tucker, "War," in *World Environmental History*, vol. 3, ed. Shepard Krech III, J. R. McNeill, and Carolyn Merchant (New York: Routledge, 2004); J. R. McNeill, "Woods and Warfare in World History," *Environmental History* 9, no. 3 (July 2004): 388–410.

20 P. Coleman, *Cost of the War: World War II Resource Guide* (Gardena, CA: American War Library, 1999).

21 Veli-Pekka Lehtola, *Saamelainen evakko: Rauhan kansa sodan jaloissa* (Helsinki: City sámit, 1994), discusses the reactions of Sámi refuges during World War II in an unfamiliar culture and environment. Heikki Lindroos, *Sotametsän perintö ja otteita eräästä metsäpäiväkirjasta* (Helsinki: Teollisuuden metsänhoitajat, 1997), examines the heritage of wartime forest. Mauri Soikkanen, *Sotilaamme erämiehinä Itä-Karjalassa 1941–1944* (Helsinki: Gummerus, 1999), investigates the practices and impacts of wartime fishing and hunting on the eastern front. Rauno Lahtinen and Timo Vuorisalo, " 'It's War and Everyone Can Do as They Please!' An Environmental History of a Finnish City in Wartime," *Environmental History* 9, no. 4 (2004): 675–96; Simo Laakkonen, "Warfare—An Ecological Alternative for Peacetime? The Indirect Impacts of the Second World War on the Finnish Environment," in Russell and Tucker, *Natural Enemy*, 175–94. See also note 22 below.

22 Laakkonen and Vuorisalo, *Sodan ekologia*; Charles Closmann, ed., *War and the Environment: Military Destruction in the Modern Age* (College Station: Texas A&M University Press, 2009).

23 See Christopher D. Stone, "The Environment and Wartime: An Overview," and Asit K. Biswas, "Scientific Assessment of the Long-Term Environmental Consequences of War," in Austin and Bruch, *Environmental Consequences of War*; Jürgen Brauer, *War and Nature* (Lanham, MD: Altamira Press, 2009).

24 For books, see Joachim Radkau and Frank Uekötter, eds., *Naturschutz und Nationalsozialismus: Geschichte des Natur- und Umweltschutzes* (Frankfurt: Campus Verlag, 2003); Franz-Josef Brüggemeier, Mark Cioc, and Thomas Zeller, eds., *How Green Were the Nazis? Nature, Environment, and Nation in the Third Reich* (Athens: Ohio University Press, 2005). Other works include Edmund Russell, *War and Nature: Fighting Humans and Insects with Chemicals from World War I to Silent Spring* (New York: Cambridge University Press, 2001); Russell and Tucker, *Natural Enemy*; Laakkonen and Vuorisalo, *Sodan ekologia*.

25 Coleman, *Cost of the War*.

26 Martin Gutmann, "The Nature of Total War: Grasping the Global Environmental Dimensions of World War II," *History Compass* 13, no. 5 (May 2015): 251–61.

27 Perhaps the best recent example of all the three aspects of this phenomenon is the Japanese government's costly efforts to detect and destroy at least some of the estimated 200,000 to 400,000 chemical weapons that were deployed in China during World War II. For background, see Hongmei Deng and Peter O'Meara Evans, "Social and Environmental Aspects of Abandoned Chemical Weapons in China," *Nonproliferation Review* (Spring–Summer 1997): 101–8. For the current situation, see the official website for the Organization for the Prohibition of Chemical Weapons, accessed October 20, 2016, https://www.opcw.org/.

28 For a good overview of studies and concepts, see Chris Pearson, "Researching Militarized Landscapes: A Literature Review on War and the Militarization of the Environment," *Landscape Research* 37, no. 1 (February 2012): 115–33; and Rachel Woodward, "Military Landscapes: Agendas and Approaches for Future Research," *Progress in Human Geography* 38, no. 1 (February 2014): 40–61.

29 Chris Pearson, Peter Coates, and Tim Cole, eds., *Militarized Landscapes: From Gettysburg to Salisbury Plain* (London: Continuum, 2010); Edwin A. Martini, ed., *Proving Grounds: Militarized Landscapes, Weapons Testing, and the Environmental Impact of U.S. Bases* (Seattle: University of Washington Press, 2015); Simo Laakkonen, ed., "Militarized Landscapes: Environmental Histories of the Cold War," special issue, *Cold War History* 16, no. 4 (November 2016).

30 The "biodiversity footprint of wars" is a neglected area of study. Traces of military actions can sometimes be observed in flora even centuries after actual events. In Finland, bracken clones flourish in sites of historical battlefields exposed to fire, and clone sizes have been used to locate and time battles that occurred between the fifteenth and nineteenth centuries. Eino Oinonen, "The Correlation between the Size of Finnish Bracken (*Pteridium aquilinum* (L.) Kuhm.) Clones and Certain Periods of Site History," *Acta Forestalia Fennica* 83 (1967): 1–51, https://helda.helsinki.fi/handle/10138/17688.

31 Raymond Dumett, "Africa's Strategic Minerals during the Second World War," *Journal of African History* 26, no. 4 (October 1985): 381–408.

32 There is a rapidly growing literature on sites of memory and memory politics of World War II. See, for example, George L. Mosse, *Fallen Soldiers: Reshaping the Memory of World Wars* (Oxford: Oxford University Press, 1990); Peter Reichel, *Politik mit der Erinnerung: Gedächnisorte im Streit um die nationalsozialistische Vergangenheit* (Munich: Carl Hanser Verlag, 1995); Siobhan Kattago, *Memory and Representation in Contemporary Europe: The Persistence of the Past* (Burlington, VT: Ashgate, 2012).

33 See, for example, Mark Harrison, ed., *The Economics of World War II: Six Great Powers in International Comparison* (Cambridge: Cambridge University Press, 1998); Gerald D. Nash, *World War II and the West: Reshaping the Economy* (Lincoln: University of Nebraska Press, 1990); James T. Sparrow, *Warfare State: World War II Americans and the Age of Big Government* (New York: Oxford University Press, 2011).

Chapter 2
Polemosphere
The War, Society, and the Environment

SIMO LAAKKONEN

In contemplating the cost that the most consuming of all wars extracted from the natural world, my thoughts turn to the question of how that war is situated in the history of mass conflict—it is as long as the history of organized societies. In this essay, I examine warfare's environmental and societal effects as a hypothetical entity that I refer to as the *polemosphere,* a term derived from the Greek war deity Polemos.[1] It refers to those aspects of the environment and society that have been affected by warfare. This essay considers the general characteristics of that dynamic relationship, and the specific consequences of warfare's transition from the limited impacts of the preindustrial era to the full-blown devastation of the century we have survived, the time of "total war." The polemosphere is not in this context a clearly defined interpretational frame, but rather a starting point for reflection; my examination thus results in questions rather than answers. This inquiry is set on a general level to complement the empirical studies of World War II that follow.

HOW DO WARTIME AND PEACETIME INTERACT?

War is collective violence for a collective purpose.[2] As social behavior, war is not abnormal, but rather characteristic of most societies throughout history.[3] Warfare is guided by the same factors that direct our other social behavior.[4] In the end, however, people tire of war, and peacetime has historically been the prevailing societal state. Wars have been primarily fought by men, so warfare is linked particularly to masculinity and the history of patriarchies, those societies that have been overwhelmingly dominated by men.[5] Most wars are in principle well planned, but because of their irrational course, wars often turn against their original objective. War cannot be controlled, and this is why the course and results of wars are unpredictable. In sum, war is human and rational activity, the results of which are inhuman and irrational.

CULTURAL HIERARCHIES OF WAR

Societal structures have shaped the ways we understand wars, even why the environmental history of war has been so little studied. There are various causes, but one fundamental reason can be pinpointed in the political norms created by militarism and warfare. These norms are created or intensified by war and steer wartime and peacetime cultural production and consumption and, through them, our conceptions of warfare. As a reflection of these norms, most visible depictions of war initially present those characteristics of warfare that reflect the interests of those in power. Still today, presidents, secretaries of state, and generals dominate public communication of ongoing wars, even in democratic societies. It is only with time that the perspectives contained in studies of war expand to allow for critical viewpoints. In this presentation I call this historical phenomenon, both its reasons and outcomes, the cultural hierarchies of war.

There are three types of cultural hierarchies of war: military, strategic, and normative. The military hierarchy starts from the upper levels of the military status scale, from the military command and upper officers, and progresses from there to junior officers, and last of all to the soldiers. Similarly, this process progresses from the frontline battle and command units to those units that support combat, and last of all to the maintenance and support troops. In terms of functions, studies focus primarily on battles, continue to those military functions that support combat, and end with the everyday life of soldiers, which nevertheless makes up the bulk of wartime existence. Correspondingly, military depictions concentrate on preparations for war and war per se; the return of society to peacetime conditions is depicted less frequently.

The strategic hierarchy affects not only war zones, but also societies as a whole. The strategic hierarchy is revealed in that military depictions focus first on those societal functions that support the armed forces, and only afterward on the home front and civilian life. In addition, this strategic hierarchy is gendered: the interest focuses primarily on able-bodied men and only secondarily on women. Other societal groups—other men and women, the elderly and children, not to mention foreigners and prisoners of war—come last in order of priority.

I. Military hierarchies

Rank	*Units*	*Phase of war*
Command	Frontline units	Battles
Senior officers	Combat support units	Preparing for war
Junior officers	Artillery	Return to peacetime
Soldiers	Maintenance	Postwar period

II. Strategic hierarchies

Relevance for warfare	*Physical condition*	*Nature*
Military economy	Able-bodied men	War horses and dogs
Home front	Able-bodied women	Production animals and plants
Everyday life	Children	Microbes
	Elderly	Wildlife
		Other nature

III. Normative hierarchies

Outcome	*Advantage*	*Genre*	*Emotions*
Victorious wars	Useful wars	Heroic wars	Pride
Lost wars	Useless wars	Nonheroic wars	Shame

Figure 2.1. The cultural hierarchies of war.

Finally, the strategic hierarchy perceives that, aside from humans, the living beings discussed are primarily those animals that directly support combat efforts, especially war horses and dogs, followed by the production of animals and plants that are useful in the wartime economy. Next, organisms that hamper war efforts—above all contagious diseases and their vectors—are addressed. Humans' relations with nature in its fullest scope, as biological communities, come last in the strategic hierarchy, if at all.

Under these hierarchies, wars are valuated on the basis of their end results. War depictions focus primarily on wars and battles that were won and only secondarily on those that were lost. In addition, depictions focus on those dimensions of a given war that are defined as heroic. The pure motives, objectives, and instruments of war are dealt with first, as well as the moral unity of the military leadership, the troops, and the populace. Only gradually can these ideals or rituals be questioned and the real experiences of soldiers and other groups finally dealt with. Feelings of fear, shame, and various traumas related to wartime are the last issues to be publicly aired.

Historical circumstances have changed drastically over the past century, yet cultural hierarchies persist in the whole range or dictatorships and democracies. The objective of these hierarchies is to create unity in the society at war and its army, in other words, to ensure discipline and combat capability. National consciousness is imbued with cultural hierarchies through the nation-state ideology and the institutions that realize it, especially media and educational and military institutions, as well as through warfare. In the end, cultural hierarchies of war are instilled into citizens' awareness by the state's propaganda and surveillance machine, ultimately by force of violence, prison, and death sentences.

Cultural hierarchies form the ways war shapes society; they are at the heart of the societal polemosphere. The uppermost levels of these three hierarchies are reserved for those in power who create and sustain the ideological ideals of war's objectives: the heroism of combat and national unity. Yet the broader the perspective from which these issues are examined, the more critical perspectives and themes emerge. Environmental issues, which are positioned on the lowest steps of the cultural hierarchies, thus tend to question the ideals of war; they are part of war's socioecological counterculture.

FROM PREINDUSTRIAL TO INDUSTRIAL WARFARE

The extent of war's socioecological sphere of influence—the polemosphere—depends on the degree of development of warfare, on an accelerating scale from preindustrial warfare to industrial warfare. This division clarifies how and why the polemosphere intensified during the latter half of the nineteenth century, and is also why this book focuses on the warfare of the industrial era in its culmination in the Second World War.

The main opponent in preindustrial war was nature, as natural conditions, epidemics, seasons, weather, and time of day did more to determine the course and outcome of war than enemy actions. Like the surrounding agricultural society, preindustrial warfare was based on renewable sources of energy: the muscle power of humans and animals, firewood, and, at sea, the wind. Battles were fought primarily during the summertime and daytime, in clearly delineated, open battlefields with so-called cold weapons of combat, meaning clubs, stones, axes, swords, spears, bows, and arrows. Wars were decided in chaotic battles that lasted one day, at most a couple of days, whose share of the total duration of the military campaign was brief.[6] The majority of time counted as waging war went to the assembling of troops and their movement, encampment, and maintenance, which was based primarily on exploiting local resources. Preindustrial warfare was subordinate to nature, labor intensive, and inefficient.

The main elements of industrial warfare started to converge after the French Revolution (1789) and expanded into the Industrial Revolution. It was based politically on new ideologies, culturally on nation-states, organizationally on modern state machines, and materially on industrial production. Advances in public health finally made it possible to maintain the long-term battle preparedness of large and healthy mass armies: for the first time, weapons killed more soldiers than epidemics did. In addition to muscle power, nonrenewable fossil fuels (coal, oil, gas) and later water power and uranium were gradually adopted as military resources, as were various engines. As a result, large amounts of troops and supplies could be transported fast and far. New scientific-technological innovations and advances in educational levels vastly increased the destructive power of war. Battles could be fought on land, at sea, and in the air almost around the clock and year-round. Warfare became more efficient as the proportion of battles to the overall duration of war grew, and more killing and destruction were achieved in a shorter time period and across a broader area than before. The preindustrial battles that were local, occasional, labor intensive, and based on close combat transformed into continuous capital- and information-intensive large-scale remote destruction—the world now had efficient industrial warfare.

THE INCREASING IMPACT OF WARFARE ON THE SOCIETAL POLEMOSPHERE

Industrialization started to transform the polemosphere in the mid-nineteenth century. The expansion of the scale and the effects of industrial warfare were most obvious in war zones. Whereas, on average, tens of thousands of men participated in eighteenth-century battles, the number of troops rose in the nineteenth century to hundreds of thousands of soldiers, and in the twentieth century to millions. The land area covered by warfare correspondingly multiplied, reflecting the increase in the range and destructive power of weapons. Preindustrial wars were concentrated in an area of at most a few square kilometers: in the Napoleonic Wars, 100,000 men fought in an area of perhaps ten to twenty square kilometers, whereas in World War I, the corresponding number of men was spread across 240 square kilometers, and during World War II, across an area of over 3,000 square kilometers. Today, the battle zone of a division of about 10,000–15,000 men might be several tens of thousands of square kilometers.[7] Correspondingly, peacetime armed forces and military industry also came to require increasingly large areas in which to function. Since World War II, military bases, testing and training sites, and various security zones have taken over huge areas around the world.[8]

Warfare between industrialized states became increasingly capital intensive. The average cost of one US Army explosive per one dead enemy solider increased sixfold between World War II and the Korean War, and eighteenfold between World War II and the Vietnam War.[9] War also became more intensive in terms of streams of material and energy. The consumption of material multiplied many times during the industrial warfare of the twentieth century. During World War I, a normal infantry division used about one hundred tons of supplies a day. The majority of this was animal feed, a part of which was acquired in war zones. At the beginning of World War II, the daily consumption of a German tank division was about three hundred tons of supplies. A modern division from the 1980s consumed an estimated 1,000–1,500 tons of supplies.[10] Estimated simply on this basis, the ecological footprint of the armed forces expanded in the past century at least tenfold. This corresponds roughly with the average growth of production and consumption in industrialized countries during the twentieth century.

Loss of human lives in wars increased exponentially as well. Whereas a worldwide average of 15,000 people had died annually in the wars of the eighteenth century, and an average of 13,000 people died annually of warfare during the relatively peaceful nineteenth century, in the twentieth century, the annual average was as high as 458,000 people. Whereas an average of 10.8 people out of every million died in the wars of the nineteenth century, by the twentieth century, victims already numbered 183.2 people per million, in other words, almost twenty times more. In World War I alone, more people died than in all the wars of the nineteenth century put together. The number of people who died in World War II is half of what is estimated as having died in all previous wars combined.[11]

The most tragic consequence of the destructive power of industrial warfare was the escalation of civilian casualties. This was a fundamental dimension of "total war": erasing the distinction between combatants and civilians. Using the new technologies, both sides in major combat attacked entire populations and even ecosystems, a practice now known as "environmental warfare." None of this was entirely new in history, but the scale, intensity, and scope of combat now engulfed entire populations.[12]

Whereas during World War I, 80 percent of victims were soldiers, as many as 50 percent of the victims of the Spanish Civil War in the 1930s and World War II were civilians. Of the victims of the Vietnam War, 80 percent were civilians. And in recent years, as many as 90 percent of the victims of war have been civilians. Today, almost as many women die in war as men, and more children die than soldiers.[13] The conclusion is harsh: during the twentieth century, industrializing war has become relatively safer for male soldiers, but at the same time it has become increasingly deadly for civilians—women, children, and the elderly.

INDUSTRIAL WAR AND THE SOCIETAL POLEMOSPHERE

Navigating the polemosphere takes us to the home front, as a demographic, material, economic, political, and cultural phenomenon. War causes demographic shifts. War divides populations by gender and age. It changes population densities and sparks three types of population movements: those of troops, immigrants, and refugees. As a result of the movements of troops, large population centers consisting of young and middle-aged men appear in often completely uninhabited or sparsely inhabited war zones.[14] The age and gender structures of the civilian populace correspondingly shift, with increased emphasis on the young, the old, and women. As a consequence of population movements caused by war, the populations of certain areas may drop and habitation may even completely vanish, in which case the natural environment in these areas is able to recover.[15] War temporarily causes high mortality rates and low birth rates, but generally these population shifts even out after the war, when birth rates skyrocket: at the onset of World War II, there were two billion people in the world; today the figure is greater than seven billion. Wars have not, then, significantly slowed the world's population growth in the short term, let alone from a long-term perspective.[16]

War results in a double economy as states try to extend control over private production and consumption. The result is an uneasy coexistence of command economy and black market.[17] Wartime economies are geographically divided, as war zones produce little and the male population in military service shifts in its prime from being producers to becoming pure consumers. War zones live off the home front—to a large extent the work of women, children, young people, and the elderly. The productivity of households is decreased not only by the absence of young men but also by the increased difficulty of importing the outputs of production, the confiscation of vehicles and domestic animals, and the material destruction of war. Still, responsibility for the soldiers as well as for other large, economically almost nonproductive groups—such as the wounded, refugees, and prisoners of war—is thrust onto the shoulders of those on the home front. Attempts have been made to rationalize economic production during wartime, but in the twentieth century the gross national product of countries in whose territories war was fought has generally dropped. Those nations with territory free from war can benefit significantly, and in them the gross national product has at times risen markedly.[18]

Warfare causes changes in the material streams of societies. Industrial warfare is extremely dependent on renewable and nonrenewable raw materials, instruments of production, labor, and waste management services used by the civilian society. In order to engage in warfare, material structures of production, consumption, and waste management are modified to ensure sufficient access to streams of strategic raw materials and energy.[19] Military operations are directed at the destruction of the opponent's strategically important industrial-urban infrastructure. Together these actions may temporarily and even significantly decrease natural resource consumption, amounts of waste, and emissions.[20]

The world wars profoundly shaped the scale of resources that were available for warfare and welfare. Military costs increased dramatically during the world wars. While military expenditures dropped rapidly following World War I to a normal level of 3–4 percent, the drop of military expenditure was much smaller after World War II, remaining instead at a level of over 10 percent throughout the Cold War.[21] Today, over $1 trillion is spent annually on armed forces and warfare. The armed forces of several rich and numerous poor countries sustain an unjust world order, the resulting costs of which for the populace and the environment can only be guessed.[22]

The national unity, or discipline, that is demanded by the authorities during wartime is in conflict with democratic principles. The experiences in

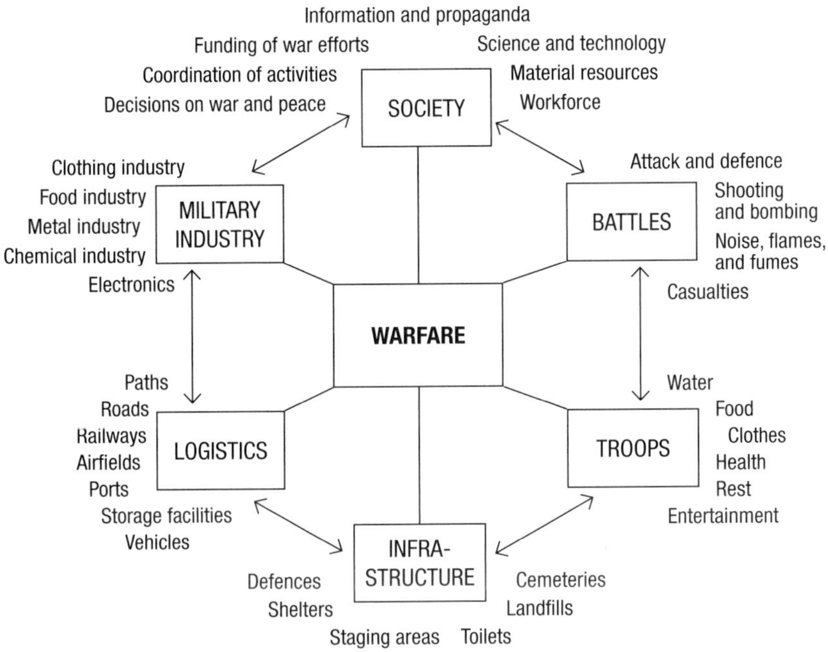

Figure 2.2. The cartwheel of war, showing the main elements of warfare.

democratic societies naturally differ from authoritarian and totalitarian societies, but war is seldom a democratic phenomenon; that is, the deeper the crisis, the fewer people are involved in decision making. The activity levels of political parties and citizen organizations tend to drop off during wartime, either voluntarily or involuntarily. In contrast, activities of voluntary organizations supporting war efforts increase.

Culture—a society's prevailing way of behavior, interaction, and understanding—has numerous effects on war, and wars have numerous effects on culture. War can raise certain ethnic cultures to a dominant status while destroying others.[23] This has a damaging effect on the cultural diversity of the world, which is one of the fundamental elements of the multipurpose use of nature. The values of individuals change, as during wartime survival becomes the primary goal. In addition to comradeship, solidarity, and equality, war elicits negative thinking and behavioral models—such as killing, violence, and concentrations of power—that are not normally accepted in societies. War hinders advances in education, mass communication, and the arts.[24]

On the other hand, lost wars in particular can dismantle established cultural structures, which may open possibilities for fresh ideas and actions. Wartime cultural changes can easily affect an entire society's relationship with nature and environmental awareness for generations to come.

INDUSTRIAL WAR AND THE ENVIRONMENTAL POLEMOSPHERE

The environmental destruction caused by industrial war increased in step with the destructive power of industrial warfare on land use, rural areas, societies, waste management, the soil, forests, bodies of water, and air.[25] Warfare has affected rural areas by the destruction of farms, domestic animals, crops, and harvests, as a result of which rare domestic animal and agricultural crop species were sometimes destroyed.[26] As one part of the wartime economy, agriculture suffered from overgrazing or depletion of the soil.[27] Rural populaces often end up the target of forced migration or experience mass mortality, whereby traditionally diverse forms of agriculture might disappear, resulting in changes in the ecosystems of entire countries, or at the extreme even entire continents.[28] During military campaigns, abundant numbers of domestic animals and pests as well as crops and weeds have traveled along with soldiers and civilians, expanding their range. Animals generally face hardship during wartime, with significant proportions on the front dying or becoming wounded. At home, domestic animals are worked to the bone and unexploitable pets are killed or abandoned, in which case they may turn feral.[29]

Warfare affects land use. Significant areas are under the control of the armed forces even during peacetime. During wartime, the areas controlled by the armed forces often expand, meaning shortages of recreational and residential areas, especially in the vicinity of large centers of population. As the boundaries of states shift, the use of new areas can undergo fundamental change. Changes in borders mean the reorganization of land use in the societies involved, which might have an effect on the use and condition of the environment for long periods, even centuries. On the other hand, frontier zones have also shielded nature from overuse throughout the ages.

War transforms the surface of the earth. Armed forces use a large amount of the earth's energy and raw material resources, which is why warfare results in large-scale military and mining operations aimed at securing strategic resources.[30] Fortress systems are among the largest structures built by mankind. Shelters and caves of various types are continuously built deeper and deeper into the earth. In war zones, the terrain is modified for military operations and structures. Efforts are correspondingly made to destroy the shelter nature offers to the enemy, which leads to the cratering of the earth's surface as well as to the mixing and polluting of the earth's surface layers. Restoration of such habitats is often slow and expensive.

Wars produce waste. Military operations leave their mark in the remnants of both parties' soldiers and civilians: corpses, body parts, gear, and personal effects. Animal carcasses, wreckage, weapons, vehicles, fuels and lubricants, vessels, residences, and entire societies are left behind. Abandoned war zones, fronts, and bases are in practice hazardous dumping grounds.[31] The military industry of the major powers has been the world's largest producer of toxic wastes. In addition to the nuclear weapons industry, the more than one hundred million land mines slowly disintegrating in the soil are perhaps the world's most pernicious waste problem.[32]

Warfare consumes forests. As preparations are made for war, forests are cut down for reasons of industrial production as well as to fund warfare.[33] The armed forces utilize forests for military purposes: forests are felled to make room for transport paths and airports, and in buffer zones and lines of fire. In addition, trees are used to build and camouflage transportation connections, positions, and breaches.[34] Soldiers use wood for construction, heating, and cooking. Vegetation is trampled by troops, and it is destroyed in bombardments and the resulting forest fires and by insect infestations.[35] In those areas abandoned by civilians fleeing war, forests might be saved, but in the areas seeing influxes of populations, the need for wood increases as new buildings are erected and land is cleared for crops. War often leads to illegal tree felling and also endangers natural conservation areas.[36]

Warfare changes the populations of natural species. Military operations scare off timid species and entice adaptable ones; they destroy existing biotic communities while creating new ones that benefit some species but harm others, in particular those that are already threatened.[37] Attempts may be made to exploit those wild species that are beneficial for military efforts and to correspondingly destroy those that are considered pests. Large mammals in particular are hunted as supplemental sources of food, while rare or harmful species are hunted in hopes of additional income. Those historically predominant wartime killers—pathogens and their carriers such as pests and rats, as well as pesticides—claim a special status in wartime conditions.[38] War continues to be a constant battle on two fronts, against people on the one hand and nature on the other.

Military operations also affect the number and quality of water bodies. Watercourses have been changed, water bodies drained, and dry land correspondingly flooded.[39] Poisoning or destruction of sources of drinking water—such as wells, oases, and springs—continues to plague the dry regions of the world. Corpses, animal carcasses, and organic waste pollute larger bodies of water, resulting in the risk of epidemics. The damaging of waste and water maintenance networks in the vicinity of population centers, the destruction of water treatment plants, and the increasing difficulty of acquiring chemicals and spare parts also pollute drinking water. Explosives, oil, chemicals, and poison gases contained in stockpiles or sunken ships cause postwar risks for water bodies, wildlife, and humans.[40] Armed conflict in the industrial era has even had effects on the seas and oceans, although in ways that we do not yet fully understand.[41]

War pollutes even the air. Noise can be a severe problem around military bases and on the home front. Noise may even be intentionally caused on the war front by military operations and weapons.[42] The equipment and energy sources used during wartime are often low grade, even toxic. Artillery fire and bombings are in and of themselves polluting, and in addition dangerous substances are readily emitted by destroyed facilities and structures.[43] Industrial warfare consumes disproportionate amounts of materials and energy, thus also playing a role in climate change.[44] But the greatest threat factor to the quality of air is the development, production, storage, and possible use of nuclear weapons and of the chemical and biological agents used in weapons of mass destruction.

CYCLES, SCALES, AND SEGMENTS OF THE POLEMOSPHERE

During the industrial era, wars have expanded in both scale and intensity around the world, intensifying the polemosphere's links between society and environment. Yet the characteristics of the polemosphere vary considerably.

The cycle of war is composed of five main stages that affect the polemosphere in different ways. During peacetime, the impact of the military complex is naturally smallest but not insignificant. Preparations for war expand the sphere of influence, which reaches its peak during wartime. Even when battles are over, the reconstruction period may have deep impacts on the environment. When societies return to peacetime, they do not return to the same conditions that prevailed before the war, because war reformulates the structures of peacetime societies. Therefore the next cycle of war starts from a new stage of the new peacetime society.

The relationship between industrial warfare and the environment in different societies can be described using a five-level mutually interactive pyramid. The foundation on which everything stands is, of course, the natural economy with its ecosystem services. The next, narrower, level of the pyramid is the human economy that sustains the basic functions of society: everyday production and consumption. The military or wartime economy comes next, in particular, production and consumption that sustains warfare in general. This is followed by the staging areas for the armed forces and their training, proving, and operations. Last of all, at the very tip, come the battles themselves. War affects all levels of the pyramid, and these in turn affect the polemosphere, but in varying ways.

The polemosphere is also geographically segmented. Today there are at least four distinctive socioecological polemospheres on earth, because the various forms of warfare are positioned around the globe according to the degree of regional socioeconomic development. The major powers are prepared to engage in industrial and information-industrial warfare with

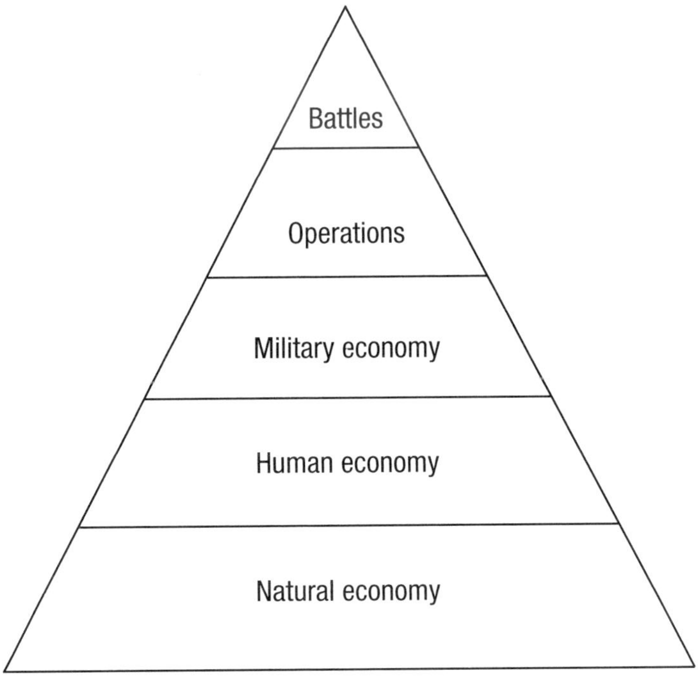

Figure 2.3. The relationship between industrial warfare and the environment can be described using a five-level mutually interactive pyramid.

weapons of mass destruction and space technologies. Wealthy industrialized nations and the most industrialized developing countries are prepared to fight advanced conventional wars with large numbers of troops on land, at sea, and in the air. The poorest developing countries are primarily capable of occasional terrestrial warfare with hand weapons and mines. In the rural regions of developing countries and the slums of both poor and rich countries, preindustrial war continues to be fought with clubs, knives, and machetes, with deadly consequences. Each of these four polemospheres has a distinct history of its own, though all four polemospheres are interlinked. International trade of weapons and related corruption and resource wars are notorious examples of such links.[45]

METHODOLOGICAL ISSUES

How can the environmental effects of war be in principle approached, depicted, and explained? Because these questions have been little discussed, I present some methodological issues that should be taken into account when investigating the polemosphere.

War is a unique subject of study. Whether the aggressor is a democracy or a dictatorship, even today wars begin with lies, rely on deceit, sustain censorship, and end in cover-ups. Further, in times of crisis, research resources are cut, the research that remains is redirected, and even basic data are lacking, especially in war zones. Gaps regularly appear around the world in sources of qualitative and quantitative data during wartime.[46] Publication of already conducted environmental studies is generally difficult during wartime, and in the later stages of war, many results are left unpublished, unarchived, or covered up for political reasons.[47] The uniqueness of war as a subject of study thus affects the number, contents, and reliability of the available data sources in the past, today, and in the future.

War is a complicated theme. Carl von Clausewitz's theses are popular, but as of yet no universally applicable theory of war exists, nor, owing to the diversity of wars, is one likely ever to emerge.[48] The subjects of the environmental history of war are two of the world's most complex agents: society and the environment. When these two agents are combined with war, which unpredictably changes both the environment and society, the result is a tripartite research frame. In this triangular model, the interaction of war and the environment can be approached from the perspectives of warfare, the environment, or society.

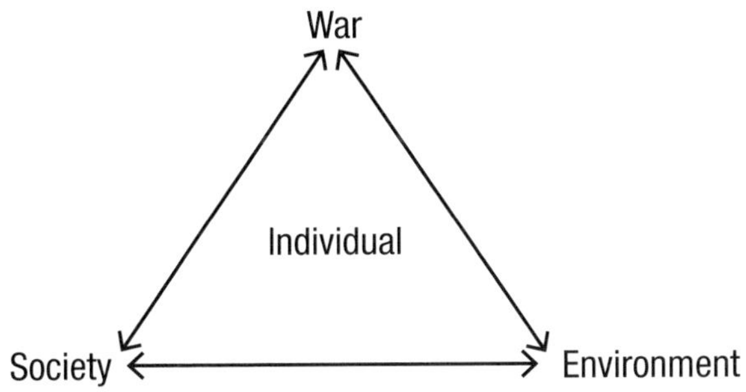

Figure 2.4. The triangular approach.

When the focus is the military perspective, the goal is to examine how the environment affects warfare. In the environmental perspective, the focus is on understanding how warfare affects the environment. From the societal perspective, on the other hand, the attempt is to understand how warfare affects society external to the war zones, and thus the environment and conceptions of it. In the following, I briefly present these three basic perspectives.

The military perspective is the most traditional. The broad strategic goal of examining the environment from a military perspective is to understand how certain natural conditions have affected the history and culture of the armed forces and warfare on a general level. The strategic influence of nature is important to remember, as the earth's seas and continents, vegetation zones, and topography continue to have a powerful effect on warfare.[49] A narrower perspective than the strategic perspective, tactical examination, investigates what operative advantage or disadvantage certain natural conditions create for individual troops or operations. These influences are not deterministic; rather, the essential thing is the environmental literacy of the troops, their ability to adopt and exploit conditions.

The traditional military viewpoint does not take into account environmental perspectives, however.[50] This is why the environmental perspective—in which the focus is on understanding how warfare has affected human living conditions, the lithosphere, the hydrosphere, the biosphere, and the atmosphere—is necessary. It is necessary to discover how the environmental effects of war and the related environmental concerns have cumulated from local concerns to issues affecting the ecological state of the entire world.

The third approach is the most complex but also the most innovative. The goal of the societal perspective is to understand how war changes the environment—and, indirectly, conceptions of it—by first changing society.

The perspective is essential because wars change societies politically, economically, socially, and culturally, and these societal changes correspondingly affect war, its environmental consequences, and our conceptions of both. The perspective can be characterized not as an institutional examination of war, but rather as an institutional examination of the environmental effects of society before, during, and after wartime.

Even though warfare is based on human interaction, negligibly little research has been conducted on the environmental history of war from the vantage point of its most natural research frame: the perspective of both parties. In examining these general reciprocal models, we should not forget the human—the individual—perspective, either. The individual functions as part of society, the environment, and war, but always forms—on the basis of his or her own tendencies, experiences, and objectives—an idiosyncratically personal view of them. Through the perspective of a single individual, we can understand how war has affected soldiers' and civilians' conceptions and memories of and personal relationship to the environment.[51]

Explaining the environmental-historical changes involved in war is no simple task. One problem is that war and peace are surprisingly difficult to separate from each other. Wars are the offspring of peace and vice versa. Warfare is based entirely on peacetime society, which produces the rhetoric, behavioral models, production structures, instruments, and staples necessary for war. At any given time, the battles of conventional wars affect a limited area of the countries at war, while the majority of the population, usually 80–99 percent of the inhabitants, does not participate in armed conflict. Thus battles and armed forces make up a surprisingly small factor in wartime societies. Further, the period of reconstruction is counted as peacetime, even though the exceptionally rapid growth in production and population it entails is brought about by war. It is easy to say when battles end—but when do the effects of war on people, societies, and the environment end?

The other problem in determining an explanation is that the main cause of environmental changes on a global scale is peacetime development: the dramatic growth of population over the past two hundred years and the related increases in agricultural and industrial production and consumption. And as the theme of peacetime environmental history has remained unexamined in most countries, how can we prove which environmental changes result from natural processes (ecosphere), from peacetime (anthroposphere), or from war (polemosphere)? In order to understand the impacts of some particular war,

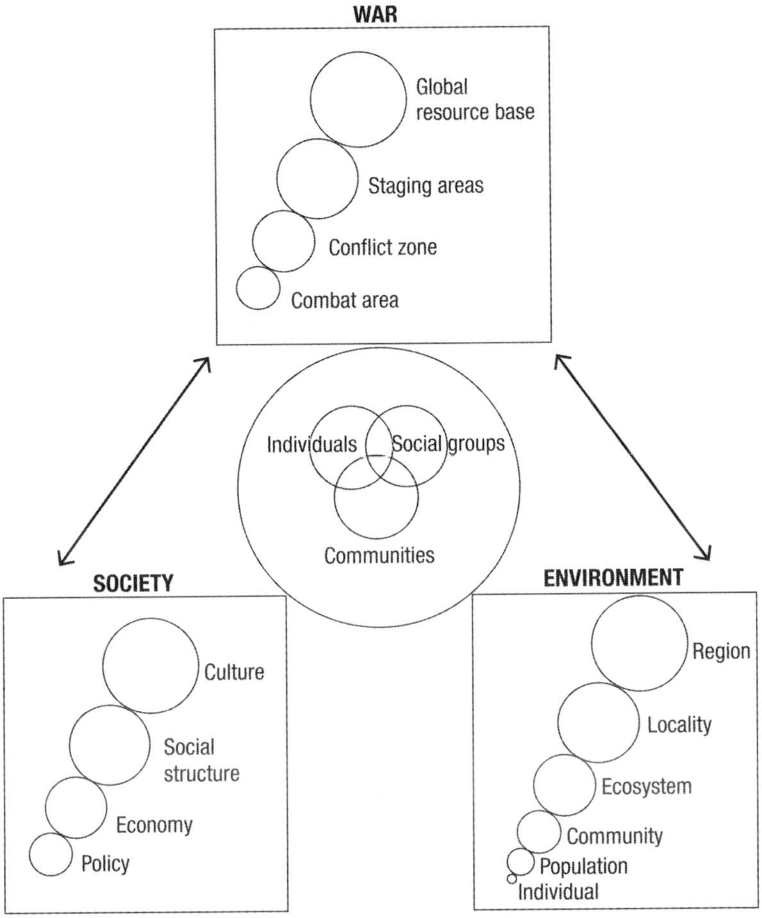

Figure 2.5. The environmental effects on the various hierarchical levels of an area's ecology. War, society, and the environment form an interactive system in which the various environmental effects of war and their comprehension are partially filtered through society. Individuals, groups, and communities can experience the ecology of war in very different ways.

we need to explore it in the framework of previous times of peace and wars and consequent cumulative impacts. Environmental cumulativity of warfare is an important dimension of the concept of the long shadows.

CONCLUSION

This chapter has explored some of the main dimensions of the socioecological polemosphere, including the impact of war on nature and society. Polemosphere as presented here is the suprahistorical and supraregional postulatory generalization in which abundant exceptions and contradictory examples can naturally be found. But it is obvious that in terms of their sphere of influence, or polemosphere, wars have expanded in both relative and absolute terms around the world. The polemosphere of the armed forces has expanded in the past century at least tenfold.

A major change in this development took place when urban-industrial revolution transformed both the social and environmental spheres of the polemosphere. Preindustrial warfare remained subordinate to nature until the mid-nineteenth century. Since then, industrial warfare started to transform local, occasional, and labor-intensive close combat into continuous capital- and information-intensive remote destruction, of which world wars, nuclear missiles, drone strikes, and cyber warfare are the latest examples.

War is an exceptionally far-reaching and multidimensional phenomenon. Since the Second World War, there have been at least four distinctive polemospheres on earth, as the various forms of warfare are positioned around the globe according to the degree of regional socioeconomic development. The relationship between industrial warfare and the environment in these different polemospheres can be described as a five-level mutually interactive pyramid. Wars affect all levels of the pyramid, and these in turn affect particular polemospheres, but in varying ways. We need multidisciplinary case studies from local to global levels to make sense of the complex environmental histories of war.

The special nature of war hampers environmental historical studies of it. The research on the environmental history of war is limited in comparison to the extensiveness of the thematics. Also, the uglier the face of war is, the more difficult it is normally to find reliable source material because of outright lies, censorship, and cover-ups. The cultural hierarchies of war are part of this problem as well. The deeper we go into history and times of crisis, the more obvious the military, strategic, and normative hierarchies become in democratic and authoritarian states, not to mention dictatorships. Therefore the environmental history of wars has been, from the start, intertwined as an element of the cultural hierarchies of war. Nature and the environment constitute one of the most recent themes that military propaganda has begun to exploit. The environmental history of wars thus constitutes a new potential arena, object, and instrument on the continuum of warfare.

Notes

1 Gaston Bouthoul, *Les Guerres: Eléments de polémologia* (Paris: Payot, 1951); Laurent Danet, "La polémosphère," *Sécurité globale* 4, no. 10 (2009): 131–49. The Roman counterpart of Polemos was Bellum.

2 A war is generally considered to be a conflict that results in more than one hundred deaths. Arto Nokkala, "The Changing Character of War," in *War or Health?*, ed. Ilkka Taipale et al. (Helsinki: Physicians for Social Responsibility, 2002), 4; Nils Petter Gleditsch, "Armed Conflict and the Environment: A Critique of the Literature," *Journal of Peace Research* 35, no. 3 (1998): 390.

3 John Keegan, *A History of Warfare* (New York: Vintage Books, 1994), 16, 29, 139, 244, 422–24.

4 David Barash, *The Whisperings Within: Evolution and the Origin of Human Nature* (Harmondsworth, UK: Penguin, 1979), 194–98. For a wider discussion, see Lawrence Keeley, *War before Civilization: The Myth of the Peaceful Savage* (New York: Oxford University Press, 1996).

5 Terry Terriff et al., *Security Studies Today* (Cambridge: Polity, 2004), 83–98.

6 Swedish historian Peter Englund has written some of the best descriptions of preindustrial warfare. See, for example, Peter Englund, *Ofredsåren: Om den svenska stormaktstiden och en man i dess mitt* (Stockholm: Atlantis, 1993).

7 Malvern Lumsden, "The Use of Raw Materials, Land and Water for Armament and War," in *War and Environment*, ed. Wendy Barnaby (Stockholm: Environmental Advisory Council, Författarna och Liber Förlag, 1981), 39.

8 Chris Pearson, Peter Coates, and Tim Cole, eds., *Militarized Landscapes: From Gettysburg to Salisbury Plain* (London: Continuum, 2010); for the American military in particular, see J. R. McNeill and David S. Painter, "The Global Environmental Footprint of the U.S. Military, 1789–2003," in *War and the Environment: Military Destruction in the Modern Age*, ed. Charles E. Closmann (College Station: Texas A&M University Press, 2009), 10–31.

9 Barry Levy, Gurinder S. Shahi, and Chen Lee, "The Environmental Consequences of War," in *The Environmental Consequences of War: Legal, Economic, and Scientific Perspectives*, ed. Jay E. Austin and Carl Bruch (Cambridge: Cambridge University Press, 2000), 53–54.

10 Martin van Creveld, *Technology and War: From 2000 B.C. to the Present* (New York: Free Press, 1989), 181.

11 Douglas Holdstock, "Morbidity and Mortality among Soldiers and Civilians," in Taipale et al., *War or Health?*, 183–97; for a critical evaluation of these concepts, see Keeley, *War before Civilization.*

12 See the penetrating discussion in Roger Chickering, "Total War: The Use and Abuse of a Concept," in *Anticipating Total War: The German and American Experiences, 1871–1914*, ed. Manfred F. Boemeke, Roger Chickering, and Stig Förster (Cambridge: Cambridge University Press, 1999), 13–28.

13 Richard M. Garfield and Alfred I. Neugut, "The Human Consequences of War," in *War and Public Health*, ed. Barry S. Levy and Victor W. Sidel (New York: Oxford University Press, 1997), 29–33.

14 Asit K. Biswas and Cecilia Tortajada-Quiroz, "Environmental Impacts of the Rwandan Refugees on Zaire," *Ambio* 25, no. 6 (1996): 403–9; John Hatton, Mia Couto, and Judy Oglethorpe, *Biodiversity and War: A Case Study of Mozambique* (Washington, DC: Biodiversity Support Program, 2001), 43–53.

15 C. R. S. Frojan and A. P. Volger, "Landmine Clearance—Its Probable Effect on Angolan Biodiversity," *Oryx* 33, no. 2 (April 1999): 89–97; "Overview A: Biodiversity, Conflict and Tropical Forests," in *Conserving the Peace: Resources, Livelihoods and Security* (Winnipeg: International Institute for Sustainable Development, 2002), http://www.iisd.org/pdf/2002/envsec_conserving_overview.pdf.

16 Colin McEvedy and Richard Jones, *Atlas of World Population History* (New York: Facts on File / Penguin Reference, 1978), 342–51; see also the national statistics of countries worldwide that contain basic historical demographic time series.

17 Robert Higgs, "Wartime Prosperity? A Reassessment of the U.S. Economy in the 1940s," *Journal of Economic History* 52, no. 1 (March 1992): 41–60. For a contemporary report, see Humeyra Pamuk and Jonathan Saul, "Insight: Syria's Economy Goes Underground as Black Market Thrives," *Reuters*, September 5, 2013, http://www.reuters.com/article/us-syria-crisis-blackmarket-insight-idUSBRE98405I20130905.

18 Angus Maddison, *Monitoring the World Economy, 1820–1992* (Paris: OECD, Development Centre Studies, 1995), 21, 66–70, fig. 1.1; Alan S. Milward, *War, Economy and Society, 1939–1945* (London: Penguin, 1977), 177–85; B. Caplan, "How Does War Shock the Economy?," *Journal of International Money and Finance* 21, no. 2 (April 2002): 145–62; Mark Harrison, ed., *The Economics of World War II: Six Great Powers in International Comparison* (Cambridge: Cambridge University Press, 1998); Karen Hagemann and Stefanie Schüler-Springorum, eds., *Home/Front: The Military, War and Gender in Twentieth Century Germany* (Oxford: Berg, 2002).

19 Matthew Evenden, "Aluminum, Commodity Chains and the Environmental History of the Second World War," *Environmental History* 16, no. 1 (January 2011): 69–93; see Simo Laakkonen, "Environmental Policies of the Third Reich," chapter 4 in this volume.

20 Simo Laakkonen, "Warfare—An Ecological Alternative for Peacetime? The Indirect Impacts of the Second World War on the Finnish Environment," in *Natural Enemy, Natural Ally: Historical Studies in War and the Environment*, ed. Richard P. Tucker and Edmund Russell (Corvallis: Oregon State University Press, 2004), 175–94.

21 Charles Feinstein, Peter Temin, and Gianni Toniolo, *The European Economy between the Wars* (Oxford: Oxford University Press, 1997), 189, table 10.1.

22 Stephen Kinzer, *Overthrow: America's Century of Regime Change from Hawaii to Iraq* (New York: Times Books, 2006); Richard Tucker, *Insatiable Appetite: The United States and the Ecological Degradation of the Tropical World* (Lanham, MD: Rowman & Littlefield, 2007).

23 Jussi Raumolin, "Metsän suojassa: Poliittisen antropologian ja geopolitiikan näkökulma," *Silva Fennica* 21, no. 4 (1987): 342–50; Paul Collier, Anke Hoeffler, and Måns Söderbom, "On the Duration of Civil War" (development research group discussion paper, World Bank, Washington, DC, 2001).

24 On the effects of war on the press, see Rauno Lahtinen and Timo Vuorisalo, "In Search of the Roots of Environmental Concern: Water Management and Animal Welfare Issues in the Finnish Local Press in 1890–1950," *Scandinavian Journal of History* 30, no. 2 (June 2005): 197–99.

25 This division partially follows Dahl's framework of examination. Arne Willy Dahl, "Environmental Destruction in War," *Disarmament* 15, no. 2 (1992): 113–15. On other frameworks for examining the environmental effects of war, see the Swedish Ministry of Defence, *Meeting on Military Activities and the Environment, 27–30 June 1995, Linköping, Sweden* (Stockholm: Ministry of Defence, 1995); Johan J. Holst, *Conflict and Environmental Degradation*, NUPI Notat 325 (Oslo: Nupi, 1985), 8 pp.; Levy et al., "Environmental Consequences of War," 51–64.

26 Arthur H. Westing, *Warfare in a Fragile World: Military Impact on the Human Environment* (Stockholm: SIPRI, 1980), 56; Jilly Cooper, *Animals in War* (Guilford, CT: Lyons Press, 1983); "The Silent Casualties of War," *UNESCO Courier* 52, no. 7/8 (July/August 1999): 12.

27 Carol R. Ember and Melvin Ember, "Resource Unpredictability, Mistrust, and War," *Journal of Conflict Resolution* 36, no. 2 (June 1992): 242–62; Daniel R. Mortensen, "The Deterioration of Forest Grazing Land: A Wider Context for the Effects of World War I," *Journal of Forest History* 22 (October 1978): 224–25; Susan D. Lanier-Graham, *The Ecology of War: Environmental Impacts of Weaponry and Warfare* (New York: Walker, 1993), 20.

28 On continental changes, see Alfred W. Crosby, *Ecological Imperialism: The Biological Expansion of Europe, 900–1900* (Cambridge: Cambridge University Press, 1986).

29 Rauno Lahtinen, "Kaupunkimaatalous ja ympäristö maailmansotien aikana" [Urban agriculture and environment during world wars] in *Sodan ekologia*, 340–74.

30 Jeffrey McNeely, "War and Biodiversity: An Assessment of Impacts," in Austin and Bruch, *Environmental Consequences of War*, 360; Raymond Dumett, "Africa's Strategic Minerals during the Second World War," in "World War II and Africa," special issue, *Journal of African History* 26, no. 4 (1985): 381–408.

31 Donovan Webster, *Aftermath: The Remnants of War* (New York: Pantheon Books, 1996); Ferenc M. Szasz, "The Impact of World War II on the Land: Gruinard Island, Scotland, and Trinity Site, New Mexico as Case Studies," *Environmental History Review* 19, no. 4 (Winter 1995): 15–30; Vesa-Pekka Herva, "Haunting Heritage in an Enchanted Land: Magic, Materiality and Second World War German Material Heritage in Finnish Lapland," *Journal of Contemporary Archaeology* 1, no. 2 (2014): 297–321.

32 Stephen I. Schwartz, ed., *Atomic Audit: The Costs and Consequences of U.S. Nuclear Weapons since 1940* (Washington, DC: Brookings Institution Press, 1998); Bruce Gray, "Landmines: The Most Toxic and Widespread Pollution Facing Mankind," accessed September 18, 2012, http://fn2.freenet.edmonton.ab.ca/~puppydog/bgray.htm; Arthur Westing, *Explosive Remnants of War* (Stockholm: SIPRI, 1985), 33–37.

33 For the most part, the demand for wood increased during wars and more forests were felled than during peacetime. See Westing, *Warfare in a Fragile World*, 51–53; Antti Erkkilä and Harri Siiskonen, "Sodan ympäristövaikutukset Namibiassa" [Environmental impacts of war in Namibia], in Laakkonen and Vuorisalo, *Sodan ekologia*, 644–73; Thaddeus Sunseri, "World War II and the Transformation of the Tanzanian Forests," in *Africa and World War II*, ed. Judith Byfield, Carolyn Brown, Timothy Parsons, and Ahmad Sikainga (Cambridge: Cambridge University Press, 2015), 238–56.

34 For an interesting alternative approach, see Tim Newark and Quentin Borsarello, *Brassey's Book of Camouflage: History of Camouflage Uniforms* (Herndon, VA: Brassey's, 1996); Henrietta Goodden, *Camouflage and Art: Design for Deception in World War 2* (London: Unicorn Press, 2007); Isla McLean Forsyth, "From Dazzle to the Desert: A Cultural-Historical Geography of Camouflage" (PhD thesis, University of Glasgow, 2012), http://theses.gla.ac.uk/3445/.

35 John R. McNeill, "Woods and Warfare in World History," *Environmental History* 9, no. 3 (July 2004): 388–410; Chris Pearson, "'The Age of Wood': Fuel and Fighting in French Forests, 1940–1944," *Environmental History* 11, no. 4 (October 2006): 775–803; Bernard Nietschamann, "Battlefields of Ashes and Mud," *Natural History* 99, no. 11 (November 1990): 35.

36 Joseph P. Dudley et al., "Conservation Biology: The Effects of War and Civil Strife on Wildlife and Wildlife Habitats," *Conservation Biology* 16 (2002): 323; Jeyamalar Kathirithamby-Wells, "The Environmental Impacts of Japan's Occupation of West Malaysia (1942–45) and Its Socio-Economic Implications," in "World War II, the Cold War, and Natural Resources," ed. Simo Laakkonen and Richard Tucker, special issue, *Global Environment: A Journal of History and Natural and Social Sciences* 10 (2012): 92–115.

37 Rajab Abdullah, "Protected Areas during and after Conflict: Nimule National Park, a Case Study," in *Parks for Peace Conference Proceedings* (Gland, Switzerland: International Union for the Protection of Nature, 1997), 195–99; Malcolm Brown et al., "War and the Environment," *Audubon* 93, no. 5 (1991): 88–99.

38 Westing, *Warfare in a Fragile World*, 57; Hatton, Couto, and Oglethorpe, *Biodiversity and War*, 43–53; Dudley et al., "Conservation Biology"; Mari Pohja-Mykrä and Sakari Mykrä, "Luonnonvaraiset eläimet sodassa ja sodan kohteina," in Laakkonen and Vuorisalo, *Sodan ekologia*, 143–92. See Helene Laurent, "The Great Louse War: Control of Typhus Fever," chapter 9 in this volume.

39 On flooding, see Levy et al., "Environmental Consequences of *War*," 53; and on draining, Emma Nicholson and Peter Clark, eds., *The Iraqi Marshlands: A Human and Environmental Study* (London: Politico's Publishing, 2002).

40 Lanier-Graham, *Ecology of War*, 58; Westing, *Warfare in a Fragile World*, 152–69; Webster, *Aftermath*; Geoff Plunkett, *Chemical Warfare Agent (CWA) Sea Dumping off Australia* (Canberra, ACT: Department of Defence Report, 2003).

41 Jacob Darwin Hamblin, *Poison in the Well: Radioactive Waste in the Oceans at the Dawn of the Nuclear Age* (New Brunswick, NJ: Rutgers, 2008); Kurkpatrick Dorsey, *Whales and Nations: Environmental Diplomacy on the High Seas* (Seattle: University of Washington Press, 2013); Simo Laakkonen and Tuomas Räsänen, "Science Diplomacy in the Baltic Sea Region: Beginnings of East-West Cooperation in Marine Protection during the Cold War, in *Northern Europe in the Cold War: East-West Interactions of Security, Culture, and Technology*, ed. Ann-Marie Ekengren, Rasmus Mariager, and Poul Villaume (Helsinki: Aleksanteri Institute, 2016), 25–48.

42 Outi Ampuja, "Ääni ja melu modernissa sodankäynnissä," in Laakkonen and Vuorisalo, *Sodan ekologia*, 304–39. For an innovative study on militarized soundscape, see Carolyn Birdsall, *Nazi Soundscapes: Sound, Technology and Urban Space in Germany, 1933–1945* (Amsterdam: Amsterdam University Press, 2012).

43 Levy et al., "Environmental Consequences of War," 53.

44 James Fleming, *Fixing the Sky: The Checkered History of Weather and Climate Control* (New York: Columbia University Press, 2010).

45 On current research, see Thomas Homer-Dixon and Jessica Blitt, eds., *Ecoviolence: Links among Environment, Populations and Security* (Oxford: Rowman and Littlefield, 1998); Thomas Homer-Dixon, "On the Threshold: Environmental Changes as Causes of Acute Conflict," *International Security* 2 (1991): 76–116; Philippe Le Billon, "The Political Ecology of War: Natural Resources and Armed Conflicts," *Political Geography* 20, no. 5 (June 2001): 561–84; Simon Dalby, "Ecological Politics, Violence, and the Theme of Empire," *Global Environmental Politics* 4, no. 2 (May 2004): 11; Tom H. Hastings, *Ecology of War and Peace: Counting the Costs of Conflict* (Lanham, MD: University Press of America, 2000); Gleditsch, "Armed Conflict and the Environment," 381–400.

46 A classic example is the lack of catch statistics during the world wars. See, for example, Poul Holm, "World War II and the 'Great Acceleration' of North Atlantic Fisheries," in "World War II, the Cold War, and Natural Resources," ed. Simo Laakkonen and Richard Tucker, special issue, *Global Environment: A Journal of History and Natural and Social Sciences* 10 (2012): 66–91.

47 For a discussion on censorship and publicity, see Dorothee Brantz, "Landscapes of Destruction: Capturing Images and Creating Memory through Photography," in The Cambridge History of the Second World War, vol. 3, *Total War: Economy, Society and Culture*, ed. Michael Greyer and Adam Tooze (Cambridge: Cambridge University Press, 2015), 725–48; Tuomas Räsänen, "Keep It Secret: The Environmental Effects of Nuclear Armament in the Soviet Union and the United States," in *Nuclear Portraits: People, Communities and the Environment*, ed. Laurel Macdowell (Toronto: University of Toronto Press, forthcoming).

48 Martin Shaw, *The Dialectics of War: An Essay in the Social Theory of Total War and Peace* (Worcester: Pluto Press, 1988), 1–16; see the first reference in Keegan, *History of Warfare*.

49 Keegan, *History of Warfare, 84–96*.

50 See, for example, Manus Midlarsky, ed., *Handbook of War Studies II* (Ann Arbor: University of Michigan Press, 2000).

51 See Ferdinand Braudel, *On History* (Chicago: University of Chicago Press, 1980); and Leena Rossi, "Yksilö ympäristöhistoriaan," in *Monta tietä menneisyyteen*, ed. Leena Rossi and Hanne Koivisto (Turku: Turun yliopisto, 1995).

Chapter 3
World War II
A Global Perspective

EVAN MAWDSLEY

This chapter faces two challenges. First is how, within the confines of a short contribution, to provide a global perspective for an event as vast as World War II. Second is how to relate such an overview to a fresh approach of environmental history. This essay is not intended to be a survey of the literature on the environmental history of the war, however.[1]

Turning first to environmental history, I am comfortable with J. R. McNeill's definition of the subfield as "the history of the mutual relations between humankind and the rest of nature."[2] But this definition is broad and could certainly take in much of economic history and perhaps much of what I would think of as "history" in general. Above all, it seems to me important to take "mutual relations"—from McNeill's definition—to be a two-way street. Much of the focus in the environmental history literature might be described as "impact studies"—examining the impact of human activities on the environment—in this case the impact of global military preparations, deployments, and operations on the natural world during the period of World War II. This is the thrust of most of the contributions to the present volume.[3] But it is also appropriate to see mutual relations the other way around, how "the rest of nature"—natural conditions and natural resources—affected the conduct of wartime operations and even the motives of the states involved in World War II.

I start by outlining what seem to be some distinctive characteristics of World War II, looking at the way the natural conditions influenced the conduct of the operations, and finally arguing that to a considerable degree the essence of the war was competition for natural resources.

THE NATURE OF WORLD WAR II

My overall conception of World War II has developed over several decades but was especially affected by attempting to produce a one-volume history of the conflict. I had just completed an operational/strategic history of the war on the eastern front and was invited to write a new and relatively compact operational/strategic history of World War II as a whole.[4]

I attempted to avoid the narrative of British popular culture, which stressed the ups and downs of the war in Western Europe—Dunkirk, the Battle of Britain, D-Day. The popular American narrative is also rather narrow, although it brings in the Pacific War; one of the misleading highlights is the Battle of the Bulge in the winter of 1944–45.[5] Neither narrative, partly as a result of the Cold War, makes much of the Russian involvement in the war, and I felt that struggle on the eastern front must receive appropriate attention.

The Russian-German front has not been neglected in the specialized professional literature. The same could not be said about the war in China, even during the first decade of the current century, when I was writing the book. The more I thought about a "new" history, the more important became the war in the Far East, the fighting between the British Empire and Japan, and especially that between China and Japan.[6] The war in China did not go well for the Chinese or for the Allied cause, sources are hard to access, and the language barrier is formidable. Nevertheless, this war was vastly important, not least for the effect it had on China's later political and social transformation.[7]

China in 1937 was the most populous single country on the planet; the estimated population numbered over 410 million people. The best recent work accepts an estimate of Chinese military deaths numbering ten million, with civilian losses in the ten to twenty million range. Losses were comparable to those of the wartime Soviet Union and much larger than the death toll of 350,000 to 400,000 (over 90 percent military rather than civilian) suffered by each of Britain and the United States.[8] In many accounts of World War II—which begin in Poland in September 1939—the Second Sino-Japanese War of 1937 is treated as a "flashback," after the attack on Pearl Harbor. The carrier battles and island assaults of the Pacific War are fascinating to a naval historian, and instrumental in forcing Japan's surrender, but they were much smaller in human scale and in a sense less important than the war in China. I made the decision that World War II, in my "new history," began in the summer of 1937, rather than on September 3, 1939.[9]

Privileging the war in China had significant consequences for my approach. If World War II is taken to start in 1937, then it is hard to see the conflict as exclusively part of the German wars or to base it simply on the Nuremberg narrative of warmongering by Hitler and the Nazis—although, of course, that element cannot be ignored. I also argue below that folding the Sino-Japanese War into World War II makes it difficult to generalize about the nature of combat at the operational and tactical level.

Another feature that I tried to bring out in a new history, apart from an emphasis on Russia and a "rediscovery" of China, was the importance of Great Britain and the British Empire. Britain played a critical political role in holding the Germans at bay after the defeat of France, at a time when both the Soviet Union and the United States were neutral. British overseas territories were also of singular importance as the object of the ambitions of a number of other powers—not so much the ambitions of the Germans as of the Japanese and the Italians, and arguably (and indirectly) also the Americans. It seemed to me that Britain's role had been obscured by the relative decline of the United Kingdom after 1945. This might appear to contradict my earlier point about British popular history, but the Empire is not a fashionable part of this narrative.[10]

The British Empire was by a considerable margin the largest political and geographical entity on the planet, with a population of over 530 million and an expanse of territory much larger than that of China. It had a powerful integrated economy, and its assets included the world's largest merchant marine and the financial influence of the City of London. The role of the "white" Dominions is relatively well known, but India—with a population of 350 million—was especially important as a military and economic factor. Troops from the subcontinent—mainly the extraordinary Indian Army—made up most of the British troops fighting the Japanese. Although India was not successfully invaded, parts of its population felt the war's economic impact, most tragically in the Bengal famine of 1943.

Moving away from the new history, what were the distinctive features of World War II? One was its geographical breadth. This was not the first global war. The world wars of the eighteenth century—especially the Seven Years' War—were global and determined the contours of the future colonial world, that is, that it would be exploited more by the British than by the French. But those wars did not involve large military deployments or reach far inland (or have much immediate environmental impact). World War I was also global, involving battles as far apart as the Falkland Islands, East Africa, and Tsingtao

(Qingdao) in China; the social, economic, and environmental impact of the 1914–18 struggle was important in terms of mobilizing global resources.[11] The seat of World War I was always Europe, however, and the mainland fighting was largely confined to the quadrilateral bounded by Flanders, Poland, the Balkans, and the eastern Alps. The main naval combat was limited to the North Sea. Significant fighting did not reach Scandinavia, North Africa, most of France and Italy, and most of European Russia, let alone the world beyond Europe.[12]

In contrast, between 1937 and 1941, two simultaneous full-scale regional conflicts were raging, one in East Asia and the other in most of Europe as well as North Africa. The Axis invasion of the Soviet Union in June 1941 brought in a new zone of fighting in the European regional conflict. This zone was six hundred or seven hundred miles (970 or 1,130 kilometers) deep and with a breadth of 1,500 miles (2,400 kilometers); it stretched from the Caucasus to the Barents Sea. These two regional conflicts—European/African and East Asian—were merged into a genuine global war in December 1941, when Japan attacked Britain's Asian colonies and the United States entered the war in the Pacific Ocean.

A second distinctive feature of the war was that it involved a huge and protracted mobilization of human and other resources, with the leading participants operating from a large industrial base. This was similar to World War I, although that war, to use J. R. McNeill's terminology, was based in the advanced countries on the coal, steel, and railways of the "Coketown cluster," while World War II was based—again in the advanced countries—on the electricity, petroleum, assembly lines, and motor vehicles of the "Motown cluster."[13] World War II was also different from the Cold War, where the planning assumptions did not include prolonged industrial mobilization once the shooting had started.[14]

World War II certainly involved more extensive mobilization than World War I, including as it did the populations and territory of Japan and China. Although the heaviest fighting, even for Germany, lasted slightly less than four years, the period of "mobilisation stress" lasted for a decade (from the mid-1930s up to 1945) in the case of Germany, the Soviet Union, and Japan.[15] The resources and production of the civilian rear in World War II were also under direct attack (aerial bombardment), which had not been the case in previous conflicts, including World War I, although homelands had certainly been under indirect attack then (through naval blockade and economic warfare).

Another distinctive feature of World War II was that it was a total (rather than a limited) war on all fronts. As far as the major Allied and Axis powers were concerned, the only outcome was the overthrow of the entire enemy system and in many cases the occupation of its territory, rather than the agreed transfer of limited territory, or a negotiated peace, where both sides ended the war intact.[16] Such limited outcomes had been characteristic of the wars of the eighteenth and nineteenth centuries. Even World War I ended with an armistice that left Germany without the Kaiser but politically intact and mostly unoccupied.

World War II was large scale in geographical terms, involved a large number of participants, and was conceived by the participants as "total." But it would be unwise to generalize too much about the nature of fighting in World War II as "modern" or "industrial." It was a global war with global variety. To be sure, the war included modern elements, and all the powers except perhaps China aspired to fight a "war of machines." In the end, however, Japan and Italy fought the war with what were essentially infantry armies of the World War I type. This was also true for nearly all the smaller participants. And the reality of the war in China has aptly been compared to the Thirty Years' War of the early seventeenth century.[17] Just to add to the complexity, while the war on land in Asia was arguably a throwback to World War I or before, the war at sea between the United States and Japan was in many respects more modern than that between Britain and Italy, and arguably more modern than the war at sea between Britain and Germany. So although the war was global, it was in many senses fragmented.

Even the more industrialized states displayed fundamental differences of approach to warfare, based on levels of development and geographical position. The Russian forces were labor intensive, while those of Britain and the United States were capital intensive. This meant that British and American armies were almost entirely motorized and had a much higher ratio of support/service troops to combat troops. Some military leaders in London and Washington even maintained that the war would be won by air power alone. The British and Americans built about 50,000 four-engine "strategic" bombers, the Germans a few thousand, and the Soviet Union ninety-one. Meanwhile, the German Navy was a large and complex organization built (by midwar) around the U-boat. It was totally different from the British Royal Navy, which was in turn different from the US Navy and the Japanese Navy.

World War II was, for the most part, a much more mobile war than the war of 1914–18, although, again, we need to be careful about generalization. The war on the western front in World War I was memorably static, creating its infamous "environments of death."[18] The campaigns in Eastern Europe in 1914–18 were far more fluid. Mobility per se is not synonymous with modernity—think of Genghis Khan and the Mongols—but mass mobility probably is. Those armies of World War II that were most modern (those of Germany, the Soviet Union, the United States, and Britain) were mostly still dependent on railways or shipping, but the availability of motor transport meant they could operate much farther from railheads and from beachheads and ports. The British and Americans were also able to carry out amphibious operations far more successfully than had been the case in 1914–18. Nevertheless, while the tank was perhaps the iconic weapon of World War II, it was unable to bring about operational breakthroughs on the Asian mainland, except in two spectacular but brief episodes: the Japanese Malayan campaign of 1941–42 and the Soviet invasion of Manchuria in August 1945.

The question of the impact of World War II on the environment is not one that I intend to discuss here. The chains of effect (the "long shadow") brought out in various chapters of this book, and elsewhere in the published literature, are fascinating. But I also find a considerable degree of truth in the proposition that the most deadly war in history (in human terms) had only limited environmental impact, especially if seen in comparative terms. "Most of the environmental changes wrought in combat proved fleeting," J. R. McNeill has argued. McNeill suggested that "preparing and mobilizing for industrial warfare" wrought more serious changes, and this aptly highlights the point that the impact of World War II has to be seen relative to other developments and other times.[19] For reasons of space, I do not pursue this argument here.

NATURAL CONDITIONS AND OPERATIONS IN WORLD WAR II

Let us return to the other side of the mutual relationship, the impact of the natural environment on the wars of humankind. First of all, how did the limits and conditions set by nature channel, complicate, or even prevent the conduct of warfare? Nature, the natural environment, as any military historian will tell you, has always had a crucial effect on operations.[20] In operational/strategic terms, climate determines the campaigning season. Physical terrain and vegetation affect mobility and the balance of advantage between attacking and defending forces.

Two of the best-known geostrategists argued, among other things, that world power was to be achieved through the ability to overcome environmental limitations. They were talking about power in a general sense, but that power related among other things to military operations. The famous American navalist Alfred Thayer Mahan (1840–1914) wrote that the ability to dominate traffic over the seas was crucial; he wrote most famously about an era (the late seventeenth and eighteenth centuries) when natural conditions and the state of technology made it difficult to move bulk goods overland.[21] The sea was the main artery of trade, and command of the seas gave victory and world power. World power went to maritime states rather than continental ones.

The British geographer Halford Mackinder (1861–1947), working on a comparable historical scale, argued that development in the mid-nineteenth century of land transport, especially the railways, had brought an end to the centuries-long "Columbian Age." The railways could now overcome natural obstacles. Resources and armies could now be moved rapidly overland, and states that held central (inland/continental) locations could both deal effectively with peripheral attacks and extend their power outward.[22] Hitler was influenced by Mackinder (via Karl Haushofer) and concluded that modern roads and railways made possible an inward-facing, self-contained, German "empire" in the European "heartland" (*Mitteleuropa*).

One of the remarkable features of World War II (potentially in all theatres) was the availability of new technology to overcome natural obstacles on both land and sea—although there would be many limitations. It is not by chance that three of the "iconic" instruments of the conflict were the tank, the four-wheel-drive Jeep, and the airplane. Each in its own way had a novel ability to overcome natural conditions, whether a few miles of rough ground, hundreds of miles of open steppe, or the English Channel. Less obvious, but as important in terms of vital logistics, were two-and-a-half-ton trucks with all-wheel drive, and roll-on/roll-off landing ships. At an even more basic physical level, radar overcame the limitations of human eyesight, distance, darkness, and clouds; as a result, this electronic technology radically changed the odds in air defense and antisubmarine warfare. Medicines and pesticides enabled "outside" troops to operate in the tropics.

Numerous specific examples could be given about how the natural conditions affected operations in World War II. I present only a few. From the time of the first great campaign in the West, weather and terrain were potential limiting factors, as they had been in previous wars. Hitler was eager for a full-scale offensive immediately after the conquest of Poland, but the strike was repeatedly postponed; the vagaries of the weather limited the availability of essential air support for the advancing German troops. But when the German invasion did finally take place, in early May 1940, surprise was achieved thanks to the incorrect French assumption that the Wehrmacht could not mount a motorized advance through the wooded hills of the Ardennes in Belgium and Luxembourg.

The timing of the invasion of Russia in 1941, and the undoing of the German campaign in the East in the winters of 1941–42 and 1942–43, were closely related to the climate—not only the extreme winter frost but also the spring and autumn mud of the *rasputitsa*.[23] After the first season of the Russian campaign, the conflict separated mainly into fighting in the forest belt and fighting in the steppe belt. By midwinter 1941–42 and up to the summer of 1944, the former zone was static, while fluid back-and-forth fighting was carried on in the south. Both the Soviet Union and China benefited from the inability of enemy land power, even in the twentieth century, to fully overcome natural conditions over vast distances, especially beyond the termini of the railway network. A similar situation existed in much of North Africa. One reason the Red Army in 1944 was finally able to break the German defensive system in Belorussia (in the forest belt) was the supply under the Lend-Lease Act of large American Studebaker 6 × 6 trucks; these made it possible to maintain mobility year-round and support a sustained ground offensive.

The feasibility of the critical cross-Channel landings—both the planned German one in 1940 (Operation Sea Lion) and the projected Allied operations in 1943 and 1944 (Roundup and Overlord)—was determined seasonally by the weather window of roughly May to September and tactically by the tides and phases of the moon. Both sides recognized that unpredictable storms could make landing or reinforcement impossible and swamp many of the landing vessels; as it was, the freak storm of June 19–21, 1944, wrecked one of the ingenious artificial harbors at Normandy. Good weather was also required to provide air support for the attacking forces; one of the reasons for the early success of Hitler's Ardennes Offensive in December 1944 was that bad weather temporarily reduced the ground-support superiority of the Allied tactical air forces. Assault landing operations, in Normandy and elsewhere, were also driven by the nature of the coastline and the quality of the beaches, the approaches to them, and possible exit routes into the hinterland.

The campaign in Libya and Egypt in 1940–43, somewhat misleadingly called the Desert War, was a remarkable case of overcoming natural conditions with motorized forces, but the limited railway coverage led to acute supply problems. Neither side could hold ground. Even when the motorized columns dashed madly from east to west (or west to east), they could not penetrate far south from the Mediterranean. At the critical point, in eastern Libya and western Egypt, the escarpment on the northern edge of the Qattara Depression greatly narrowed the front and strengthened the position of the defending forces.

The unexpected and crucial defeat of British Empire troops in Malaya in the winter of 1941–42, climaxing in the fall of Singapore, resulted in part from faulty assumptions by the defending British commanders regarding natural conditions, the supposed impassability of the "jungle" terrain of Malaya, and the supposed impracticality of amphibious landings on the east coast of Thailand and Malaya during the annual "northeast monsoon."

The acclimatization of outside troops—Japanese as well as European and American—fighting in tropical regions was an important factor in military success or failure. The Japanese initially approached this in a more methodical way than the Allies, which greatly aided their conquest of Southeast Asia.

Natural conditions also affected naval warfare. This was no longer the era of the Spanish Armada; ship technology had reached the point in the mid-twentieth century where safe navigation was possible in most oceans and at most times of year. (An unusual example of large-scale damage caused by natural conditions, rather than enemy action, was the experience of the American Task Force 38 in a typhoon off the Philippines in December 1944.) Nevertheless, extreme conditions would be met in the northern latitudes; there the likelihood of violent storms, the relative length of day and night, and even the proximity of ice fields were directly related to the time of year. The extreme case was the Arctic convoys to North Russia, which were highly vulnerable to air and submarine attacks in the Barents Sea during summer, when the sun shone all day long. Thermoclines also had a large seasonal effect on the effectiveness of sonar in submarine and antisubmarine operations in both the Mediterranean and the Arctic. The sciences of both meteorology and oceanography were much developed by wartime necessity.

In general, the warships of World War II were more like the sailing ships of Admiral Horatio Nelson's time than the coal-burning dreadnought battleships of 1914–18. They repeatedly carried out long-range oceanic operations, not because they were sailing ships, but because they were larger, fueled by oil, and capable of being refueled and resupplied while under way at sea.

Regional and seasonal weather greatly influenced the air war. The Battle of Britain came to a halt partly because of the deteriorating weather in the autumn of 1940. Although conventional wisdom holds that the American B-17s and B-24s carried out "pinpoint" bombing during the day, and the Royal Air Force (RAF) carried out "indiscriminate" area bombing at night, the unfavorable European weather meant that the Americans often had to bomb blindly "through the clouds." In the air war in the Far East, the jet stream unexpectedly reduced the effectiveness of unescorted high-altitude B-29 bombing raids against Japan in 1944–45.

What was actually important was not so much the ability of aircraft to fly at higher and higher altitudes (although that was the theory behind the pressurized B-29 Superfortress in 1944–45). More significant was the remarkable extension of aircraft range in the 1930s. Longer-range flights determined much of the nature of the fighting, which was about seizing airfields or denying them to the enemy. Hitler's diversion into the Balkans in the spring of 1941 was partly driven by the need to overrun bases from which the RAF could bomb Axis oil wells and refineries in southern Europe. In the Pacific, the ability to quickly build airstrips on coral atolls was an important ingredient. Perforated steel Marston matting for runways was a significant example of a material widely used by Allied forces to overcome natural conditions.

Successful partisan or guerrilla warfare was dependent on, among other things, natural conditions—vegetation, topography, and climate. The most successful partisan campaigns were fought in the forests of Belarus or the rough country of Yugoslavia or northern China. The rural guerrilla bands in France were named after the thickets (*maquis*) of the southeast. It has also been suggested that some of the severity of Wehrmacht occupation policy was based on the exposure of German troops to the extreme conditions of the Russian environment.[24] Another important factor in creating partisan strongholds in both Europe and Asia was remoteness from roads and railways.

NATURAL RESOURCES AND THE AIMS OF WAR

Some historians have discussed changes in the environment as a source of war. Charles E. Closmann noted that "military conflict is often a . . . consequence of environmental decline," giving as an example the effect of climate change and desertification in modern-day Darfur.[25] Environmental decline was not evidently a cause of World War II, but what was the role of natural resources in planning, carrying out, and ending it?

A large element in the motivation of World War II was a global struggle for natural resources, especially food and key minerals.[26] It was not the first war to be so motivated, of course, but the spectrum of resources was especially broad, programs came closer to reality, and the approach adopted was ruthlessly systematic.

The Axis states, which initiated the war, always stressed that their cause was a "new order," redressing an existing world imbalance. As the preamble to the Tripartite Pact, the September 1940 treaty that created the Axis, put it:

> The governments of Germany, Italy and Japan, consider it as a prerequisite of a lasting peace that each nation of the world would receive the space (*Raum*) to which it is entitled. They have, therefore, decided to stand by and co-operate with one another in regard to their efforts in the greater East Asia space and in the European region respectively wherein it is their prime purpose to establish and maintain a new order of things (*eine neue Ordnung*) calculated to promote the mutual prosperity and welfare of the peoples concerned.

Raum meant not simply greater living space, but also access to food and fertile land, and minerals and other raw materials. The Germans' ambitions were intensified by the experience of 1914–18, when their defeat was seen to be caused by lack of food and other resource shortages.

The grand strategic aim of the Axis powers was to extend their zones of control and broaden their resource base. The Germans were concerned with the exploitation of the natural resources of European Russia (including the Caucasus). The Japanese intended to exploit the natural resources of northeastern China (Manchuria) and Southeast Asia and formulated the structure of the Greater East Asian Co-Prosperity Sphere.[27] Mussolini coined the phrase "war of continents," and fascist ideology made much of Italy's *spazio vitale.*[28] The cause of the three major Axis powers was not world domination; indeed, they were fundamentally antiglobalist. They strove to establish—for each country individually—regional resource autarky in central Europe, in the Mediterranean basin, and in the Far East. Germany's drive for *Weltmacht*, for example, meant for many in the twentieth-century German elite becoming a world power, not becoming *the* world power.

Conversely, the Allied powers wished to maintain their existing physical or economic control over natural assets or, in the case of the Americans, to avoid the further subdivision of world resources. This partly explains the surprising partnership of the democratic British Empire and the United States alongside the Stalinist Soviet Union and the authoritarian and anticolonialist nationalist China.[29]

The British and French governments (in 1939) had aimed at least to maintain the status quo. They wanted to keep control, in the face of European and extra-European challenges, over the share of global resources that they had gained in preceding centuries, especially the overseas empires. Unfortunately for them, their metropolitan territories were relatively small geographically and demographically and relatively vulnerable. A prolonged war also threatened them with bankruptcy.

One of the foreign policy objectives of the United States was to gain free access to world resources, the so-called Open Door. This was not why the United States allied itself strongly with Britain (and opposed Germany and Italy) in 1940–41, although it was one reason the American government had come into increasing conflict with an expansionist Japan. This was also not why the United States entered the war; much public and political opinion in America was opposed to direct involvement in the "European war," and the country was thrown into the fighting by the surprise Japanese attack on Pearl Harbor. After December 1941, however, it was evident that access to resources continued to be important to the US government, and that this access was to be achieved at the expense of allies (the British and French empires) as much as of enemies. The quid pro quo of American aid to the United Kingdom under the Lend-Lease Act was access to British imperial resources and markets, and the Roosevelt administration was committed to the self-determination of the European colonies.

The Soviet Union and China were in a somewhat different position. Given their huge size and level of economic development, neither state was especially motivated by a desire to gain additional resources. The Soviet government had since the mid-1920s been self-consciously attempting to control the environment, and specifically to extract resources from remote parts of its territory. Unlike the United States, neither country was actively involved in overseas trade in the 1920s and 1930s. The Chinese nationalists were eager to eliminate the external colonial presence in China. Stalin's government in the Kremlin would undoubtedly pursue a policy of expanding control over

Eastern Europe, but this was primarily driven by political and military considerations rather than by a desire for further resources. Moscow was essentially defending a large resource base. The Russians had a resource and industrial base located in the deep rear. This was the opposite of the Chinese situation, where the coastal resource/industrial base had been overrun by Japanese forces in 1937–39.

The outcome of World War II as a whole, based on the relative populations and resources of the two coalitions, was not so surprising. The Axis operated from two regions situated at opposite ends of the Eurasian land mass. After June 1941, they had almost no physical contact aside from a handful of submarines and blockade runners; there was only one minor attempt by the Italians to open an air link to the Far East. The Allies from the end of 1941 controlled the rest of the world; the Russians dominated the core of Eurasia, and the British and Americans dominated the world's oceans. China possessed too many people and too much land for the Japanese to digest. This imbalance—the combined productive, resource, and demographic advantage of the British Empire, the Soviet Union, the United States, and China—was stressed in public speeches by the leaders of the new United Nations in December 1941, when the global alliance setups were confirmed.[30]

How could the governments of the Axis powers ever have thought they would win such an uneven struggle? One explanation would be that the leaders of Japan in 1937, and Germany in 1939, underestimated the resolve of their local opponents. Mussolini's government in 1940 assumed that the balance of power in Europe had already fundamentally shifted after the defeat of France, and that Italy would not actually have to fight anyone. The Germans in 1941 extrapolated the lessons of the French campaign and underestimated the military abilities and political cohesiveness of their Soviet opponents and the environmental operational challenges involved; they did not expect to fight a long war in Russia.

The Japanese civil and military authorities spent much time in 1940–41 slowly weighing the pros and cons of escalating the China war by attacking two superpowers. They exaggerated racial ("spiritual") and military factors, and exaggerated the closeness of Britain and Soviet Russia to defeat. They also assumed that the non-Asian powers—especially the United States—having lost control of their Asian/Pacific resources and bases, would lack the will to reestablish their influence. Put another way, Japan's leaders underestimated the ability of the United States to overcome environmental challenges and make the long, island-hopping journey across the Pacific to Tokyo Bay.

CONCLUSION

World War II had a powerful impact on the environment, especially striking in remote regions that had not previously been exposed to advanced technology. The conflict was global and involved states in quite different stages of development. The fighting perhaps had more direct environmental effect in China, while in the more economically developed parts of the world the environmental impact came from industrial production and from the indirect results of social change that occurred during the conflict.

The natural conditions and resources, in all their variety, had a great influence on the nature of the war both operationally and in terms of overall national objectives. World War II was fought between two global coalitions. One coalition had access to nearly unlimited natural resources, as well as the capability to move them anywhere, more or less at will. The other coalition did not have that capability. The "have-nots" had attempted to redress this prewar inferiority; because they started from a position of disadvantage, they failed.

Notes

1 For such a survey, see Jacob D. Hamblin, "Environmental Dimensions of World War II," in *A Companion to World War II*, vol. 2, ed. Thomas W. Zeiler and Daniel M. Dubois (Malden, MA: Wiley-Blackwell, 2013), 698–716.

2 J. R. McNeill, "Observations on the Nature and Culture of Environmental History," *History and Theory* 42, no. 4 (December 2003): 6.

3 This is also the emphasis of most of the essays in Richard P. Tucker and Edmund Russell, eds., *Natural Enemy, Natural Ally: Toward an Environmental History of War* (Corvallis: Oregon State University Press, 2004), and in Charles E. Closmann, ed., *War and the Environment: Military Destruction in the Modern Age* (College Station: Texas A&M University Press, 2009).

4 Evan Mawdsley, *Thunder in the East: The Nazi-Soviet War 1941–1945* (London: Hodder Arnold, 2005); idem, *World War II: A New History* (Cambridge: Cambridge University Press, 2009).

5 An extreme version of this may be seen in *Beyond All Boundaries*, a film shown at the National World War II Museum in New Orleans.

6 The Far East (China and Southeast Asia) and the Pacific (Oceania) are treated in my *World War II* as two distinct regions of conflict, which works at least up to 1944.

7 In the last five years, our knowledge of the "China War" has been much widened by the articles in Mark Peattie, Edward Drea, and Hans Van de Venn, eds., *The Battle for China: Essays on the Military History of the Sino-Japanese War of 1937–1945* (Stanford, CA: Stanford University Press, 2011), and by Rana Mitter, *China's War with Japan, 1937–1945: The Struggle for Survival* (London: Allen Lane, 2013). Also valuable in the environmental context are the

essays in Diana Lary and Stephen MacKinnon, eds., *Scars of War: The Impact of Warfare on Modern China* (Vancouver: University of British Columbia Press, 2001).

8 Peattie et al., *Battle for China*, 46.

9 For the present purposes, there is no need to spend too much time defining the exact parameters of World War II. I discuss this in *World War II*, 67.

10 A useful corrective, with an appropriate title, is Christopher Bayly and Tim Harper, *Forgotten Wars: The End of Britain's Asian Empire* (London: Allen Lane, 2007). See also Ashley Jackson, *The British Empire and the Second World War* (London: Hambledon Continuum, 2006); and David Edgerton, *Britain's War Machine: Weapons, Resources, and Experts in the Second World War* (New York: Oxford University Press, 2011).

11 This global dimension is a major theme of Jay Winter, ed., *The Cambridge History of the First World War*, 3 vols. (Cambridge: Cambridge University Press, 2013).

12 The same could be said of the Napoleonic Wars. They took in more of European territory than did World War I, although not all at the same time.

13 J. R. McNeill, *Something New under the Sun: An Environmental History of the Twentieth Century* (New York: W. W. Norton, 2000), 296–97.

14 The Cold War is a different topic, but it went through several phases. The extreme thermonuclear phase was in the late 1950s and early 1960s. Before and after that decade, there was a strong interest in building up conventional forces, especially in central Europe, and for fighting "brushfire wars" in the Third World.

15 For insights, see Joseph Maiolo, *Cry Havoc: The Arms Race and the Second World War, 1931–1941* (London: John Murray, 2011).

16 One interesting exception was the the fate of Finland in 1944.

17 Ronald Spector, "The Sino-Japanese War in the Context of World History," in Peattie, *Battle for China*, 467. Spector seems to find the Peninsular War (1808–14) a better analogy because it involved two different kinds of armies.

18 Dorothee Brantz, "Environments of Death: Trench Warfare on the Western Front," in Closmann, *War and the Environment*, 68–91.

19 McNeill, *Something New under the Sun*, 344–47. This quote was about the twentieth century in general, but all of McNeill's examples relate to World War II or later.

20 Judith Bennett stated that, in addition to failing to address environmental consequences, military historians did not often "portray the environment as an active agent." *Natives and Exotics: World War II and Environment in the Southern Pacific* (Honolulu: University of Hawai'i Press, 2009), xxi.

21 Jon T. Sumida, *Inventing Grand Strategy and Teaching Command: The Classic Works of Alfred Thayer Mahan Reconsidered* (Baltimore: Johns Hopkins University Press, 1997).

22 Brian W. Blouet, *Global Geostrategy: Mackinder and the Defence of the West* (New York: Frank Cass, 2005).

23 See, for example, R. H. Stolfi, "Chance in History: The Russian Winter of 1941–1942," *History* 65, no. 214 (June 1980): 214–28.

24 Omer Bartov, *The Eastern Front, 1941–45: German Troops and the Barbarization of Warfare* (New York: St. Martin's Press, 1985).

25 Closmann, *War and the Environment*, 1. For an important discussion of future prospects, see Thomas F. Homer-Dixon, *Environment, Scarcity, and Violence* (Princeton, NJ: Princeton University Press, 1999).

26 Three recent works that bring this out well are Adam Tooze, *The Wages of Destruction: The Making and Breaking of the Nazi Economy* (London: Allen Lane, 2006); Alex J. Kay, *Exploitation, Resettlement, Mass Murder: Political and Economic Planning for German Occupation Policy in the Soviet Union, 1940–1941* (Oxford: Berghahn Books, 2006); and Lizzie Collingham, *The Taste of War: World War Two and the Battle for Food* (London: Allen Lane, 2011). "Food," Collingham observes, "was at the heart of many of the policies which set Germany, Italy and Japan on the path to war in the 1930s" (16). Interestingly, she argues that,

unlike World War I, the outcome of World War II in Europe "was not determined by food," because the Germans had enough food in the later stages of the war (at the expense of the occupied territories), while the Soviet population had an "appalling food hardship" (464).

27 Michael A. Barnhart, *Japan Prepares for Total War: The Search for Economic Security, 1919–1941* (Ithaca, NY: Cornell University Press, 1987); Louise Young, *Japan's Total Empire: Manchuria and the Culture of Wartime Imperialism* (Berkeley: University of California Press, 1998).

28 David Rodogno, *Fascism's European Empire: Italian Occupation during the Second World War* (Cambridge: Cambridge University Press, 2006).

29 This is an extreme simplification of the nature and dynamics of the conflict, which had complex political and historical roots. Nevertheless, it does correspond to one aspect of World War II.

30 Aside from Italy and some smaller Axis allies dropping out in 1943–44, the final change was the Soviet entry into the war with Japan in August 1945.

Part II

SOCIAL AND ENVIRONMENTAL IMPACTS OF THE WAR

Chapter 4
Environmental Policies of the Third Reich

SIMO LAAKKONEN

Associating the Third Reich with environmental policies, let alone deep-seated ecological ideals, may be difficult. What on earth does a fascist dictatorship have in common with nature conservation and environmental protection? Historian Milan Hauner aroused discussion on the theme by claiming that the National Socialists were the first rulers in the world to be radical environmentalists,[1] an assertion that sparked intense academic debate. According to Anna Bramwell, who has studied the ideological history of ecology, the National Socialist government actively strove to promote nature conservation.[2] Technological historian Peter Staudenmaier, on the other hand, criticizes Bramwell for her uncritical assessment of the National Socialists.[3] Environmental historian Thomas Lekan sees researchers as having concentrated overly on the National Socialist worldview and ignoring military-political consequences.[4] Others like Joachim Radkau and Franz Uekötter have argued that the National Socialist period did not signify increasingly environmentally friendly policies, but rather intensified industrialization and destruction of nature.[5] This discussion between peacetime ideals and the industrialization demanded by warfare offers a backdrop to this essay.

Historians have studied different elements of the environmental policies of National Socialist Germany, which is why they have seen the achievements of the policies in divergent ways. Whereas some researchers have emphasized the idealism of the policies, others have addressed the racist motives of the government and exposed the minimal results achieved. In addition, while the majority of existing research focuses on the history of nature conservation, the history of environmental protection has so far been much less studied. Consequently, the prevailing overall picture of the environmental views of the National Socialists and how they were applied in the Reich's policies is highly fragmented and biased. The main aim of this essay is to provide a synthesis of current discussion, including the main sectors of the environment in the Third Reich.

Environment and environmental policies are used here as an ahistorical umbrella term covering societal discussion of and decision making related to nature conservation (conservation of indigenous nature and landscapes) and environmental protection (in this case, primarily reducing pollution of air, water, and soil). To provide an overview of these issues is a challenging task. The twelve-year period (1933–45) when the National Socialists were in power was an exceptional time, starting from the economic crisis, continuing through the police state, and ending with seven years of societal discussion limited by war. On the other hand, National Socialism was not some unified ideology, nor was the Reich administration a monolithic machine. Each sector of environmental policy has its own dynamics and history, which is why it is difficult to present comprehensive answers to questions related to developments in the field as a whole. It is particularly difficult to demonstrate to what extent National Socialist environmental policies were applied via practical measures, and how these measures affected the state of the environment in or outside Germany, so such questions will not be investigated here.

Understanding and explaining environmental policies of the Third Reich entail adopting a holistic framework. This essay argues that one should take into account at least three different yet deeply connected factors and phases—that is, ideals, tools, and goals—of the Nazi regime. Ideals were significant in the Third Reich because the country was a one-party state, whereby only views that were supported by National Socialists could be realized.[6] The impact of National Socialist ideology concerning nature and race was considerable during the first years, when the Nazis were in parliament and then in power (1933–36). Tools in addition to political dictatorship included autarky, a policy for economic self-sufficiency, which was tied to the rearmament policy as the chronic shortage of foreign currency and several important raw materials set obvious limits to the future capabilities of the German armed forces. Autarkic economic policies were adopted gradually at first and then in full force during the first Four Year Plan (1936–39).[7] From the beginning, the primary political goal of Adolf Hitler and the National Socialist ideology and autarky was imperialist-racist war (1939–45) in order to gain *Lebensraum* for the German people.

This unholy trinity—ideals, tools, and goals—and the related three main phases of the Nazi regime formed a powerful continuum that should be taken into account when exploring the different sectors of environmental policy of the Third Reich from 1933 to 1945. This essay explores this issue in six significant sectors: rural policies, forestry, nature conservation, animal rights, water protection, and waste management. Even a brief exploration of each theme helps to make sense of the environmental policy of the Third Reich.

BLOOD AND SOIL: RURAL POLICIES

The National Socialists adopted a nature-based ideology that emerged from the combination of ultranationalist views, evolutionary theory, and racial theories. According to this thinking, nature gave birth to different nations, assigning them different demands and tasks. Nature meant the right of more powerful races and peoples to expand their territory by force, at the expense of those who were weaker.[8] This conception of nature, based on the right of might, played a central role in National Socialism.

The nature-based ideology of the Third Reich was applied in practice in its various spheres of government. The Third Reich's agricultural policies—or, more accurately, extensive rural policies—were led by the agronomist Richard Walther Darré, who was a zealous advocate of rural life and represented the ultranationalist tradition of mystifying the Aryan countryside.[9] Darré, minister of agriculture, was the first to adopt the slogan that later became famous: *Blut und Boden*, or "blood and soil." With this phrase, Darré meant that Germans had a special tie to the blood (race) and soil (rural and nature) of their homeland. Other nations, especially rootless Jews, did not enjoy a correspondingly powerful bond. Darré said, "The notions of soil and blood award us the moral right to restore to our care as much land from the east as is necessary to achieve harmony between our nation and the geopolitical situation." The Third Reich was to restore the unity of blood by cleansing the racial map of Europe.[10]

Darré was one of the few ideologues of the Nazi party. He combined the Nazis' various, even contradictory, anticivilization, antiurbanization, anticapitalist, and antiliberalist views into a potent, mystical notion of the rural as a revivifying force. As a result, a culture of rural life in harmony with nature became one of the ideological pillars of National Socialism. The German people represented the organic culture rising from the soil, not the soulless and materialistic civilizations of cities of the rest of Europe.[11]

Figure 4.1. Argentinian-born Walther Darré rose to the position of Minister of Agriculture in the National Socialist regime. An opponent of industrialization, he used the slogan *"Blut und Boden"* to champion racially pure agricultural communities that relied on natural methods to till the soil. Source: Bundesarchiv, Bild 183-H1215-0503-009, Zentralbild, 1937.

In addition to ideological influence, Darré's status was exceptionally strong in the labyrinthine power structures of the National Socialists. As a visionary and expert on rural life, he enjoyed strong support in rural areas as far back as the early 1930s; indeed, the National Socialists rose to power in large part thanks to votes Darré garnered from the countryside. From 1933 to 1942, he held the status of minister of agriculture and Reich peasant leader. The Ministry of Agriculture had the fourth-largest budget in the Reich.[12] From these positions of power, Darré, whose views were also supported by Reich Chancellor Rudolf Hess, could carry out his visionary rural policies in Germany.

Darré's most significant achievement was the adoption of natural agricultural methods in Germany. The goal was to till the land in accordance with the laws of nature. Darré took the model for these reforms from Rudolf Steiner's anthroposophy and the biodynamic farming methods it espoused.[13] But Darré's societal goals were even more radical than the naturalization of German agriculture. Darré's long-term strategy was to get rid of industrial society altogether and replace it with peasant communities. According to him, the capitalism and industrialism that had shown their true colors in the turmoil of the Depression would rapidly disappear, along with the urbanization and mechanization of society. Many of those hurt by the Depression shared these views, but Darré aimed at promoting this transformation: the country's sick industrial areas should be stifled by isolating them from the vital countryside, and a new "aristocracy of the German soil" would arise to lead the country. Once capitalism fell, Germany and all of Europe could develop anew their characteristic, indigenous, and ideologically healthy agricultural society: racially diseased industrial-urban Europe had to be ruralized and thereby purified.[14]

The practices of the Third Reich proved otherwise. The growing production of the war industry continually needed more muscles to work its machines. In the 1930s, over a million Germans moved from the country to work in the city. During the war years, 7.6 million foreign, forced, and slave laborers sustained production. In agriculture, half of the workers were foreigners.[15] Darré's visionary ideas about the creation of a new, rural National Socialist society were not warmly received at the highest levels of the Third Reich. Walther Darré was forced to resign in 1942. The man named as the new minister of agriculture, Herbert Backe, was an opponent of Darré who understood the immediate needs of the military economy.[16]

FORESTRY

During the interwar period, the German forest was perceived through a powerfully vital myth as having ascended from the primeval forests that once covered Europe and their free tribes, which not even the Roman Empire could vanquish. Such cherished myths did not have anything to do with German forests or forest policies, however, because contemporary capitalist-industrial forestry policy was developed in the late 1800s in Germany.[17] The goal of the new German school of thought was to achieve the maximum annual profit possible from a forest by turning the country's rather diverse forest types into so-called normalized forests. In practice, this meant modifying the existing mixed forests through clear-cuts and forest cultivation into production forests containing trees of a uniform species and age.[18]

National Socialist forest policies sharply diverged from these mainstream German forest policies, despite the fact that both capitalist and National Socialist reformers of forest policy shared the same goal: to maximize the production of the country's forests for economic purposes. But the practical methods each school adopted to increase production differed significantly. One of the most important goals of National Socialist forest policy was to move away from the "liberal-capitalist" forest economy that emerged in the late 1800s toward an "idealist-nationalist" forest economy.[19]

In practice, the National Socialist forestry policy favored mixed-growth over single-species forests. This policy would permit a primeval forest landscape to replace the monotonous capitalist "tree plantations," called production forests by the National Socialists. Mixed forests were assumed to effectively utilize the resources of variable locations and soils. Variable age structure of the growing timber stock was also considered good for a forest's growth. The new forest program opposed clear-cutting and lowering the age of trees at harvest, in other words, the felling of young trees that would hinder the production of high-quality timber. And instead of artificial reforestation, the National Socialist forest program recommended the natural regeneration of forests because it saved on tree cultivation costs and led to the appearance of naturally healthy mixed forests with diverse age structures.[20] These views were grounded in the nature ideal of the National Socialists, in which foresters should give up the cold, calculating liberal-capitalist production-forest economy (*Forstwirtschaft*) and move toward an idealist-nationalist forest-based economy (*Waldwirtschaft*) that considered the forest as a living organism.[21]

Figure 4.2. Adolf Hitler focused on leading Germany's racist-imperialist war policy while Hermann Göring also addressed forest and animal protection issues before the beginning of the new world war. Source: Bundesarchiv, Bild 141-1917.

This nature-based forestry (*Dauerwaldlehre*) became the official management program for the nation's forests in 1934.[22]

As the military economy demanded ever-increasing amounts of wood, the new forestry policy ran into problems. There was also local-level skepticism regarding the new national policy, so it was carried out in a neglectful fashion, which led to its failure.[23] In 1937, Hermann Göring was forced to name a subservient member of the Schutzstaffel (SS) as the new leader of the forestry authority and repeal the earlier regulations. The new official goal of the forestry policy was a "natural production forest," accurately reflecting the compromise made between conservation and economic interests. In 1933, the National Socialists had begun to renew national forestry legislation, but this work could not be brought to completion, initially owing to internal conflicts and later because of the world war.[24] After the war, the ideology of soft forestry continued only in Switzerland, which was spared from war, from where it returned to Germany in the 1980s.[25]

NATURE AND LANDSCAPE CONSERVATION

The German nature conservation movement was made up of extremely disparate groups and viewpoints. Generally speaking, however, nature conservationists viewed the National Socialists' rise to power with cautious optimism. This was because nature conservationists of various schools, the National Socialists, and the traditional political right all shared an admiration for nature. The hopes of conservationists were particularly roused by the fact that the National Socialists had proposed setting the common good above the individual, and they looked unkindly on the privileges of the church, the nobility, and the bourgeoisie. Although the Weimar Republic that preceded the Third Reich had in principle been amenable to nature conservation, one conservation initiative after another had run into hurdles, including an overemphasis on the protection of private property.[26]

The number of members in the four largest German nature organizations grew from 40,000 in 1925 to almost 50,000 by 1935, 70,000 by 1939, about 90,000 by 1940, and continued to grow during wartime. In part, these figures can be viewed as illustrating the growth of significance of nature and nature conservation under the new regime. Once the war ended, the combined number of members of these nature organizations collapsed by half, not reaching wartime levels again until the mid-1970s.[27]

The main achievement of the National Socialists in the field of nature conservation was the establishment of a national Nature Conservation Act in 1935. This act was a significant reform in Germany, as it was the country's first federal, rather than state-specific, nature conservation law. Targets of protection under the law included plants, nongame animals, natural monuments, and those elements of the landscape whose conservation was consistent with the common good. The law went so far as to require that any officials with concerns about projects that could transform the rural environment be heard out before work could begin. Such notification obligation was progressive in its comprehensiveness, but this specific clause was least followed in practice.[28]

In the period 1934–35, the National Socialists established several statutes on the protection of wild animals. In their eyes, real hunters not only hunted, but also had a duty to nurture and protect wild animals so that older and stronger animals could reproduce and be conserved. Göring in particular was concerned about the endangered species of greater Germany, such as the brown bear and the European bison, and attempted to create for them breeding programs as well as conservation laws and payments. Of those species that were extinct in Germany, the wolf was protected in the Third Reich, and there was a desire to bring back moose to the region. The hunting laws enacted during Göring's time are for the most part still in force.[29]

Relying on the new nature conservation law, Göring established a total of four so-called Reich nature reserves. Three of the parks were located in Germany; the fourth was established in what is currently Lithuania, near the harbor town of Memel (now Klaipeda), which Hitler annexed to Germany by threat of force in March 1939.[30] As a result of the reforms, by 1940 there were about eight hundred different nature reserves, whose total land area exceeded five thousand square kilometers.[31] This jump was also to a large extent a result of administrative codification and the registering of already-existing reserves, the first of which were established already in the early nineteenth century.[32]

The most radical goal of the National Socialist nature conservationists—the conservation of nature in all spheres of the Third Reich's activities—was a particular demand of Fritz Todt. Todt was one of the most influential National Socialists, the minister who answered for the Reich's technological and industrial policies. Todt was an engineer and an ardent lover of nature who strove to harmoniously combine technology and nature wherever possible. Todt became known for aiming to take "the laws of nature" into consideration during the construction of the *Autobahn* system, the enormous

Figure 4.3. Reichminister Dr. Fritz Todt demanded that environmental conservation be taken into consideration in all Reich activities. After Todt's accidental death, the more flexible Albert Speer was named as his successor. Source: Bundesarchiv, Bild 146-1969-146-01, Röhn, 1940.

motorway construction project that was his responsibility. Todt impressed upon the engineers that the National Socialist network of motorways should be designed "according to nature's terms, not in contradiction to them." By this, he meant that motorways should be designed using innovative curving shapes suited to the landscape, local materials should be used in their construction, and plants should be planted in between lanes and along the shoulders. Billboards were not allowed. In Fritz Todt's opinion, the National Socialist motorway engineers represented the middle way, a specifically German school of technology that would simultaneously save technology from the soulless materialism of capitalism and Marxism as well as repudiate the unfounded opposition between nature and culture.[33]

In the mid-1930s, Todt and his assistant in the motorway construction program, Reich Landscape Commissioner Alwin Seifert, enthusiastically championed a comprehensive "Reich-level law for protecting Mother Earth," so that "the decline of the irreplaceable foundation of all life might be prevented."[34] The Mother Earth law was never realized, but at least the Reich Nature Conservation Act established in 1935 included progressive viewpoints.

When Todt died in an accident, Albert Speer, who was considered more flexible than Todt in the eyes of the powers, was named as his successor.[35] But a lack of funds and workforce proved to be a problem for the administrators responsible for carrying out the law. The bolstering of military institutions took precedence over enforcement of the act, as did the construction of roads and waterways, job creation, soil conditioning, and exploitation of natural resources.[36] By 1939, almost four thousand kilometers of cement-paved motorway had been built in Germany, and the number of automobiles had grown exponentially. Motorways and other such major projects of the Third Reich, no matter how much attention had been paid to nature conservation during their execution, inevitably caused massive changes in and destruction of the environment.[37]

ANIMAL RIGHTS POLICIES

In Europe, the animal rights movement emerged in the early nineteenth century, drawing attention to the brutal treatment of domestic and production animals. Animal rights were also an important issue for the National Socialists. In the Prussian Parliament in 1927, the National Socialist deputy demanded action against animal cruelty and the kosher slaughter practiced by Jews. These demands contained nothing new per se; in the early 1900s, animal rights activists in other countries also harshly criticized kosher slaughter, in which the jugular vein is severed and the animal allowed to bleed to death.[38] The unified appeal for animal rights was justified in different ways, however. National Socialists depicted Jews as a cold, calculating race that had a tendency to animal abuse, in contrast to the northern race that nurtured an organic relationship between humans and animals.[39]

In 1932, the National Socialist German Workers' Party (NSDAP) once more proposed prohibition of animal testing at the behest of SS leader Heinrich Himmler, who was like many other leading National Socialists an ardent animal lover.[40] The National Socialists approved the law in April 1933, immediately upon coming to power. In August 1933, Hermann Göring delivered a later-famous radio speech in which he focused on the proposed prohibition of vivisection that he had presented a couple of weeks earlier. In this context, Göring was presumably referring to experimental surgery on living animals. He harshly criticized the torture of animals through pointless animal testing, presenting the National Socialist relationship to nature as an alternative. The Reich marshal also used history and the unique affinity between the

German race and animals to justify his prohibition of animal testing: animals and Aryans had shared home and field for millennia and together defended the land against conquerors. Göring stressed that a hard fate awaited violators of animal rights laws. Lawbreakers who treated animals like lifeless chattel would be arrested or locked away in concentration camps.[41]

The Third Reich strove to advance animal rights in many ways. During the period 1933–43, decrees on a number of activities, from shoeing horses to handling fish to using anesthesia in veterinary work, were issued. One decree focused on the most humane, painless way of killing crayfish and lobsters; it recommended dropping them and other shellfish one by one into boiling water. The decrees enacted by the National Socialists were often preceded by a lively debate, taken part in by researchers and animal rights activists. Animal rights became a subject at schools and universities, and animal psychology was introduced as a subject of study at the latter.[42]

But there was a significant ethical gap between National Socialist animal rights activists and those representing other political views. The National Socialists justified their views through the inequality between different groups of humans and animals. In the worldview of the National Socialists, racially pure Aryans stood at the top of the natural hierarchy and then came animals, of which the most respected were powerful predators that subjugated others through their strength. These were followed by other animals and finally subhumans, including mentally ill or crippled Germans, Gypsies, Jews, Slavs, and in general people whom Nazis did not like. The Nazis drew the judicial line between animals and subhumans, which is why the strict animal rights laws did not apply to those beings classified as subhumans.[43]

Animal rights were doubtless an important issue for National Socialists, for personal, ideological, and worldview reasons. It was necessary to temper strict animal rights laws in practice, however, because doctors were not eager to replace test animals with human guinea pigs. In addition, the German Army alone had almost three million warhorses and mules[44] that were sacrificed without a second thought to the demands of the military economy and war efforts. Naturally, among those who worked with animals, there was a great sense of loss.

AIR AND WATER PROTECTION

Next the National Socialist era is examined from the perspective of industrial-urban environmental issues. The Third Reich did not bring notable differences to Germany's anti–air pollution policies; progress in one sphere of the polycentric administration was shadowed by lack of it in another.[45] The greatest changes took place in water protection policies.

At the beginning of the 1930s, the Great Depression once more put near-total halt to the construction of treatment facilities and the development of treatment methods. The rise of the National Socialists to power did not really affect the technical-scientific development of treatment methods, but it changed the aims of wastewater policies.[46] In 1935, absolute priority in wastewater treatment was given to their economic exploitation, as the goal of the military economy was autarky. The Four Year Plan of 1936 outlined four main goals for the economic utilization of wastewater: the recovery of raw materials from industrial wastewater, the exploitation of the gas (methane) resulting from decomposition in wastewater, the separation of oils and fats from wastewater and sludge, and the utilization of wastewater in agriculture.[47]

Of these four goals, the utilization of wastewater in agriculture had priority. There was an attempt to boost food production by using the nutrients contained in wastewater to improve Germany's self-reliance in foodstuffs. The state strove to promote the use of wastewater from communities, cities, and to some extent also industry in agriculture for the irrigation and fertilization of crops. Porous underground hoses, open channels, and movable sprinkler systems were developed to conduct wastewater to crops. The activity was organized by obligating those who emitted wastewater to join with farmers to form wastewater-use associations that were then supervised by state authorities; the costs of these associations were at least in principle determined according in relation to the benefits.[48] The state supported the utilization of wastewater by covering as much as half of the costs of the projects.[49] Thus the use of wastewater in agriculture served not only the economy but also the ideological ideals of National Socialism of vital countryside.

Industry and cities strove to make economic use of the wastewater by separating out the sludge, fats, oil, petroleum, and gas it contained for reuse. Laboratory tests were used to attempt to recover and utilize acetone, organic acids, carbon, alcohol, and other constituents from wastewater. In practice, however, utilization of gas and fat became the only large-scale activities.[50] New treatment methods were developed, primarily for neutralizing the acid wastewater produced by industry, as its amount had greatly increased with the military economy.[51]

The National Socialist aim to economically utilize wastewater can be considered good in principle, but the directing of resources to the military economy detracted from this utilization. When it ran up against the lack of material resulting from the military economy, water protection activity stopped and even regressed because numerous industrial and municipal wastewater treatment plants were destroyed or damaged by war.[52] The work undertaken in 1932 to standardize the criteria used in the wastewater industry and reform water laws was also left unfinished. A proposal for a new, unified Reich water law was prepared in 1941, but the world war prevented its enactment.[53]

URBAN-INDUSTRIAL WASTE MANAGEMENT

States were already intensifying their efforts at waste utilization and collection during the First World War.[54] But in the Third Reich, the NSDAP took actions, almost immediately after its rise to power in 1933, that aimed to improve waste management and set it under strict party and state control. Already in 1934 the Nazi regime had launched a state-run waste material program (*Altstoffprogram*) to ameliorate the utilization of waste.[55] In addition to the development of waste collection and utilization, it involved efforts to prevent generation of waste, save raw materials, and develop synthetic substitutes for natural resources.[56] The goal of the new waste management program was to realize autarky, to achieve "raw material freedom" for the German people.[57] The German prewar economy was short of many strategic minerals and raw materials that were important for the military. For example, 95 percent of the German textile industry depended on imports.[58]

In the Weimar Republic, waste management was mainly operated by several small-scale businesses. In 1933, there were over 6,500 registered companies that collected waste products, such as scrap metal, paper, and glass. These firms were as a rule small in size, often run by an individual scrap dealer or a married couple. Also there were in the 1930s about 30,000 ragmen who earned their living by collecting used textiles and reselling them. At first, the practitioners of the waste material trade took a positive attitude toward Nazi efforts to reorganize the waste management system. They regarded the *Altstoffprogram* launched by the National Socialists as an opportunity both to increase sales and improve the reputation of the branch. But they were soon to realize that the Nazis aimed to exclude many of them.[59]

The introduction of the Four Year Plan in 1936 signified organizational restructuring of waste management on the national, regional, and industry

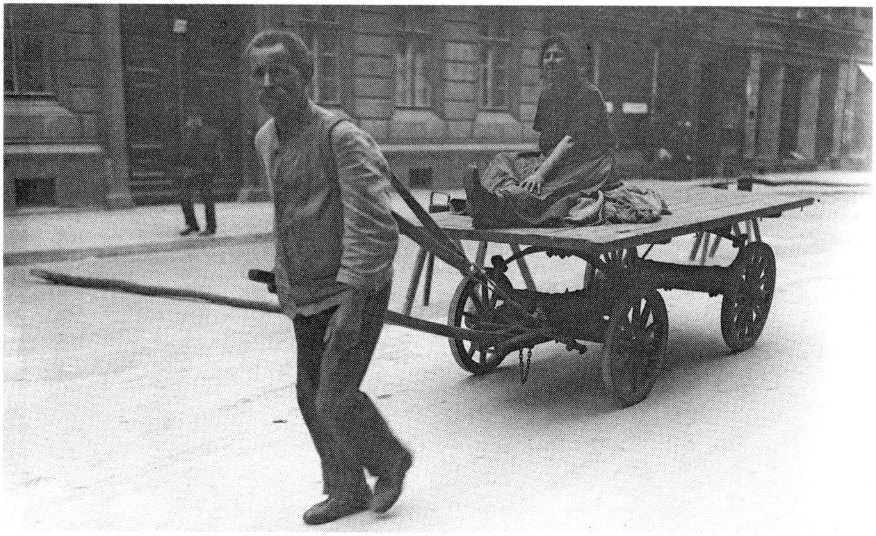

Figure 4.4. Rag collectors returning from work. Before the war, waste management in Berlin was also taken care of by individual entrepreneurs, many of whom were Jews, Gypsies, and other minorities. After these entrepreneurs were liquidated, Nazi waste policy wound up in a state of crisis in Germany. Source: Bundesarchiv, Bild 102-08773, Georg Pahl, 1929.

levels. The goal was to improve the efficiency of waste management and build up the branch on more acceptable racial grounds. The main concern was that many people taking care of waste management were "subhumans" and, above all, Jews. Measures to oust the "improper elements" and Aryanize the branch began in 1936, and by the end of 1938 this purge was completed.[60] But the forced exclusion of Jews, Gypsies, and others became a self-inflicted problem for the regime. Already in spring 1939, Willy Ziegler, the first Reich commissary for scrap recycling, complained that a labor shortage reduced the efficiency of waste management in the whole Reich.[61] The Nazi officials tried to remedy the problem by mechanization of the branch and by increased use of forced labor. Also, volunteers from Nationalist Socialists organizations such as Hitler-Jugend or Deutsche Arbeitsfront launched waste collection campaigns.[62]

As the demand and competition for labor and raw materials aggravated during the war, so did the already-strained relations between different branches of Nazi administration. The needs of waste management were always seen as secondary to the military and armament industry. In the continuously worsening situation, the new Reichskommissar Hans Heck warned in 1941 that the utilization of waste products would soon become impossible. There were some indications of satisfactory results in waste management, although the publicly proclaimed goal of *restlose Erfassung* (total collection) was illusory at best.[63] In fact, raw material problems persisted during the whole duration of the National Socialist rule—but this was the situation in most belligerent countries as well. In the Third Reich, however, waste management suffered from a particular split. On the one hand, authorities sought to improve waste utilization procedures and recycling, but on the other hand, these efforts were blurred by organizational disorder, ideological extremism, and genocide.

CONCLUSION

In some aspects of environmental policy in the Europe of its time, National Socialist Germany was progressive. Yet National Socialists' thoughts on environmental policies were seldom of their own invention. Also, the motives and outcomes varied markedly from one field to another owing to the contradictory impact of their ideals and economic and political goals. Nazi ideals of nature were important in forming rural policies and issues related to forestry, nature protection, and animal rights. The racial aspects were mostly marked in rural policies and animal rights issues. Nature-based ideals, however, rarely addressed air pollution, water pollution, or waste management because urban-industrial areas had no particular status in the National Socialist ideology. The striving autarky had instead a deep impact on waste management and a considerable impact on water protection, but almost none on air pollution policy. Racial ideology had no impact on air and water protection owing to their highly mechanized nature, while it effectively crippled waste management. To sum up, Nazi ideals affected above all nature conservation, while policies of economic autarky had a deeper impact on environmental protection.

These examples indicate clearly that the National Socialists' environmental policies and the administration as a whole were marked by an ideological-political paradox. This paradox meant that National Socialist Germany looked simultaneously to the past and the future. It rejected rationalism of the Enlightenment while fully embracing modern technology. Consequently, the National Socialist worldview combined the magician of the mythic world, *homo magus*, and the technician of the modern world, *homo faber*.[64] Environmental policy was realized as a part of this paradoxical worldview, and therefore also National Socialist environmental policies turned into an ideological-political paradox. The mythical-ideological world of the National Socialists saw Germanic nature and Aryan society as having lived in organic interaction. This ideal lifted environmental politics to the peak of the totalitarian state, but even the first Four Year Plan for the military economy stifled conservation initiatives, and the ideals finally foundered in the world conflagration. The rise and fall of German environmental policies are thus both explained by the inherent paradox of the National Socialist worldview in which nature and society acted in accordance to the same "laws of nature," as part of a world based on the right of might. National Socialist environmental policy arose out of the war-seeking ideal of blood and soil, and for the same reason, it sank in blood and soil.

The development of current environmental policies is difficult to understand without taking into consideration the period of the Third Reich. Because of the defeat, Germany's powerful influence in European and international environmental research, technology, and policy diminished after World War II. The impact of this decline was multiplied by the almost complete substitution of the German language by English in the postwar world. Generally speaking, destruction and division brought about by the world war and the consequent cold war weakened considerably the role of European institutions and languages in global environmental policy making. This opened up doors to the arrival of new winds in environmental issues from the victors, particularly the United States. It is hardly a coincidence that in the postwar period most influential environmental figures and publications came first from the United States and that the main language of international cooperation was English. The global directional shift in the international history of environmental policy and politics was thus also closely bound to the outcome of World War II.

This postwar directional shift in environmental policy had a deep impact on Europe, at least until the late 1960s or early 1970s, when European voices again spoke up and were subsequently heard around the world. The gradual recovery of Germany, the largest country in the region, and its environmental administration and policies, this time in the spirit of democracy, restrengthened Europe's environmental influence considerably. Especially in the European Union, Germany's status as a leader that recognizes the significance of protecting the environment has been a prerequisite for advances in environmental policy on the continent as a whole. Since World War II, however, Europe has struggled to become a true pioneer and leader in global environmental policy. The political shadow cast by the Third Reich has been both wide and long.

Notes

The author gives warm thanks to Professor Thomas Lekan, who commented on the draft of this manuscript. The author also thanks researcher Paula Schönach, Heikki Länsisalo, and Pekka Niemelä for advice, and MSc Lasse Salonen for collecting material for expanding this essay, which is based on Simo Laakkonen and Hendrik Seeger, "Kolmannen valtakunnan ympäristöpolitiikka," in *Sodan ekologia: Nykyaikaisen sodankäynnin ympäristöhistoriaa*, ed. Simo Laakkonen and Timo Vuorisalo (Helsinki: Suomalaisen Kirjallisuuden Seura, 2007), 433–73.

1 Milan L. Hauner, "A German Racial Revolution?," *Journal of Contemporary History* 19, no. 3 (July 1984): 685; see also Gert Gröning and Joachim Wolschke, "Naturschutz und Ökologie im Nationalsozialismus" [Nature protection and ecology in National Socialism], *Die Alte Stadt: Vierteljahreszeitschrift für Stadtgeschichte, Stadtsoziologie, Denkmalpflege und Stadtentwicklung* 10, no. 1 (1983): 1–17.

2 Anna Bramwell, *Ecology in the 20th Century: A History* (New Haven, CT: Yale University Press, 1989).

3 Peter Staudenmaier, "Fascist Ecology: The 'Green Wing' of the Nazi Party and Its Historical Antecedents," accessed August 15, 2015, http://www.spunk.org/library/places/germany/sp001630/peter.html. This online article is the first section of Peter Staudenmaier and Janet Biehl, *Ecofascism: Lessons from the German Experience* (Edinburgh: AK Press, 1996).

4 Thomas Lekan, *Imagining the Nation in Nature: Landscape Preservation and German Identity, 1885–1945* (Cambridge, MA: Harvard University Press, 2004).

5 Joachim Radkau and Franz Uekötter, eds., *Naturschutz und Nationalsozialismus: Geschichte des Natur- und Umweltschutzes* (Frankfurt: Campus Verlag, 2003); Franz-Josef Brüggemeier, Mark Cioc, and Thomas Zeller, eds., *How Green Were the Nazis? Nature, Environment, and Nation in the Third Reich* (Athens: Ohio University Press, 2005); Franz Uekötter, *The Green and the Brown: A History of Conservation in Nazi Germany* (Cambridge: Cambridge University Press, 2006); for the periods preceding and following National Socialism, see Thomas Lekan and Thomas Zeller, eds., *Germany's Nature, Cultural Landscapes and Environmental History* (New Brunswick, NJ: Rutgers University Press, 2005).

6 Volker Ludwig, "Die Entstehung des Naturschutzgebietes Hohenstoffel," *Hegau* 42 (1997/98): 153–90.

7 Friedrich Huchting, "Abfallwirtschaft im Dritten Reich," *Technikgeschichte* 48, no. 3 (1981): 252–54.

8 George L. Mosse, *The Crisis of German Ideology: Intellectual Origins of the Third Reich* (London: Weidenfeld and Nicolson, 1966), 1–5; Daniel Gasman, *The Scientific Origins of National Socialism: Social Darwinism in Ernst Haeckel and the German Monist League* (London: Macdonald, 1971), 30–34; Adolf Hitler, *Mein Kampf* (Munich: Franz-Eher-*Verlag*, 1925), 287.

9 For more on Darré's personal history, see Anna Bramwell, *Blood and Soil: Walther Darré and Hitler's "Green Party"* (Bourne End: Kensall Press, 1985).

10 Staudenmaier, "Fascist Ecology"; Bramwell, *Ecology in the 20th Century*, 191.

11 Radkau and Uekötter, *Naturschutz und Nationalsozialismus*, 43.

12 Bramwell, *Ecology in the 20th Century*, 203.

13 Ibid., 197–206; Peter Staudenmaier, "Organic Farming in Nazi Germany: The Politics of Biodynamic Agriculture, 1933–1945," *Environmental History* 18, no. 2 (2013): 383–411.

14 Anna Bramwell, "Was This Man the Father of the Greens?," *History Today* 34, no. 9 (September 1984): 8; See also David Schoenbaum, *Hitler's Social Revolution: Class and Status in Nazi Germany, 1933–1939* (New York: Doubleday, 1966).

15 Norbert Frei, *National Socialist Rule in Germany: The Führer State 1933–1945* (Cornwall: Blackwell, 1995), 139.

16 Heinz Haushofer, *Ideengeschichte der Agrarwirtschaft und Agrarpolitik im deutschen Sprachgebiet, Band II* (Munich: Bayerischer Landwirtschaftsverlag, 1958), 271. Cited in Staudenmaier and Biehl, *Ecofascism*.

17 Jussi Raumolin, *The Problem of Forest-Based Development as Illustrated by the Development Discussion, 1850–1918*, Research Reports 4 (Helsinki: University of Helsinki Department of Social Policy, 1990), 49.

18 Ibid., 47.

19 Pekka Niemelä and Leena Laiho, "Kansallissosialismi—Euroopan ensimmäinen vihreä liike?" [National Socialism—Europe's original Green movement?], in *Aiivan, mutta oletko ajatellut, että . . . Ystäväkirja Antero Jyrängille hänen täyttäessään 70 vuotta*, ed. Kari Immonen and Pekka Niemelä (Joensuu: Joensuun yliopistopaino, 2003), 117–18.

20 Ibid, 118–23. The original source is "Deutsche Waldwirtschaft im Dritten Reich," *Der Deutsche Waldwirtschaft* 15, no. 89 (1933).

21 Niemelä and Laiho, "Kansallissosialismi," 123–25.

22 Andreas Gautschi, *Der Reichsjägermeister: Fakten und Legende um Hermann Göring* (Suderburg: Nimrod-Verlag, 1999), 63. See Michael Imort, "'Eternal Forest—Eternal Volk': The Rhetoric and Reality of National Socialist Forest Policy," in Brüggemeier et al., *How Green Were the Nazis?*, 43.

23 Gautschi, *Der Reichsjägermeister*, 64.

24 Niemelä and Laiho, "Kansallissosialismi," 117, 123–24; Joachim Radkau, *Natur und Macht* (Munich: Verlag C. H. Beck, 2000), 296; Edeltraud Klueting, "Die gesetzlichen Regelungen der national-sozialistischen Reichsregierung für den Tierschutz, den Naturschutz und den Umweltschutz," in Radkau and Uekötter, *Naturschutz und Nationalsozialismus*, 92; Gautschi, *Der Reichsjägermeister*, 65.

25 Anto Leikola, "Bärenthoren—Kestometsätalouden ihannemetsä" [Bärenthoren—the ideal forest of the sustainable forest economy], *Metsänhoitaja* 1 (1993): 69–73.

26 Raymond H. Dominick III, *The Environmental Movement in Germany: Prophets and Pioneers, 1871–1971* (Bloomington: Indiana University Press, 1992), 101, 114. On the *Heimatschutz* movement, see Gautschi, *Der Reichsjägermeister*, 62.

27 Dominick, *Environmental Movement in Germany*, 84, 122.

28 Ibid., 107–8; Radkau, *Natur und Macht*, 295; Karl Ditt, "Die Anfänge der Naturschutzgesetzgebung in Deutschland und England 1935–49," in Radkau and Uekötter, *Naturschutz und Nationalsozialismus*, 114–15; Lekan, *Imagining the Nation in Nature*, 168–73.

29 Arthur Arluke and Boria Sax, "The Nazi Treatment of Animals and People," in *Reinventing Biology: Respect for Life and the Creation of Knowledge*, ed. Lynda Birke and Ruth Hubbard (Bloomington: Indiana University Press, 1995), 230; Uekötter, *The Green and the Brown*, 99–108.

30 Klueting, "Die gesetzlichen Regelungen," 95–96; Franz Uekötter, "Natur- und Landschaftsschutz im Dritten Reich: Ein Literaturbericht," in Radkau and Uekötter, *Naturschutz und Nationalsozialismus*, 470–71.

31 Klueting, "Die gesetzlichen Regelungen," 100.

32 Michael Wettengel, "Staat und Naturschutz 1906–1945: Zur Geschichte der Staatlichen Stelle für Naturdenkmalpflege in Preußen und der Reichsstelle für Naturschutz," *Historische Zeitschrift* 257 (1993): 463. The lack of clarity regarding the number of nature reserves indicates that the history of the German reserves continues to be poorly known.

33 Lekan, *Imagining the Nation in Nature*, 212–14, 227–29.

34 Staudenmaier and Biehl, *Ecofascism*.

35 Eduar Schönleben, *Fritz Todt: Der Mensch, der Ingenieur, der Nationalsozialist* (Oldenburg: Verlag Gerhard Stalling, 1943), 13, 71–72.

36 Klueting, "Die gesetzlichen Regelungen," 99.

37 Lekan, *Imagining the Nation in Nature*, 234–51; William H. Rollins, "Whose Landscape? Technology, Fascism, and Environmentalism on the National Socialist Autobahn," *Annals of the Association of American Geographers* 85, no. 3 (September 1995): 494–520.

38 Arluke and Sax, "Nazi Treatment of Animals and People," 229.

39 Boria Sax, *Animals in the Third Reich: Pets, Scapegoats, and the Holocaust* (New York: Continuum, 2000), 106. For instance, the famous propaganda film *The Eternal Jew* comes to a climax in an extremely brutal scene of kosher slaughter. "The Eternal Jew (*Der ewige Jude*)," YouTube video, 1:05:19, a 1940 Nazi propaganda film directed by Fritz Hippler at the behest of Joseph Goebbels and the Reich Ministry of Public Enlightenment and Propaganda, posted by Elliot Jones, January 15, 2013, https://www.youtube.com/watch?v=CIQp31Oyn70.

40 Bramwell, *Ecology in the 20th Century*, 204; Arluke and Sax, "Nazi Treatment of Animals and People," 240–41.

41 Arluke and Sax, "Nazi Treatment of Animals and People," 229, 244–45.

42 Ibid., 230, 232. Examples of the discussion include the 1933 pamphlet on the animal and plant conservation act, in which the statute was commented on by leading experts. *Tier- und Pflanzenschutzverordnung vom 10. März 1933 mit Einführungserlass vom 15. Mai 1933* (Berlin: Walter de Gruyter, 1933).

43 Arluke and Sax, "Nazi Treatment of Animals and People," 234–39.

44 Richard L. DiNardo, *Mechanized Juggernaut or Military Anachronism: Horses and the German Army of WWII* (Mechanicsburg, PA: Stackpole, 2008), 10–13.

45 Franz Uekötter, *Von der Rauchplage zur ökologischen Revolution* (Essen: Klartext Verlag, 2003), 206, 215–16, 282; idem, "Polycentrism in Full Swing: Air Pollution Control in Nazi Germany," in *How Green Were the Nazis?*, 102.

46 Hendrik Seeger, "The History of German Waste Water Treatment," in "The History of Urban Water Management in the Baltic Sea Region," vol. 2, ed. Simo Laakkonen and Sari Laurila, special issue, *European Water Management* 2, no. 5 (October 1999): 51–56.

47 Adolf Heilmann, "Städtereinigung und Vierjahresplan," *Gesundheits-Ingenieur* (henceforth *Ges-Ing*) 60 (1937).

48 Angerer, "Neues esfolgreiches Verfahren der wirtschaftlichen Abwasserbeseitigung und -verwertung durch Abwasserverwertung," *Ges-Ing* 54 (1931); M. Grevemayer, "Landtwirtschaftliche Abwasserverwertung durch unterirdische Bewässerung," *Ges-Ing* 60 (1937): 367; Radkau, *Natur und Macht*, 297.

49 Grevemayer, "Lanswirtscahfliche Abwasserververtung," 367; Karl Imhoff, "Die Verwertung der Abwasserstoffe in der Landtwirtschaft," *Ges-Ing* 68 (1947).

50 Heilmann, "Städtereinigung und Vierjahresplan," and "Über die biologischen Grenzen der landswirtschaftlichen Verwertung städtischer Abwässer," *Ges-Ing* 64 (1941). For Finnish and Swedish views, see "Biprodukterna vid spillvattenrening," *Hufvudstadsbladet*, December 15, 1938; "Nyttiga produkter vid vattenrening," *Dagens Nyheter*, December 18, 1938.

51 Herbert Rohde, "Eine einfache Neutralisationsanlage für säurehaltiges Abwässer," *Ges-Ing* 60 (1937); F.-M. Schlehte, "Über die Reinigung säurehaltiger Abwässer," *Ges-Ing* 60 (1937); Heinrich Keppner, "Entsäuerungsanlagen für Abwasser," *Ges-Ing* 62 (1939); Karl Imhoff, "Behelfsbauten und Dauerbauten zur Reinigung von Abwasser," *Ges-Ing* 65 (1942).

52 Walther Kunze, "Einfache Bauweisen für Kläranlagen auch nach dem Kriege," *Ges-Ing* 65 (1942). On attitudes, see "Generalinspektor für Wasser und Energie, Zurückstellung des Baues von Kläranlagen," *Ges-Ing* 65 (1942); M. Prüss, "Entwicklung, derzeitiger Stand und zukünftige Aufgaben der Abwasserreinigung in Deutschland," *Ges-Ing* 69 (1948), 261. See Seeger, "History of German Waste Water Treatment," 51–56.

53 Klueting, "Die gesetzlichen Regelungen," 103.

54 Markus Ossberger, "Geschichte der Abfallwirtschaft in Österreich," (master's thesis, Technische Universität, Vienna, 1997), http://publik.tuwien.ac.at/files/PubDat_143783.pdf, 42.

55 Susanne Köstering, "'Pioniere der Rohstoffbeschaffung': Lumpensammler im Nationalsozialismus, 1934–1939," in *WerkstattGeschichte* 17 (1997), 49.

56 Friedrich Huchting, *Prüfung alter Verwertungstechnologien aus Mangel- und Kriegszeiten: Umweltforschungplan des Bundesministers des Innern* (Düsseldorf: Verein Deutscher Ingenieure, 1979), 21; Susanne Köstering, "'Millionen im Müll?' Altmaterialverwertung nach dem

Vierjahresplan," in *Müll von Gestern? Eine Umweltgeschichtliche Erkundung in Berlin und Brandenburg*, ed. Susanne Köstering and Renate Rüb (Münster: Waxmann Verlag, 2003), 139.

57 Ossberger, "Geschichte der Abfallwirtschaft in Österreich," 42.

58 Friedrich Huchting, "Abfallwirtschaft im Dritten Reich," *Technikgeschichte* 48, no. 3 (1981): 254.

59 Köstering, *Pioniere der Rohstoffbeschaffung*, 46–49, 55.

60 Huchting, "Abfallwirtschaft im Dritten Reich," 139, 260–61.

61 Köstering, "Millionen im Müll?," 47, 140–41. See also Heike Weber, "*Towards Total Recycling: Women, Waste* and *Food Waste Recovery* in *Germany, 1914–1939*," in "Recycling and Re-Use in the Twentieth Century," ed. Ruth Oldenziel and Heike Weber, special issue, *Contemporary European History* 22, no. 3 (August 2013): 371–97.

62 Huchting, "Abfallwirtschaft im Dritten Reich," 264–65. Finnish and Swedish visitors reported the widespread use of synthetic materials, extensive mechanization of waste management, and use of forced labor in sorting waste. "Romunkerääjämme opintomatkalla," *Helsingin Sanomat*, July 19, 1938; "Gott utbyte av studiebesöket i Ruhrdistriktet," *Borås Tidning*, January 24, 1938.

63 See Huchting, "Abfallwirtschaft im Dritten Reich," 257, 268–70, and also the attached organization scheme; Köstering, "Millionen im Müll?," 143.

64 Ernst Cassirer, *Myten om staten* (Stockholm: Natur och kultur, 1948), 302; on the cultural paradox of National Socialism, see also Jeffrey Herf, *Reactionary Modernism: Technology, Culture and Politics in Weimar and the Third Reich* (Cambridge: Cambridge University Press, 1986).

Chapter 5
The Costs of the War for the Soviet Union

PAUL JOSEPHSON

The German armies attacked the Soviet Union on June 22, 1941. They rapidly overwhelmed Soviet forces, destroyed much of the existing Soviet Air Force on the ground, and captured millions of soldiers as the Blitzkrieg moved toward Leningrad, Moscow, the Don River, and the Caucasus. Ultimately they overran approximately 850,000 square kilometers of territory. As the Soviet armies retreated early in the war, they adopted a scorched-earth policy, leaving behind fields of stubble and smoldering forests, and they dynamited buildings and factories. The German soldiers continued the war by stripping what was left, surrounding and starving Leningrad, approaching the outskirts of Moscow and reaching Stalingrad, where a horrific battle of block-to-block fighting left only rubble. Virtually no trees remained. Toward the end of the war, their supply lines destroyed, those Nazi armies that managed to flee the pursuing Red Army stripped bare what was left in the war zone. World War II had long-term, wide-ranging impacts across a vast swath of the European Soviet Union.

This essay evaluates the extensive environmental costs of World War II for the Soviet Union in two ways. First I examine the physical costs of World War II for Soviet territory. In the concluding section, I consider the long shadow of World War II—and the Cold War—over the environment of the Soviet Union. Granted, on many levels it is impossible to precisely evaluate these environmental costs. We have rough approximations of territory conquered, extensive descriptions of battles, and chapter and verse on materiel, artillery, tanks, and airplanes. Yet no complete literature exists on total numbers of shells, explosives, mines, and so on employed during the four-year Great Fatherland Patriotic War. Eyewitness accounts and archival materials focus not on nature as such, but on the battles and soldiers. Because of the extent of the devastation of the environment, it will suffice in this environmental history to focus on military and social issues to understand the impact on flora and fauna, and on rivers, lakes, and streams.

Building on information about engagements, battles, tactics, and urban warfare in the western Soviet Union, where the bloodiest battles of the war occurred, it is possible to create a terrifying account of the environmental costs of World War II for the Soviet Union, including on the built environment. This will broaden the definition of environmental history to consider the farms, villages, towns, and cities. Consider alone that the Nazi armies conquered nearly all of Ukraine, Belarus, and Moldova. The Wehrmacht reached the Don River and came within two hundred kilometers of the shores of the Caspian Sea, made it as far north as Vologda, and to the outskirts of Moscow.[1] Millions of soldiers engaged in pillaging, plunder, and pitched battles, and they engaged in this violent conflict in the middle of the most productive agricultural lands of the nation and many urban centers. Farmland and forest, rivers and lakes, and industrial and agricultural objects in those ecosystems were destroyed. Discussion of the Leningrad Blockade, the conquest of Belarus, and the greatest tank battle in history at Kursk in summer 1943 also enables us to gauge the scale of environmental degradation during the war.

THE NAZI INVASION OF THE SOVIET UNION

The Stalinist economic system was a constant wartime economy. The performance of this economy, with its inefficient and undervalued extraction of natural and mineral resources, contributed to extensive environmental change in taiga, tundra, and steppe, in both burgeoning cities and farm villages from the Arctic to Central Asia, in the lead-up to what Soviet leaders considered the "inevitable" war between the two competing world systems: socialist and capitalist. Party officials, economic planners, and engineers sought to master the empire's extensive natural resources with the goals of economic self-sufficiency and military strength. They ordered armies of workers to construct giant dams and reservoirs on the major European rivers, the Don, Dniepr, and Volga. The workers were thrown into building massive chemical combines, metal smelters, and oil refineries in both European and Siberian parts of the country. The Bolsheviks established the gulag labor enterprise that had its own great environmental and human costs. Armies of slave laborers built dams, canals, roads and railroads, and forestry enterprises from Karelia to Central Asia with little concern for nature.

The environmental costs of the Stalinist wartime economy were so great because of the speed and scope of the effort, the low level of concern for the environment, and the productivist ethos that prevailed among party officials and planners. Leaders were determined to leap to the level of the advanced capitalist nations, and especially the United States, in producing iron and steel, extracting coal and ore, pumping oil and gas, felling timber, and creating and expanding networks of roads and railroads. They created entire production cities dedicated to iron, concrete, asbestos, and mining (e.g., Asbest, founded in the Urals region in 1933). In the process, they refused to tolerate deviation from plans even though the plans were in reality far-fetched; targets did not reflect real possibilities but symbolized determination to transform the nation into an industrial powerhouse.

Similarly, in the pursuit of collectivized agriculture, Soviet planners and political officials established targets for the destruction of individual peasant holdings and the creation of larger cooperative farms that pushed boundaries to the horizon, eliminated fallow and meadows, destroyed edge effects, and, through the improper use of machinery and overuse of chemicals, destroyed soils. Undergirding this entire economic revolution was a view of nature as an object to be manipulated according to plan. Nature itself, in many ways, came to be equated to an enemy to be subjugated to the positive forces of the Communist Party to reshape it in a socialist fashion. Rivers and forests would operate according to plan; ore would be mined with minimal concern about pollution.[2] In spite of being proclaimed as "supremely rational," the planning system was therefore anything but rational in terms of resource use and production processes. Nor did it fully prepare the nation for the war effort.

The Soviet people and country suffered miserably during World War II. In spite of the obvious environmental costs of war—fires, explosions, shells, haphazard and intentional disposal of fuels and ordnance, the blowing up of factories and oil fields, the spread of chemicals, and so on—little has been written in a systematic fashion about these costs. Bombardment from airplanes and tanks left rubble and craters. Chemical weapons led to despoliation, defoliation, and toxic pollution.[3] The destruction of urban infrastructure led to the displacement of millions of people in the Soviet Union. Armies targeted transport and irrigation systems; dams were destroyed, levees and embankments were dynamited, ports "were clogged with unexploded ordnance and sunken ships." Soldiers decimated forests and farms, and dumped oil and gas into rivers and lakes in order to deprive enemy troops of cover, shelter, food, and water. People denuded the cities and countryside of anything they could use, including animals, as they fled eastward or hid in the forests. The dead bodies of humans and animals decomposed in the open sun. Mines and other unexploded ordnance remained a threat for decades. There is no question that battles and scorched-earth policies devastated the Soviet landscape, especially in Belarus and Ukraine. Worst of all, by 1945, twenty-five million people perished and another twenty-five million people had no shelter.[4]

The Nazi force that invaded the Soviet Union comprised "the greatest army ever assembled." More than three million men in 134 divisions with three army groups, 7,184 artillery pieces, 3,350 tanks, 2,770 aircraft, and 600,000 vehicles joined in the Blitzkrieg toward Leningrad, Minsk, Moscow, Kiev, and Baku. More than two million Soviet soldiers in 132 divisions including thirty-four armored divisions with 10,000 tanks and 2,300 planes opposed them. The invasion on June 22, 1941, caught no one but Joseph Stalin by surprise; Stalin disappeared from public view for ten days, likely out of shock, briefly requiring Viacheslav Molotov and other high party officials to lead the nation. Before the end of the third week of the attack, 300,000 Russians had been captured, and 2,500 tanks, 1,400 artillery guns, and 250 aircraft had been captured or destroyed. Hitler's armies moved so rapidly through the countryside that they compromised their own supply and communication lines. This led them to madly strip the countryside of anything that remained behind in the wake of Stalin's scorched-earth policy.[5]

Stalin and the military leaders were determined to prevent any resources from falling into Nazi hands. They ordered the evacuation of all industrial and military resources, all rolling stock, all engines and wagons, all fuel and grain, all farm animals, and any other raw materials. Almost 32,000 industrial enterprises were destroyed, although the Soviets succeeded in evacuating some 1,500 key factories to the east, in particular to the Ural, Siberia, Central Asia, and Volga regions.[6] What they could not withdraw they destroyed with explosives or fire: buildings, public records, monuments, homes, dams, locks, and factories. Also destroyed was roughly 2.5 million kilowatts of generating capacity—including the Dniepr hydroelectric power station—so that less than 300,000 kilowatts of power remained intact. Soviet demolition was so effective that it prevented the Nazis from increasing capacity to 630,000 kilowatts until two years later. Basic industry—coal, iron ore, steel, electricity, cement—was for all practical purposes totally destroyed; mines were flooded.[7]

Figure 5.1. Block to block and building to building fighting between the Wehrmacht and Red Army in Stalingrad between August 1942 and February 1943 led to two million casualties and is the bloodiest battle in world history. Source: *Stalingrad—Hero City*, Vadim Volikov, 1943. Gamborg Gallery, Moscow, Russia.

The Soviet scorched-earth strategy included the deportation of twelve to thirteen million men, women, and children—largely of Polish, Ukrainian, and Jewish descent—along the German-Russian frontier who were forced from the borderlands, many of them to Siberia. As the war burned through the countryside from west to east, almost no cattle or other farm animals, grains, gasoline, or any other supplies remained. The Soviets busily harvested the fields as they retreated. According to some estimates, the so-called Occupied Eastern Territories produced forty-three million tons of grain under Soviet rule in 1940. Under German administration, the harvest in 1941 was perhaps thirteen million tons and dropped to 11.7 million tons in 1942.

By November 1941, the German Army had occupied territory in which 40 percent of Soviet citizens lived; 63 percent of coal, 68 percent of iron, 58 percent of steel, 60 percent of aluminum, 38 percent of grain, and 84 percent of sugar were produced; and 38 percent of cattle, 60 percent of pigs, and 41 percent of railroads were located. Agricultural production fell by 40 percent.[8] The German administration succeeded in seeding no more than three-quarters of the prewar acreage. Fertilizer was practically unavailable. Compared to the average yields per hectare of approximately 2,200 pounds (fourteen bushels/acre) in the Ukraine in the late 1930s, the Germans managed to obtain just 1,500 pounds (ten bushels/acre) in 1942. Furthermore, the Soviet scorched-earth policy now began to show its full effects: the use of seed grains to relieve the worst hunger in the cities, the increasing partisan menace, and the dearth of personnel and machinery ruined the harvest.[9]

The Germans enslaved or imprisoned 8.7 million people.[10] The millions of civilians left behind were both at the mercy of the Nazis and also essentially starved. The Nazis singled out Jews, Gypsies, and Slavs for enslavement and genocide. At least 160 Nazi concentration and labor camps, some holding tens of thousands of prisoners, were established in Ukraine alone. The major goal was to conquer the Soviet Union and destroy Bolshevism, but the Germans intended the Occupied Eastern Territories as *Lebensraum*, or living space, for German-run agricultural settlements using Slavic populations as slaves. Many Ukrainians were also sent to Auschwitz and other death camps in Poland. Through the camps, disease, and starvation, Ukraine lost more people in World War II than any other European country—if we consider the Baltic states and Ukraine European countries—perhaps a quarter of its population,[11] although according to some sources one-sixth.

THE LENINGRAD BLOCKADE

A good example of the Nazi intentions for the decimation of Leningrad concerned Tsarskoe Tselo (formerly called Pushkin)—the great palace that Catherine the Great built, nearby to Pavlovsk—which was Tsar Paul's classical castle and sculpted gardens. At the beginning of the war, some of the treasures from Pavlovsk were evacuated to Siberia or hidden in the cellars underground, but the Soviets were unable to protect either site. Exquisite fragile panels that could not be removed were hidden under wallpaper, but the Nazis discovered and removed them in less than thirty-six hours. Pavlovsk, which was used as the Nazis' headquarters outside of Leningrad, was occupied for two and a half years until 1944, when they burned it down.[12] Similarly, Pushkin was destroyed.

In summer 1941, the German North Army of 500,000 men under General Field Marshal Wilhelm Ritter von Leeb commenced an operation designed to destroy the Red Army in the north, taking Baltic bases and eventually Leningrad. By July 9, von Leeb was in Pskov and on the July 10 in Luga. Only 180 kilometers remained to Leningrad. By the end of August, the North Army had secured the remaining rail lines into Leningrad. The Red Army hung tough, preventing further progress, so that Hitler determined to starve Leningrad to death. On September 8, 1941, the 871-day blockade of Leningrad, with its great human losses and environmental costs, commenced.[13]

At the beginning of the blockade there were 2.6 million people in Leningrad, including 400,000 children, and in the surrounding regions another 343,000 people.[14] On September 8, at 6:55 p.m., the Nazis began bombing the city, dropping more than 6,327 incendiary bombs. More than 178 fires broke out. The damage to storage facilities initially was not significant, but the fact of the fires led to rumors that central stores had burned, and this was the reason for the subsequent famine. Documents relating to the events of fires and losses have only recently been declassified. Inadequate storage facilities, interrupted transport, and poor management by the authorities made matters worse.[15]

During the entire blockade, the Nazis fired 150,000 heavy artillery shells and 107,000 incendiary and other bombs on the city.[16] In addition to housing, factories, and infrastructure, this destroyed salt, vegetables, salt-preserving equipment, candies, and baked goods. On September 10 and 11, the authorities carried out another survey of farm animals, birds, grain, and so on, and determined that they had sufficient reserves for thirty-five days of flour and grain, thirty days of grits and macaroni, thirty-three for meat, oils and fats about forty-five days, and sugar and other baking supplies sixty days. Potatoes and vegetables were hardly available. The authorities ordered the killing of all livestock, the processing of the carcasses, and the use of the feed for humans. They established a bare-bones ration. From October 1, workers and engineers received ration cards for bread of four hundred grams/day, and everyone else two hundred grams/day, while all other foods sharply declined. By the end of 1941, the authorities had no more than a week of supplies of basic foods.

As things grew bleak into the winter, the mood of the people understandably took a turn for the worse, and the censors took out their pens with greater frequency and vigilance. By December, more than 20 percent of letters were being censored. Everyday "life becomes worse." "It is dreadfully cold." "I witnessed a horse collapse on the street. People ran up to it with axes and knives, began to cut it to pieces and carry it home. It was terrible."[17] Children and invalids would perish first.

The one thin hope was an ice road across Lake Ladoga. The lake had been poorly studied from a transport point of view before the war, but it now assumed critical importance. In August 1941, the State Committee on Defense determined to seek solace at the small harbor of Osinovets on Ladoga, fifty-five kilometers from Leningrad. On September 12, two barges arrived carrying 626 tons of grain and 116 tons of flour. This was the modest beginning of the "Road of Life." The authorities sent troops, shells, and fuel to the port; the railroad would carry the supplies on to Leningrad. The volume of the road was never great, but psychologically and in other ways it was a crucial factor in saving Leningrad. In the first two months, 24,000 tons of grain, flour, and grits arrived, along with 1,130 tons of meat and milk products. By mid-November the ice had frozen to 1.8 meters (about six feet). They shifted to trucks that moved at best at the speed of a horse-drawn carriage. On November 23, they delivered only nineteen tons of supplies. Deep freezes actually created a thick foundation, and so the weight of freight increased through the winter manyfold. The Germans constantly strafed the convoys, and by spring the ice road was covered with thirty to forty centimeters of water. Still, transport continued until April 24, 1942. Between November and the end of April, brave solders and drivers had managed to deliver 361,309 tons of freight.

In November 1941, the authorities cut the ration of bread for engineering workers to 300 grams, and for others to 150 grams (four slices of bread), and soon cut that miserly ration further. While most people acted with great heroism, others could not, stealing food and ration cards, waiting for a waif to pass out and then descend on the body for boots, socks, rags, and any morsels in the pockets. Famine enveloped the city. One citizen wrote, "Leningrad became a morgue. The streets became boulevards of the dead. In every house the basement was storage for the dead."[18] Cannibalism broke out. In December, there were forty-three reported cases, in January, 366; in February, 612; in March, 399; in April, 300; in May, 326; and in June, fifty-six. But factories had to work to contribute to the war effort. The Kirov Elektrosila Factory was at constant threat from artillery in the open, but workers—emaciated men, women, boys, and girls—stood at presses and drills, turning out shells and other equipment.

The pace of evacuation picked up, reaching a massive scale in 1942. At first, children, mothers, the sick, the wounded, and invalids had priority, followed by scientific workers, students, and workers of evacuated factories. But the death rates far exceeded the evacuation effort, with 96,751 people dying in January 1942 (the period of the highest death rate); 36,606 people died in the first ten days of February, and 34,852 in the second ten days.

Thousands of homes, factories, and other buildings were destroyed: schools, libraries, and government buildings, and parts of the Hermitage, Russian Museum, and Senate. The city's infrastructure deteriorated, disease-carrying rats spread, and trash piled up everywhere. The Leningrad City Council therefore ordered at the end of March 1942 that all able-bodied individuals be mobilized to clean courtyards, remove filthy snow and ice, and bodies—or else epidemics would surely break out. By April 15, Leningraders had cleaned up more than 12,000 courtyards and over three million square meters of streets, squares, and parkways along the canals and river, and carted away one million tons of garbage and snow. The brigades constructed 4,100 pillboxes and dugouts and built more than thirty-five kilometers of barricades and antitank barriers out of rubble. Signs of normalcy returned in 1943 with resumption of a few tramlines and the opening of several factories, schools, and movie theaters. Most homes were reattached with water and sewerage.

The Soviet leadership in Leningrad under Marshal K. E. Voroshilov determined that they must be prepared for Nazi entry into the city. They drew up a plan to strip and then blow up the largest Leningrad factories, power stations, roadways, bridges, and the Baltic Fleet. They set 325,000 kilograms of explosives in the foundations of enterprises to be ready to turn them into rubble at a moment's notice. This plan would have led simultaneously to the destruction of a thousand buildings, all related to transport, electrical distribution, cables, and railroad depots, and telegraph and telephone stations.

After the war, of course, Stalin and the party leadership established Moscow and Leningrad, where most of them lived, as priority areas for reconstruction. While the centers of these cities were quickly cleaned and rebuilt, the outlying regions looked like war zones until 1950. To make matters worse for cleanup and recovery, Stalin devoted investment to such projects as the seven "skyscrapers" in wedding-cake style that dot Moscow's skyline as symbols of his and Soviet power, at the same time as millions of people lived underground in holes and rubble (*zemlianki*). Stalin focused most available resources on rebuilding industry. Hence the farms, the landscape, and the forests suffered longer, as there was little investment in planting grass, bushes, trees, and eventually flowers. The rebuilding of priorities Leningrad and Moscow took years to complete; imagine the rest of the nation. Imagine Belarus, which was at the center of World War II.

THE DEVASTATION OF BELARUS

Operation Barbarossa and the Nazi conquest of the Soviet Union involved the effort to destroy the Slavic people (or to enslave the survivors as agricultural workers); seize all natural, personal, and state wealth; and subjugate the survivors to harsh rule. The Nazis looked east for *Lebensraum*. Belarus and Ukraine may have suffered more than any other republics of the former Soviet Union. According to official Belarusian estimates, the total material loss of the Nazi occupation was seventy-five billion rubles (in 1941 prices), or thirty-five times the entire budget of the republic in 1940. In Belarus, the Germans looted and burned 209 of 270 cities and regional centers (80 to 90 percent of Minsk, Gomel, and Vitebsk were destroyed), and 9,200 villages. Industry and energy production dropped to pre-1913 levels; the Germans leveled 100,465 buildings connected with industry and stripped as many as 10,000 enterprises of machinery, including 85 percent of large electrical

Figure 5.2. At the end of World War II some 1,500 cities were left in shambles across the western Soviet Union. Farmland and forest had also been devastated. Source: *Stalingrad in 1944*, Natalia Gippius, 1944. Gamborg Gallery, Moscow, Russia.

power stations. The losses to agriculture were equally great, as 10,000 collective farms, ninety-two state farms, 316 machine tractor stations, and 1.2 million agricultural structures—including sheds, barns, huts, and houses—were destroyed. Museums, libraries, theaters, clubs, scientific research institutes, and 8,825 of 12,294 schools were also destroyed by fire.[19]

A horrible symbol of this devastation was the village of Khatyn, which was burned to the ground with all its inhabitants, a tactic repeated in another 618 towns and villages. On March 23, 1943, in response to the shooting of a German officer on the road near the town, German soldiers, determined to punish Khatyn's inhabitants, rounded up all the residents, locked them in a shed, and set it on fire. Some people managed to break through the burning doors, only to be killed by machine guns. In all, 149 people including seventy-five children were slaughtered.[20]

The Central Nazi Army rolled through Belarus with fifty divisions, two motorized brigades, 823 tanks, 11,027 artillery, and 1,677 fighter planes. German airplanes quickly turned many villages to ruins. The authorities attempted to save what they could, organizing evacuation of archives and banks, and enabling the escape of perhaps as many as 1.5 million citizens, although more likely only 750,000 people, farm animals, and 124 large industrial enterprises. In spite of fierce fighting, by June 28, 1941, Minsk was occupied, and perhaps 350,000 people had died, and by the beginning of September all of Belarus with its eight million inhabitants had fallen to the Nazis while 900,000 Soviet soldiers had died. The Nazis organized something like six hundred concentration camps. Murder and genocide of the Jews followed. Among the roughly 2.3 million inhabitants of Belarus killed during the war were 715,000 Jews.[21]

The retaking of Belarus was equally as violent. In Operation Bagration in summer 1944, the Red Army and partisans with 2.4 million men, 36,000 pieces of artillery, 5,200 tanks, and 5,300 airplanes destroyed the "center" army group consisting of 1.2 million soldiers, 9,500 pieces of artillery, 900 tanks, and 1,350 planes, and killed 145,000 German soldiers.[22]

THE INDUSTRIALIZATION OF WAR

In the twentieth century, beginning with World War I, mass production of armaments and weapons—cannon, ships, tanks, airplanes, and so on—provided the industrial foundation to war. During this time, civil populations also became the targets of industrial war. The largest industrial battle in the world was perhaps the Kursk tank battle, the largest tank battle in history, in 1943. For its attack on the Kursk salient, the German military leadership gathered 900,000 soldiers in fifty divisions, including sixteen tank and motorized divisions, 10,000 pieces of artillery, 2,700 tanks and 2,000 aircraft, or about one-third of Germany's remaining military strength in the Soviet Union. Red Army leaders focused 1.3 million soldiers, 20,000 artillery pieces, 3,600 tanks, and 2,400 planes on the salient. They established networks of antitank defenses. By June 1943, 300,000 civilians were helping the Red Army build defenses around the Kursk salient. The Soviets strengthened their troops from April to July with ten infantry divisions, ten antitank brigades, fourteen artillery regiments, eight mortar regiments, and other tank and motorized regiments. From March until July, another 5,635 artillery, 3,522 mortars, and 1,294 airplanes were given over to the battle.[23] Roughly two-thirds of the German Army was based in Russia with plans to break the Red Army. At the beginning of July, battle preparations were complete.

The Kursk battle took place over a salient south of Orel and north of Kharkiv at Kursk.[24] If the salient were not removed, the Russians would likely succeed in trapping the Germans between two Russian forces. The Germans attacked using the Luftwaffe. But Soviet leaders were equally prepared—and those preparations involved extensive remaking of the landscape. Their intelligence had alerted them to a massive German offensive; they knew where it would be, the numbers involved, and near enough to when it would start. The Soviets stationed huge numbers of men and equipment in the Kursk bulge. In addition to scores of kilometers of trenches, the defense installed thousands of mines in every square kilometer. At the beginning of June along the Kursk bulge, there were eight defensive zones to three hundred kilometers in depth.

German prisoners captured by Russian shock troops before the battle began told the Russians that the attack was to be on July 5. To preempt the attack, the Russians launched a massive artillery bombardment before dawn that day. This had an impact on the morale of the Germans, as it was clear that their plan had been compromised. After the bombardment had finished, Germany started her attack at 4:30 a.m. with an artillery barrage. A tank and infantry attack started at 5:30 a.m. once air cover had arrived. The main thrust contained five hundred tanks; heavy tanks at the front, supported behind by medium ones with infantry behind these. The Germans tried to break through on four occasions. They gained ten kilometers of land in the first twenty-four hours of fighting but lost 25,000 men killed or wounded, two hundred tanks and self-propelled guns, and two hundred aircraft. This pattern repeated over the next few days. By July 10, the German Ninth Army had lost two-thirds of its tanks including the new Tigers, whose side armor was vulnerable to antitank weapons.

This was all a prelude to the greatest tank battle in history on July 12, when 1,500 tanks engaged at Prokhorovka some eighty kilometers to the southeast of Kursk. By July 16, the Germans were on the defensive, in the process losing another 350 tanks and 10,000 men. With the strength of the Germans in the southern Kursk salient broken, the Soviets launched a major counteroffensive that led to German defeat. The Soviets launched another counteroffensive in the north of the salient to relieve Orel. They outnumbered the Germans two to one in all areas. Unable to call in reinforcements from their men fighting in the south, the Germans were unable to hold off the Russian offensive. The Russian Air Force prevented the Luftwaffe from providing support.

In August, fighting accelerated with constant bombardments, artillery fire, strafing from airplanes, and engagements of tanks. The front moved twelve to twenty-six kilometers along the bulge. By August 5, 1943, Orel was back in the hands of the Russians, while any German retreat was severely hindered by partisans who destroyed German field communications, rear-logistical operations, and rail lines, ensuring that train engines piled up at rail heads and making them an easy target for the Russian Air Force.

At the southern sector of the salient, the German Army faced a similar formidable enemy. While the Germans had 300,000 men and about six hundred tanks, the Russians had nearly one million men and many more tanks. On August 5, the Soviet forces entered Belgorod. The partisans who operated in this area derailed more than one thousand trainloads of troops in August—a major factor in preventing the Germans from moving their soldiers around. On August 13, the Soviets broke through the outer defenses of Kharkiv and ten days later liberated the city, ending the Battle of Kursk. (Kharkiv lost 300,000 residents, 160,000 were sent to labor camps in Germany—many of whom died along the way—and the Germans destroyed more than 160 million square meters of housing, five hundred industrial enterprises, and all cultural, educational, and medical institutions.)[25] What remained of the farmland and forests in the region after the battle included gouged-out craters, debris, ordnance, and rivers and streams filled with rock, soil, military materiel, fuel oil, chemicals, and bodies.

Study of treaded vehicles of any sort reveals their immediate and long-term impact on the environment. This is one reason why forestry and agricultural operations in the postwar years throughout the world turned to soft balloon-like tires for tractors and skidders, and why many manufacturers abandoned treads and hard rubber tires. Even one pass of a snowmobile skidding over the ice and snow has immediate and long-term impacts, such as compaction of the soil and destruction of flora and fauna, not to mention interrupting the life of larger mammals with noise and exhaust.[26] The thousands of tanks that engaged over scores of square kilometers at the Battle of Kursk left devastation no less than the tanks and troop carriers and other military machines that destroyed landscapes throughout the entirety of the Soviet war theater.

THE POSTWAR SHADOW OF THE WAR

In the transformation of cities into death zones, the siege of the countryside and the slaughter of innocents within it, and the large-scale industrialization of twentieth-century war, the belligerents of World War II changed forever the relationship between nature and the military. Because of the character of the conflict—as the experiences of Leningrad, Belarus, and Kursk demonstrate—and the fact that much of the battle between 1941 and 1945 took place on Soviet soil, environmental degradation reached a scale, depth, and longevity in the country that rivaled that in China, Japan, Germany, and elsewhere; indeed, it was likely greater still. And because Joseph Stalin and other Soviet leaders were determined to never again permit invasion and risk an end to the glories of the great socialist motherland, the war never ended. From Victory Day on May 9, 1945, until the collapse of the Soviet Union in the late 1980s, the leadership determined to harness the economy even more than before to military preparedness, with continued and extensive consequences for Soviet public health, ecosystems, and resource development.

The long shadow of World War II carried over the vast expanses of the environment of the Soviet Union. The extent of devastated lands ranged from Karelia through central Russia to the Don River, from blockaded Leningrad to Moscow, to Belarus and Ukraine, both of which may have suffered greater losses than any other areas of the world during the war. Over four years, an area of perhaps five million square kilometers was the constant battleground for soldiers moving and yielding, attacking, thrusting, retreating, and pushing back across fronts. Peasants stuck in these war zones were fodder for the war effort, on their own taking what they could to survive, selling, consuming, burning other things, and subject to murder, theft, and exploitation. Tens of thousands of homes, schools, enterprises and the like

Figure 5.3. With the German armies defeated, the Soviet Red Army swept westward toward Berlin and established control over nations of East Central Europe. The Red Army left rubble where anything was left standing, including in Budapest, Hungary. Source: *Budapest. Battle at the Royal Palace*, Natalia Gippius, 1945. Gamborg Gallery, Moscow, Russia.

were destroyed. Most major cities were 90 percent leveled. Sewerage, water treatment, electrical supply, and so on were either destroyed by the Soviets upon evacuation or dynamited by the Nazi soldiers. In the countryside, marauding armies stripped everything that could be used, coming and going and coming again. Trenches, unexploded ordnance, mines, shells, and tanks created what might be called, with only modest exaggeration, a moonscape. By the end of the war, one-quarter of the prewar capital stock of the Soviet Union had been destroyed, and in occupied areas as much as two-thirds of it. Oil, gas, chemicals, and wastes had been burned, spilled, and dumped far and wide.

Another factor contributing to the long shadow was the rejuvenation of the Stalinist economic system following the victory over the Third Reich. The postwar years signaled a complete return to collectivization of agriculture and rebuilding of heavy industry destroyed during the war, all to be prepared for the next full-scale military engagement. The fourth five-year plan (the first postwar plan) hence was directed toward military enterprises and activities with little concern about efficiencies of production, the environment, housing, or other public overhead capital. As before the war, investment targeted heavy industry while the government extracted capital from agriculture, and prevented sufficient funding for cleanup and reclamation.[27]

To prepare for the next war, the Soviets pursued full-scale development of Siberian, Arctic, and Ural resources on top of those of the western regions of the country with no concern about environment, indigenous people, or local residents.[28] In 1948, the Communist Party unanimously passed the Stalinist Plan to Transform Nature that included dams, canals, irrigation systems, hydropower stations, and forest defense belts, all to guarantee that nature itself functioned according to plan. They believed that "she" (the Russian word, *priroda,* is feminine) was capricious and unreliable, but could be made to support state power like industry and workers.[29]

The war's shadow fell over metallurgical combines at Monchegorsk, Vorkhuta, and Norilsk, three of the most polluted regions that provided strategic metals for the Cold War. Norilsk on the Taimyr Peninsula grew out of the Norgulag that, from 1935 to 1956, employed tens of thousands of prisoners to extract ore from the largest nickel-copper-palladium deposits in the world underneath the permafrost, leading to the deaths of thousands of those prisoners and the destruction of fragile Arctic ecosystems, owing to haphazard waste disposal and smelters operating without pollution control devices.[30]

In one last major—and equally violent—way, Soviet leaders promoted policies that contributed to significant environmental degradation. As in the United States, France, Great Britain, and China (the other four original nuclear powers), the Soviet Union adopted haphazard waste disposal practices in the production of nuclear, chemical, and biological weapons.[31] These Cold War practices destroyed so many ecosystems across such a vast landscape that costly and ongoing remediation in the twenty-first century may never catch up. Suffice it to say that, under Stalin, the enormous pressure on scientists and engineers to develop and then mass produce these weapons led them to skimp, take shortcuts, and ignore not only common sense but also common decency.

Hiding under the blanket of "national security," Soviet authorities ordered the dumping of high- and low-level nuclear and other toxic wastes across the landscape, in lakes, rivers, and even in the oceans. Engineers developed reactors that used "once-through" cooling, so that water from a lake or river that cooled the reactor was piped back into the lake or river contaminated with radioisotopes and at a higher temperature that destroyed water ecosystems. In developing weapons production and test sites ("nuclear polygons"), the authorities moved people out of their homelands, especially the Nentsy from Novaia Zemlia and Kazakhs in Kazakhstan. Yet, like the Bikinians and Polynesians in the South Pacific, forced from their homes by the Americans and French, respectively, people were displaced to areas where they might still suffer the consequences of radioactive fallout. In a number of cases, the nuclear waste spread far and wide, contaminating thousands of square kilometers of land and requiring tens of thousands of people to be evacuated, although with some delay. The regions most affected by Stalinist Cold War weapons of mass destruction were the Southern Urals, the region around Semipalatinsk in Kazakhstan, and the Arctic. In their production of plutonium for bombs the Soviets—and the Americans—paid little attention to the grotesque environmental costs of their efforts. They destroyed entire ecosystems. They lied to the people toiling in the towns and regions about the impact of the pollution of the highly dangerous production facilities and haphazard waste disposal practices on them and their children.[32] The impacts on the environment and public health would be felt well beyond the breakup of the Soviet Union in 1991.[33]

In the twentieth century, war became an industrial process, with the indiscriminate devastation of "strategic" military targets and "civilian" ones. The development of entire economies geared predominantly toward war—primarily those of the Soviet Union and National Socialist Germany—led inexorably not only to war, but also to direct and indirect, massive, and seemingly irreversible environmental change. As Erich Maria Remarque wrote in *All Quiet on the Western Front* (1929) about a different war, World War I,

> We come to the communication-trench and then to the open fields. The little wood reappears; we know every foot of ground here. There's the cemetery with the mounds and the black crosses. That moment it breaks out behind us, swells, roars, and thunders. We duck down—a cloud of flame shoots up a hundred yards ahead of us. The next minute under a second explosion part of the wood rises slowly in the air, three or four trees sail up and then crash to pieces. The shells begin to hiss like safety-valves-heavy fire—
>
> "Take cover!" yells somebody—"Cover!" . . . Not a moment too soon. The dark goes mad. It heaves and raves. Darknesses blacker than the night rush on us with giant strides, over us and away. The

> flames of the explosions light up the graveyard. There is no escape anywhere. By the light of the shells I try to get a view of the fields. They are a surging sea, daggers of flame from the explosions leap up like fountains. It is impossible for anyone to break through it. . . . The earth bursts before us. It rains clods. . . . A hole is torn up in front of me. Shells hardly ever land in the same hole twice, I'll get into it. With one bound I fling myself down and lie on the earth as flat as a fish; there it whistles again, quickly I crouch together, claw for cover, feel something on the left, shove in beside it, it gives way, I groan, the earth leaps, the blast thunders in my ears, I creep under the yielding thing, cover myself with it, draw it over me, it is wood, cloth, cover, cover, miserable cover against the whizzing splinters.

CONCLUSION

Soviet efforts to control the economy and subjugate nature to economic planning mechanisms failed to bring preparedness for that inevitable war with capitalism and fascist Germany. But even before the Nazi invasion, the Winter War with Finland in 1939–40 revealed the Soviet soldier to be capable and patriotic, but even with superior numbers and firepower unable to defeat the Finns decisively, short on materiel and without capable leadership (out of paranoia, Stalin had viciously ordered the murder of 40,000 Red Army Corps officers during the Purges as enemies of the people). The absence of leadership surely handicapped the Soviet response to the June 1941 invasion of the Wehrmacht and the significant losses of hardware, soldiers, and citizens. The natural and built environments, as we have seen, also were devastated.

Several interrelated political, economic, and ideological factors distinguished the war on nature carried out during nearly three-quarters of a century of Soviet power, an assault that accelerated during the Cold War years. The first was the almost fetishistic embrace of planning mechanisms with certainty that the "plan" would enable the socialist economy to avoid the costs of industrialization—squalor, dingy and dangerous workplaces, smoke-belching factories—that capitalist nations and their workers had endured. Similarly, through scientific application of plans, nature herself could be made to function in a machinelike fashion. The second was the centralized bureaucratic nature of Bolshevik rule that led almost inevitably to the pursuit of large-scale, centralized development projects, each of which acquired nearly unstoppable technological momentum. Soviet ideologists and officials argued that massive facilities—dams, hydroelectric stations, irrigation systems, smelting and petrochemical factories, and thermal and later nuclear power stations—permitted them to achieve economies of scale, but overlooked these projects' inefficiency and human and environmental costs. The so-called hero projects of Soviet power—massive technologies and entire cities devoted to industrial production—had ideological and political purposes that forestalled apprehension of the costs. They enabled party spokesmen to claim that the socialist system was capable of achieving size, scale, and production on a level impossible in the capitalist West, while simultaneously providing a forum where nearly illiterate peasants could be transformed into politically conscious workers devoted to industrialization.

Last, the constant struggle against enemies, against hostile capitalist encirclement even before World War II, then against fascist Germany, and against the United States and its allies during the Cold War was always accompanied by war against capricious nature itself. Those who advocated a moderate approach to resource development, consideration of the social factors in production,[34] or the need to create nature preserves were ignored, hounded for their lack of Communist conviction, or at times even labeled as "wreckers" in the life-and-death struggle with the West. The language of war on nature was not unique to the Soviet Union,[35] but it was ubiquitous.

Two issues in the Soviet case make it difficult to distinguish between environmental effects that happened during the Cold War and those that occurred because of the Cold War. First, a constant, impatient, heroic, and violent effort to transform the nation—its people, institutions, nature, and resources, including primarily, as the Marxists would say, its productive forces and productive relations—prevailed throughout Soviet history. Beyond the sine qua non of Soviet existence—increasing industrial production by leaps and bounds—this effort reflected the belief among Soviet leaders that they must remain vigilant against enemies at home and abroad who threatened that transformation.

Second, and similarly, Soviet industry was tempered at the forge of Stalinism, and maintained its major features of large-scale enterprises hitched to the desiderata of production until the breakup of the Soviet Union. In many respects, political commissars had embraced the militarization of production from the first steps of Soviet socialism under war Communism when Leon Trotsky introduced it in a futile effort to stem anarchy in industry after the Bolshevik seizure of power. Many of the features of militarization—organization of workers into brigades, harsh punishments for "violation of labor discipline" (tardiness, absenteeism, drunkenness) and pilfering, the appointment of Bolshevik managers who had learned to secure victory at any cost at the front in the Red Army, not at the forge in the factory—carried through to the Gorbachev era.

Conceptually, significant changes in worldview accompanied Cold War Soviet resource management and exploitation practices. These changes went beyond those common to Enlightenment thinking about the desirability of reshaping nature to serve human needs, the inevitability of progress, and the ability of humans—and increasingly their research, financial, industrial, and state institutions as major actors in the process—to subjugate nature. After all, Bolshevik leaders were Marxists trained in the European tradition who believed that both humans and capricious nature might function according to plans and dictates; as it turned out in the Soviet case, both received serious and often coercive reeducation. The changes included the belief that nature was an enemy, and must be joined to state programs to protect the motherland. According to National Socialist thinking in Germany, the Aryan people (the *Volk*) had special status with respect to nature as the sole beneficiaries of proper nature preservation and management, and also were the only ones capable of properly tending it. Similarly, in the Soviet Union, the proletariat benefited from Bolshevik management of nature. One of the benefits for Soviet leaders and citizens was military preparedness. But the persistent, even irreversible, cost was the industrialization of war and its terrible environmental legacy.

Notes

The author thanks Raffael Scheck and Simo Laakkonen for their helpful comments on an early draft of this essay, and also the anonymous reviewer for critical commentary.

1 Among the many studies of the German invasion of the Soviet Union, see O. Bartov, *The Eastern Front, 1941–1945: German Troops and the Barbarisation of War* (London: Palgrave, 1985); H. Boog, J. Forster, J. Hoffman, E. Klink, R.-D. Müller, and G. R. Ueberschar, Germany and the Second World War, vol. 4, *The Attack on the Soviet Union* (Oxford: Oxford University Press, 1994); Alan Clark, *Barbarossa: The Russian-German Conflict 1941–45* (New York: William Morrow, 1965); and J. Erickson, *The Road to Stalingrad* (New Haven, CT: Yale University Press, 1999).

2 On the Stalinist economy and the environment, see Paul Josephson, Nicolai Dronin, Ruben Mnatsakanian, Aleh Cherp, Dmitry Efremenko, and Vladislav Larin, *An Environmental History of Russia* (Cambridge: Cambridge University Press, 2013), chap. 2.

3 For detail on the Soviet biological and chemical weapons program, and briefly their use in World War II, see Milton Leitenberg and Raymond Zilinskas, *The Soviet Biological Weapons Program: A History* (Cambridge, MA: Harvard University Press, 2012). See Edmund Russell, *War and Nature* (Cambridge: Cambridge University Press, 2001), for discussion of the use of chemicals on insects and humans from World War I until the publication of Rachel Carson's *Silent Spring* in 1962.

4 On the impact of war on the environment, see Jennifer Leaning, "Environment and Health: 5. Impact of War," *Canadian Medical Association Journal* 163, no. 9 (October 2000): 1157–61, and Walter Laqueur, *Europe since Hitler* (New York: Penguin, 1984), 1516; and J. W. Dower, *Embracing Defeat: Japan in the Wake of World War* II (New York: W. W. Norton, 1999), 33–120, both as cited in Leaning, "Environment and Health." See also "War," accessed November 18, 2014, http://www.pollutionissues.com/Ve-Z/War.html#ixzz0u9UcVYoZ.

5 C. N. Trueman, "Operation Barbarossa," The History Learning Site, accessed November 18, 2014, http://www.historylearningsite.co.uk/operation_barbarossa.htm.

6 Mark Harrison, *Accounting for War* (New York: Cambridge University Press, 1996); Denis Brand, *L'Experience Sovietique et sa Remise en Cause* (Paris: Sirey, 1993).

7 Walter Sanning, "Soviet Scorched-Earth Warfare: Facts and Consequences," Institute for Historical Review, accessed November 18, 2014, http://www.ihr.org/jhr/v06/v06p-91_Sanning.html.

8 N. A. Voznesenskii, *The Economy of the USSR during World War II* (Washington, DC: Public Affairs, 1998).

9 Alexander Dallin, *German Rule in Russia, 1941–1945* (London: MacMillan, 1981). See also Sanning, "Soviet Scorched-Earth Warfare."

10 On the human costs of Nazism and Stalinism, see Timothy Snyder, *Bloodlands: Europe between Hitler and Stalin* (New York: Basic Books, 2010).

11 Ann Lencyk Pawliczko, *Ukraine and Ukrainians throughout the World* (Toronto: University of Toronto Press, 1994), 62.

12 "The Alexander Palace," accessed November 18, 2014, http://eng.tzar.ru/museums/palaces/alex_ander.

13 German officialdom announced on September 21, 1941, "At first we will blockade Leningrad and destroy the city with artillery and aviation. . . . In the spring we will penetrate the city, and we will take everything that remains alive into the deep interior of Russia, or we will imprison it, cover Leningrad with earth and give the northern region of the Neva to Finland."

14 I have drawn on such sources as "Blokada Leningrada," accessed November 17, 2014, http://www.blocada.ru/.

15 See Dmitri Pavlov, *Leningrad 1941: The Blockade*, trans. John Adams, foreword by Harrison Salisbury (Chicago: University of Chicago Press, 1966). For recent scholarship on the human aspects of the Leningrad blockade, see Polina Barskova, "The Spectacle of the Besieged City: Repurposing Cultural Memory in Leningrad, 1941–1944," *Slavic Review* 69, no. 2 (Summer 2010): 327–55. See also Denise Youngblood, review of *The Blockade*, directed by Sergei Loznitsa, *Russian Review* 66, no. 4 (October 2007): 693–98.

16 "The Trial of German Major War Criminals," The Nizkor Project, accessed November 17, 2014, http://www.nizkor.org/hweb/imt/tgmwc/tgmwc-07/tgmwc-07-62-02.shtml.

17 See "Stroki iz Pisem, iz iatykh Voennoi Tsenzuroi (iz Arkhibnykh Dokumentov Upravleniia FSB po S.-Peterburgu i Oblasti)," at such sites as Blokada Leningrada: Dos'e, 9 May 2012, accessed November 17, 2014, http://civility.ru/blokada-leningrada-dose/.

18 Ibid.

19 "Posledstviia Velikoi Otechestvennoi Voiny dlia Belarusi," accessed November 17, 2014, http://archives.gov.by/index.php?id=697135.

20 "The Tragedy of Khatyn," Ministry of Culture of the Republic of Belarus, accessed November 17, 2014, http://khatyn.by/en/tragedy/.

21 Barbara Epstein, *The Minsk Ghetto, 1941/1943: Jewish Resistance and Soviet Internationalism* (Berkeley: University of California Press, 2008).

22 "Istoriia Voiny: Obzor Sobytii," accessed November 17, 2014, http://archives.gov.by/index.php?id=427420.

23 C. N. Trueman, "The Battle of Kursk," The History Learning Site, accessed November 17, 2014, http://www.historylearningsite.co.uk/battle_of_kursk.htm.

24 The following section is largely taken from "Kurskaia Bitva," accessed November 17, 2014, http://histrf.ru/ru/lenta-vremeni/event/view/kurskaia-bitva; see Trueman, "Battle of Kursk," and "Kurskaia Bitva," accessed November 17, 2014, http://www.otvoyna.ru/kursk.htm.

25 For a definitive study, see David Glantz and Jonathan House, *The Battle of Kursk* (Lawrence: University of Kansas Press, 1999).

26 W. J. Wanek, "Snowmobile Impacts on Vegetation, Temperature and Soil Microbes," in *Proceedings of the 1971 Snowmobile and Off-Road Vehicle Research Symposium*, Michigan State University Department of Parks and Recreation Resources Technical Report No. 8, ed. M. Chubb (East Lansing: Michigan State University, 1971), 116–29; and J. T. Stangl, "Effects of Winter Recreation on Vegetation," in *Effects of Winter Recreation on Wildlife* (Washington, DC: National Park Service, 1999), 119–21.

27 G. Khanin, *Dinamika Ekonomicheskogo Razvitiia SSSR* (Novosibirsk: Nauka, 1993). The Soviets carted off two-thirds of the Germany aviation and electronics industries, a major part of rocketry, automobile, construction, military, and other factories. See also Anthony Sutton, *Western Technology and Soviet Economic Development* (Stanford, CA: Stanford University Press, 1968).

28 See Ilmo Massa and Alla Bolotova, "Opening the Circumpolar Arctic World," chapter 15 in this volume.

29 Paul Josephson, "The Stalinist Plan for the Transformation of Nature and the East European Experience," in *Stalin's Plans for the Transformation of Nature and Their Impact in Eastern Europe*, ed. Doubravka Osakova (Oxford: Berghahn Books, 2016), 141.

30 Paul Josephson, "Industrial Deserts: Industry, Science and the Destruction of Nature in the Soviet Union," *Slavonic and East European Review* 85, no. 2 (April 2007): 294–321.

31 On the Soviet biological and chemical weapons program, see Leitenberg and Zilinskas, *Soviet Biological Weapons Program.*

32 On the environmental impacts of plutonium production in the Soviet Union, Russia, and the United States, see Kate Brown, *Plutopia: Nuclear Families in Atomic Cities and the Great Soviet and American Plutonium Disasters* (Oxford: Oxford University Press, 2013).

33 Among the dozens of excellent compilations and reports that indicate the extent of nuclear waste and other problems of the Soviet Cold War legacy, see the Bellona Foundation reports: Nils Bøhmer, Aleksandr Nikitin, Igor Kudrik, Thomas Nilsen, Andrey Zolotkov, and Michael H. McGovern, *The Arctic Nuclear Challenge* (Oslo: Bellona Foundation, 2001); and Igor Kudrik, Aleksandr Nikitin, Charles Digges, Nils Bøhmer, Vladislav Larin, and Vladimir Kuznetsov, *The Russian Nuclear Industry—The Need for Reform* (Oslo: Bellona Foundation, 2004). See also Paul Josephson, *Red Atom* (Pittsburgh: University of Pittsburgh Press, 2005); Zhores Medvedev, *Nuclear Disaster in the Urals* (New York: Angus & Robertson, 1979); "Russia: Mayak Radioactive Waste Facilities," Nuclear Threat Initiative, accessed July 31, 2006, http://www.nti.org/db/nisprofs/russia/fissmat/pumayak/nucwaste.htm; and B. I. Bulatov, *200 Iadernykh poligonov SSSR* (Novosibirsk: Tseris, 1993). For an encyclopedic compilation of the human and environmental costs of the Soviet development model, see Murray Feshbach and Alfred Friendly Jr., *Ecocide in the USSR* (New York: Basic, 1992).

34 The party met efforts to suggest a moderate approach with coercion, arrest, and execution of engineers and others, as Loren Graham explores in *The Ghost of the Executed Engineer* (Cambridge, MA: Harvard University Press, 1993).

35 Russell, *War and Nature.*

Chapter 6
Conceptualizing Wartime Flood and Famine in China

MICAH S. MUSCOLINO

During World War II, or the "Anti-Japanese War of Resistance" (1937–45) as it is known in China, North China's Henan Province endured a series of war-induced ecological disasters. The first struck in June 1938, when Chinese Nationalist armies under the command of Chiang Kai-shek breached the Yellow River's dikes in Henan in a desperate attempt to block a Japanese military advance.[1] For the next nine years, the Yellow River's waters spread to the southeast into the Huai River system via its tributaries, inundating vast quantities of land. Perhaps the single most environmentally damaging act of warfare in world history, the strategic interdiction threw long-established water control infrastructure into disarray, leading to floods that persisted until after the conflict had come to an end. Investigations carried out after 1945 estimated that the resulting floods killed over 800,000 people and made nearly four million into refugees in the provinces of Henan, Anhui, and Jiangsu. In Henan Province alone, wartime flooding killed over 325,000 people and displaced over 1,170,000.[2]

But that was not all. An even greater catastrophe struck Henan in 1942–43, when climatic anomalies associated with a global El Niño event, wartime disruption of transport, and the food energy demands of Chinese and Japanese armies stationed in Henan precipitated a famine of terrific magnitude.[3] The Henan famine led to as many deaths—approximately two million—as the famous Bengal famine that occurred in India at nearly the same time, and millions more Henan residents took to the roads and fled to escape this subsistence crisis.[4] This chapter contends that the disasters that took place in Henan during World War II resulted not from arbitrary destruction, but from the nature of war and militarization itself.

MILITARY METABOLISM

To comprehend Henan's war-induced ecological catastrophes, one can employ a conceptual approach that traces energy flows through and between societies and environments.[5] Metabolic processes transform energy and materials, enabling biological systems (whether organisms or higher-level ecosystems) to maintain life, to grow, and to reproduce. Socioeconomic systems also depend on continuous throughputs of energy and materials to maintain their internal structures. By exploiting various energy sources, human societies modify and manipulate land, water, plants, and animals to fulfill their basic needs. The concept of "social metabolism" likens this dependence to the biological metabolism of a living organism. Unlike the biological notion of metabolism, however, this socioecological concept links energy and material flows to social organization. The quantity of resource use, its material composition, and the sources of these flows are a function of socioeconomic production and consumption systems that vary greatly across time and space. This framework analyzes sociometabolic patterns at different spatial, functional, and temporal scales, while also tracing their environmental consequences.[6] By seeing human societies as embedded in larger organic systems, an energy-centered approach renders legible the connections between phenomena that historians have conventionally treated as discrete. Rather than artificially separating socioeconomic and biophysical processes, this approach enables us to highlight the seamless interrelationships and interdependencies that existed among societies, environments, and the military establishments that they support.

Like all socioeconomic systems, militaries have metabolisms. Nature's energy makes warfare possible. Fighting and preparing for war, like all work, requires appropriating and exploiting energy sources. Militaries consist of vast agglomerations of humans, animals, machines, raw materials, logistical networks, engineering works, and many other components. No military system can survive without energy inputs from the surrounding environment. They take in food, fuel, building materials, and other resources; they emit wastes. Focusing on energy and its transformations allows for a better understanding of war-environment connections than any interpretation premised on an artificial division between the "human" and "natural." The metabolism of militaries and societies shapes the choices of commanders, the fates of civilian communities, and the course of environmental change. Moreover, thinking in terms of energy offers a conceptual apparatus that can tie together the conflicts waged in vastly different times and places that, when taken together, constitute the environmental history of war.

We conventionally define energy as the capacity to do work. Work occurs when a force acts on a body, causing it to move some distance in that force's direction. Moving an object entails doing work and expending energy. The specific amount of energy depends on the size of the object, how far it moves, and the resistance it encounters. Energy assumes many forms that have the potential to do work. Capturing more of that energy and using it with greater efficiency enables more work to be done. On this planet, the primary source of energy is the sun. Solar energy drives energy conversions at all levels. Photosynthesis, the process by which plants capture and store solar energy as chemical energy, is central to life on earth. As Edmund Burke III explains, "All complex life forms have devised methods for accessing the solar energy stored in plants. Human metabolism allows us to unlock this store of energy either directly, by consuming plants, or indirectly, by consuming animals. Alone among other complex forms of life, humans have been able to devise means of storing and using solar energy."[7]

Two laws govern energy flows. The first law of thermodynamics states that energy can change from one form to another, but it cannot be created or destroyed. The same amount of energy exists before and after it is transformed.[8] The second law of thermodynamics dictates that whenever energy changes forms, part of the energy becomes heat. Energy conversion is never 100 percent efficient. Some energy will always become heat and dissipate into the environment. No energy transformations occur without some energy being degraded from a concentrated to a more dispersed form. The functioning of complex entities involves numerous energy conversions. As energy is converted to do work, some of it changes into heat. Energy transferred as heat is still energy, but it is no longer useful for doing work. The quantity of energy is fixed, but its quality is not. As energy conversion chains progress, potential for useful work steadily declines. Entropy represents this dissipation of useful energy.

All complex structures require energy inputs to maintain their organization and to keep functioning. In a closed system, energy dissipation from entropy will lead to loss of complexity, greater homogeneity, and more disorder. In actuality, however, most energy conversions happen in open systems that interact with the surrounding environment. Complex entities can temporarily defy entropy by importing and metabolizing energy. They arise in a balance between the usable free energy in the environment, which they can put to work, and the entropy they throw off. Inputs of high-quality energy make it possible for complex structures to combat decay from within. In the process, they also dissipate large amounts of energy as heat, increasing entropy overall. As complex systems, living organisms maintain continuous energy inflows and outflows. Metabolism enables organisms to avert decay and stay alive by drawing energy from their environment. But they maintain their structures at the expense of increased contribution of entropy to the surrounding environment.[9]

Grasping the environmental dimensions of war and militarization requires understanding how energy is converted for military purposes. Militaries can be thought of as organic systems that continuously interact with their environments, engaging in transfers of energy and materials. Militaries must constantly find new sources of useful energy and develop more effective mechanisms for handling extraordinarily large energy flows. As complex organisms, military systems extract energy to do work and maintain their internal organization, while at the same time releasing low-level energy via entropy.

The forms of energy that can support this "military metabolism" are strictly limited. Other complex systems—including agrarian ecosystems and hydraulic networks—stake their claims to these finite energy sources as well. As it is transferred across different spatial scales, energy changes forms. But because the total amount of energy remains constant, appropriating energy in forms that are needed to fight or prepare for war necessarily entails losing it in others. Even if war and militarization lead economies to exploit new energy forms drawn from nature, they render energy unavailable for other purposes. Militaries struggle for strategic advantage, as well as for energy sources that drive their metabolism. The better militaries are able to gather, store, and deploy energy, the greater their potential for organized violence, coercion, and destruction. Military systems exploit finite sources of useful energy to maintain themselves, do work, and expand. At the same time, they release heat, pollution, and waste. Building complex military structures and expanding their realm of operations lead to disorganization, chaos, and degradation in the environments on which they depend. This degradation occurs at the level of ecosystems, as well as in the wastage of human bodies.

We can imagine far-reaching webs woven together by transfers of energy. Various actors that strive for power occupy and move from one node to another in these webs. By straining one thread in the web of energy transfer, actors may break another thread. Enormous transfers of energy occur across vast spaces, as military and nonmilitary actors renegotiate their roles and possessions. The catastrophes that occurred in Henan during World War II derived from the impact of channeling various forms of energy for military purposes, as well as the acute disturbances that these distorted flows created.

THE YELLOW RIVER FLOOD

The same energy that propels rivers drives all human activities, including the waging of war. As Richard White has observed, labor that humans expend trying to control rivers ties them so closely to the environment that they become fundamentally inseparable from it.[10] White also reminds us that energy and work intertwine with power. Sometimes, power measures rates of energy flow and energy use. At other times, power means doing work and effecting change. In other instances, power refers to the ability to command the energy and work of others. All these different meanings of power "involve the ability to do work, to command labor. To be powerful is to be able to accomplish things, to be able to turn the energy and work of nature and humans to your own purposes."[11] Energy harnessed as power becomes a means of attaining human ends and pursuing human goals, including military ones.

By the twentieth century, exploiting China's rivers and waterways for military-strategic purposes had plenty of historical precedents. From ancient times, China's extensive irrigation systems had been a "sword of Damocles—and almost anyone could snip the thread."[12] On many occasions, armies intentionally flooded China's rivers to gain an upper hand against their opponent. Chiang Kai-shek and his subordinates perceived the Yellow River in similar strategic terms. In June 1938, Nationalist armies transformed nature's awesome energy into a weapon by breaking the Yellow River southern dike in Henan to block the Japanese Army's advance. With this decision, the Chinese Nationalist armies exploited the river's energy to counter Japanese military power. Considered another way, manipulating rivers in this fashion was a means to alter the energy needs of the enemy. Breaching the dikes greatly increased the energy required for the Japanese Army to carry out its military advance as well as other logistical requirements. Those logistical requirements, of course, were also energy requirements.

Throughout World War II, Chinese and Japanese armies in Henan engaged in hydraulic warfare as they struggled to harness the river's energy and deploy it against their military adversaries. Following the river's diversion in 1938, Chinese and Japanese armies confronted one another across its new course, making it a strategically vital frontline area. Long after the initial flood, the Yellow River remained an actor in the military conflict, with battles between natural and human forces continuing to play out as part of the Sino-Japanese War. Military actors on either side expended huge amounts of energy working with, on, and against the river to attain objectives, undertaking projects to channel and redirect its flow to fortify their positions and deploy it against their enemies. Their struggles were over power; contests over who could control the river, manipulate its energy, and secure its benefits. Manipulating the Yellow River to attain military-strategic advantage also required power in another sense—the ability to gain advantage from the labor of others. The river demanded vast expenditures of human energy to match its own energies. To carry out wartime hydraulic engineering projects, military forces and water control agencies that allied with them had to mobilize massive flows of labor and materials. But as refugees fled Henan in the wake of the 1938 flood, those energy sources became extremely difficult to obtain. The task of providing these inputs placed an even greater burden on localities that had already been devastated by warfare and flooding.

The Yellow River was not a passive object, but acted to frustrate human efforts aimed at shaping its behavior for military-strategic purposes. As in times past, "hydrological systems kept twisting free from the grip of human would-be mastery, drying out, silting up, flooding over, or changing their channels. By doing so they devoured the resources needed to keep them under control or serviceable."[13] In a time of war, when armies devoured or destroyed virtually all available resources, this cycle grew even more vicious. Given the sheer level of damage, Henan's flooded area had precious little energy to devote to hydraulic engineering, let alone for repairing ecological devastation. Military systems drained energy out of local society, making it virtually impossible to recover from war-induced disasters. Energy is finite. If it is diverted to waging war, it is unavailable for the work of reconstruction.

Competition for energy and power played out not solely between Chinese and Japanese armies — or between these military forces and the river — but also between militaries and the local communities that they exploited. Tensions emerged between central authorities who wanted to exploit limited energy sources for national defense and localities in the flooded area that struggled to retain them for their own needs. War-induced distortion of energy flows rearranged the environment physically, disrupted human-constructed hydraulic and agroecological systems, and made their repair all but impossible. The Yellow River's hydroenergy benefited the Nationalist military in the short term by slowing the Japanese invasion, but the fateful decision to divert its course left the Nationalists stuck in a quagmire that was at least partly of their own making.

THE EFFECT OF WAR ON ENVIRONMENTAL INFRASTRUCTURE

Writings on war and the environment tend to describe the myriad ways in which military conflict—or preparation for it—degrades previously healthy or unspoiled landscapes. But Henan's wartime experience points to the importance of baselines in reconstructing narratives of war's environmental effects. All agricultural landscapes are human made or "anthropogenic." From late imperial times, human labor sculpted China's agricultural landscapes to an unusually high degree. Such hyperanthropogenic landscapes required vigilant management, as well as constant investments of labor and resources. Without those inputs, rapid and costly deterioration followed. Given the vulnerability of Henan's human-constructed landscapes to neglect, wartime population loss brought rapid and severe ecological decline. To put it another way, creating and maintaining agroecological and hydraulic systems consumed huge energy flows in the form of human labor power and materials. In the 1930s and 1940s, diversion of energy to military conflict led to environmental degradation and disorder.[14]

North China's anthropogenic environments were neither purely "natural" nor entirely the product of human artifice. Rather, human-engineered agricultural landscapes inhabited the "twilight zone between Nature and Culture" characteristic of what Emmanuel Kreike calls "environmental infrastructure." Examples of environmental infrastructure include cultivated landscapes, farms, fields, and water-management systems. Humans, as environmental actors, "work *with* nature (which is at once an actor and a medium), rather than dominating nature or being dominated by nature." Any changes in how human societies interact with and maintain their environment will have implications for environmental infrastructure as well.[15]

Prior to the Japanese invasion of the late 1930s, therefore, Henan's environment did not exist in any sort of pristine or natural condition. Centuries of intensive human exploitation and management thoroughly altered the landscape, removing forests and other vegetation to make way for farms. Ecological diversity, and hence resilience, declined dramatically. Population growth pressed resources to the limit; ecological buffers were lost.[16]

By the nineteenth century, Henan and the rest of North China confronted what many historians have labeled an environmental crisis. Deforestation led to intensified erosion and caused sediments to accumulate in the Yellow River, the Huai River, and many other river systems, necessitating higher dikes and resulting in more frequent and costly flooding. Loss of vegetation cover led to critical shortages of fuel as well as building materials.[17] Maintaining a balance between hydraulic and agricultural systems demanded ever-greater amounts of energy. During the early twentieth century, in Lillian Li's assessment, "Environmental decline, especially the siltation of rivers, was historically unprecedented. . . . The cumulative effect of centuries of deforestation, intensive land use, and excessive control of rivers posed a problem of greater magnitude than had ever been experienced." The frequency and scale of flood and famine in the Republican period (1911–49) "far exceeded anything that had occurred previously."[18] Under these circumstances, World War II caused dislocations that destabilized fragile environments and upset precarious ecological balances, triggering acute shocks.

Upheavals created by the Japanese invasion and the Yellow River's strategic diversion led to large-scale population displacement that dispersed energy embodied in human labor power and made intensive management of Henan's environmental infrastructure impossible. Rapid ecological damage was the inevitable result. Attaining and maintaining order in anthropogenic agroecological and hydraulic systems, as with militaries and other complex systems, takes energy. But the energy used to maintain armies and fight wars—or energy that is dispersed as a result of military dislocations—cannot be utilized for other purposes. Without adequate inputs of labor and resources, Henan's agricultural and hydraulic systems quickly spiraled into disarray and recovery became nearly impossible.[19] Military conscription drained labor power from local societies, while the army's appetite for energy took food supplies out of the hands of civilian populations. As the direction of energy flows changed, agricultural landscapes transitioned into "warscapes" transformed by destructive human action.[20]

THE HENAN FAMINE

As the dislocations of war dealt a severe blow to production of energy surpluses through agriculture, short-term climatic fluctuations added additional shocks. In 1941–42, Henan and other parts of North China experienced unpredictable weather patterns and low rainfall levels connected with a strong El Niño event, causing agricultural output to fall precipitously.[21] As Peter Perdue has observed, "Whether or not famine happens depends on the effect of climate conditions on interactions between officials, peasants, relief agencies, and markets for labor and grain." People's ability to sell their labor and resources on markets, not simply the aggregate availability of food, determines life or death. In late imperial times, Chinese officials had effectively mobilized relief programs to deliver food, cash, and work opportunities to starving communities.[22] In the early 1940s, wartime instability combined with state security interests to rule out effective interventions to assist Henan's rural population and avert famine.

Even as harvests collapsed, the imperatives of war demanded huge concentrations of energy. As a frontline region, Henan had to supply vast amounts of resources to the military. Satisfying this appetite required seizing energy and materials from already-devastated agroecological systems. The Nationalist regime collected taxes in kind and extracted an array of other levies to secure food for the army; other armies launched grain raids for the same purpose. In these circumstances, all combatants had to compete violently for food supplies, which took on the utmost military-strategic importance. In Hans van de Ven's words, "Battles for the harvest followed between the Nationalist state, its armies, local militaries, and the population, which the Japanese also joined."[23] The Chinese Communist Party's military forces actively participated in these food struggles as well. With military and state actors trying to capture more and more energy to fight a war, the energy available for other life forms dwindled. Channeling energy to the military distorted food entitlement relations in Henan and made grain scarce or unavailable for vulnerable segments of society.

It should be stressed that in the 1930s and 1940s, Henan and the rest of the North China Plain remained overwhelmingly an "advanced organic economy" without a significant industrial sector powered by fossil fuels. No mechanisms existed to extract and transport hydrocarbon energy, or to process it into forms that were useful for modern militaries.[24] Waging war in this particular region of China required massive concentrations of people, animals, food, and fuel in battle zones. Under what J. R. McNeill terms the "somatic energy regime," in which energy resources came primarily in the form of human and animal muscle, more people and livestock meant more productive power. Human and animal populations were a "flywheel in the society's energy system" that could be mobilized regardless of the abundance or scarcity of plant crops that were the primary source of energy. The only way to achieve political dominance and military success was to direct the somatic energy regime, harness its surplus, and apply it to fuel warfare. Because muscle was the main energy source, attaining power required controlling large numbers of people and livestock.[25]

Viewed from this perspective, agriculture was "a solar-energy system controlled by humans, in which the energy output of selected plants is monopolized for human purposes." States and militaries thus regarded humans as "ambulatory solar-energy storage systems."[26] Differential access to energy sources corresponded to power differentials between groups of people. Large-scale conscription of able-bodied males in wartime Henan—both to fight and to work—kept labor power away from agriculture, making recovery difficult. Extracting energy from a war-devastated environment influenced by climatic anomalies threw ecological systems into disarray, creating an "energy crisis" in which civilian populations could not obtain caloric intake needed for their minimal subsistence.

War-induced flood and famine displaced millions of people in Henan, forcing them to adapt to unfamiliar environments and find ways to secure the energy they needed to survive. Their ecological adaptations had an impact. Many flood victims earned income by gathering and selling energy stored in the form of organic biomass as fuel for heating and cooking. But few refugees could cut down trees. In a part of China that had been deforested for centuries, and where severe fuel shortages existed since the nineteenth century, refugees took whatever they could find. They gathered brush, grass, chaff, and roots—anything that could serve as fuel. Though fuel shortages existed in northern China for decades, refugee survival strategies made them more severe. Extraction of biomass deprived soils of organic nutrients and subjected land that was already jeopardized by erosion to greater damage.

Displaced people added pressure on scarce resources in areas of in-migration and dramatically altered environments, especially in the sites where China's wartime state resettled refugees to reclaim wasteland for agricultural production. China's Nationalist and Communist regimes alike envisioned land reclamation as a way to mobilize refugee labor power to boost agricultural production and strengthen China's military resistance. Much of this "wasteland" consisted of upland areas covered with trees and other vegetation. Unlike Henan, Shaanxi Province's Huanglongshan region, site of the largest wartime resettlement project, had forests. Land reclamation converted wooded landscapes for agricultural production, simplifying ecosystems, jeopardizing wildlife habitats, and accelerating soil erosion. All of these ecological changes carried deleterious consequences for the health and well-being of most refugee populations.[27]

AFTERMATH

World War II did tremendous damage to North China's natural landscape. But war-induced environmental decline is only part of the story. Assessments of war's ecological legacies also have to consider the capacity of societies to restore war-ravaged landscapes to productivity.[28] Along with presenting a particularly graphic example of the immediate impact of military conflict on the natural landscape, which has been the focus of previous environmental histories of warfare, Henan's experience also illustrates the resilience of human societies and their capacity to restore war-ravaged landscapes into productive agroecosystems. After World War II ended—and even as China was embroiled in civil war between the Nationalist and Communist parties—Henan's most badly damaged areas experienced resettlement, socioenvironmental reconstruction, and the re-creation of humanized agricultural landscapes.

Based on his environmental history of World War II in Japan, William Tsutsui asserts "that the effects of warfare on the environment (be they favorable or detrimental) are often less lasting and less significant than we might imagine." War's environmental legacies "are complex, contingent, and often surprisingly transitory."[29] Writing of the twentieth-century world as a whole, J. R. McNeill likewise writes that "combat had its impacts on the environment, occasionally acute but usually fleeting. More serious changes arose from the desperate business of preparing and mobilizing for industrial warfare." In his view, "patient labor and the processes of nature" have usually hidden the scars of war and "assimilated into the surrounding countryside the sites of even the most ferocious battles—except where there has been conscious effort to preserve the battlefields as memorials." He notes that dryland agriculture, as practiced in Henan and on the rest of the North China Plain, "recovered quickly from war, on average in about three years."[30] These observations ring true for

Henan's flooded area, at least to an extent. The rebuilding of agroecological systems and return of agricultural productivity to prewar levels occurred in a rather short time. In line with McNeill's assertion, agricultural output in Henan's flooded area returned to pre-1937 levels about three years after full-fledged peace and stability returned with the Chinese Civil War's conclusion in 1949. Human labor and investment remade landscapes of unparalleled devastation into environmental infrastructure.

In other respects, however, Henan's postconflict experience makes it necessary to qualify these conclusions about war's transitory environmental effects. Recovery did not result from nature's powers of regeneration alone. Repairing the ecological degradation caused by war-induced neglect came about only through active human management. Recovering from the damage caused by war-induced neglect of agroecosystems demanded huge flows of energy embodied in labor and resources, as well as agencies capable of channeling and coordinating them. After 1945, large-scale external assistance came from the United Nations Relief and Rehabilitation Administration (UNRRA), which launched redevelopment programs in war-damaged areas of China in conjunction with the Nationalist regime's Chinese National Relief and Rehabilitation Administration (CNRRA).

In 1946–47, tens of thousands of laborers supervised by UNRRA-CNRRA exerted their energies to return the Yellow River to its pre-1938 course. UNRRA-CNRRA offered considerable material support to refugees who returned to their homes in Henan's flooded area and assisted them in bringing land back under cultivation. In conjunction with the labor of Henan's rural populace, processes of postconflict ecological reconstruction garnered huge energy subsidies from these transnational relief agencies. Without those external inputs, human communities could not have remade war-torn landscapes into productive agroecosystems so quickly, and the environmental scars of warfare would have persisted much longer. With this external assistance, but mostly thanks to their own exertions, farmers reclaimed nearly all of the previously flooded area by 1949, and the landscape gave little indication of the disasters that had occurred only a few years earlier.

At the same time, the rapidly changing ecological conditions in Henan's flooded areas shaped the geostrategic terrain on which the final stage of the Chinese Civil War between the Nationalists and the Chinese Communist Party (CCP) was fought. The Yellow River took on renewed strategic importance in the late 1940s, as Nationalist military leaders sought to isolate

Communist forces in eastern Henan by returning the river to its pre-1938 course. Contrary to Nationalist intentions, the river's rediversion decreased the water barrier between Nationalist and Communist armies, affording the CCP greater mobility. The Chinese Civil War of 1946–49 proved much less devastating than World War II, which enabled recovery to proceed without serious impediment. Return of human population and agricultural recovery gave the Communists and Nationalists reason to exert control over the flooded area, making it a point of military contestation.

William McNeill once conceptualized food energy consumption by armies as the "macroparasitism of military operations."[31] To sustain themselves, militaries have to rob agriculturalists of a portion of their harvests. But any parasites that kill or harm their host do not last long. Just as viruses evolve less virulent strains that exploit hosts without killing them, armies have normally recognized the importance of protecting the farmers they exploit. But such symbiotic balances have not always existed. In times of crisis, as David Christian rightly notes, able rulers often become brutal and destructive predators. Less able—or more desperate—rulers employ destructive fiscal methods that they or their advisers know will undermine the basis of their power.[32] The direction of the Chinese Civil War in Henan depended largely on which side was able to secure flows of energy and resources for its military.

Struggles over energy flows in wartime Henan intersected with twentieth-century China's most significant political transformation. At the height of World War II and into the Chinese Civil War that followed, Nationalist "military macroparasitism" damaged its host society, seizing food supplies and drafting men as soldiers and laborers. Without effective structures to channel and coordinate large energy flows, the Nationalist military position in Henan crumbled. A more symbiotic balance, by contrast, undergirded expansion of Communist influence. In the wake of the 1942–43 famine, the CCP's forces improvised relatively stable and effective methods of exploitation that benefited armies and civilian populations alike. After 1945, the CCP proved adept at attracting returning refugees—and labor power—to its base areas in Henan's flooded area, enacting land reform in conjunction with land reclamation. In large part, the Communist capacity to innovate new, more stable ways of extracting energy from ecological systems explains the CCP's consolidation of power in local society.

For the Chinese Nationalist regime, a desperate need to extract energy for the army and to remold the hydraulic environment to meet its strategic priorities gave rise to socioecological disruptions that paved the way for its eventual military collapse in North China. The Chinese Communist Party's military forces in Henan, during the 1940s at least, proved more capable of adapting to changed environmental conditions and improvising ways of capturing the energy they needed to survive. To an extent, the contrast between the ecological impact of the Chinese Nationalist and Communist militaries was derived from their tactics. The Nationalist armies engaged in positional warfare while the Communists relied primarily on guerilla warfare, which gave them greater mobility and flexibility. Different military-strategic choices necessitated by the war against Japanese thus resulted in different ecological outcomes, which also had significant political implications.

CONCLUSION

The energy-centered approach employed in this essay complements other ways of thinking about the war-environment nexus. Edmund Russell, for instance, has suggested that analyzing military supply chains as food chains "will help us uncover the indirect and hidden, but absolutely essential, links between armed forces and civilian, agricultural and natural systems." Thinking in terms of food chains, as he notes, demonstrates "that the area of militarized landscapes extends far beyond battlefields and bases, growing ever wider as the supply chain lengthens."[33] For ecologists, trophic pyramids are used to represent the roles of different organisms within food chains. In terrestrial ecosystems, plants anchor the bottom level, herbivores the next, and predators the next up from them. "Species at each level depend not only on the level immediately below them, but on *all* lower levels—though their dependence becomes less apparent as the food chain lengthens."[34]

This mode of analysis also highlights the importance of energy flows. Though Russell does not dwell on the point, it is worth stressing that trophic pyramids map energy transfers between producers and consumers at each step in the food chain. As he explains, "The width of the pyramid represents biomass (the weight of organisms). Transforming energy from one form to another always comes at the cost of lost energy, so the biomass of each level must always be less than that of the level below it."[35] Russell usefully applies the model of a trophic pyramid to militarization's ecological effects: "Starting at the bottom, we can label the levels natural systems, agricultural systems, political, economic and technological systems, and armed forces." Armed forces depend on political, economic, and technological systems for their sustenance. "Less apparently but just as much, they rely on the agricultural and natural systems that support the political and economic systems. Moreover, since each level must harvest greater biomass than itself to survive, the impact of military consumption widens as one goes down the scale. This means that militarization grows ever more pervasive as it becomes ever less visible."[36] Fully grasping the ecological impact of warfare and militarization requires investigating energy conversions at every level of the food-web pyramid.

Seeking to expand the environmental history of warfare "beyond the battlefield" to the "host of semiperipheral contexts where war etched its distant imprint on the land," Matthew Evenden analyzes commodity chains, or "the linked labor and production processes involved in the making of a commodity from production to finished good."[37] As Evenden explains in his pathbreaking research on aluminum production during World War II, "far from dividing the environmental history of the Second World War into a series of national histories, commodity chains bridge the distance between places, point up the importance and irrelevance of international boundaries, and connect social and environmental change on several spatial scales. The commodity chain thus offers a useful angle of vision to help understand the dynamics of warfare and environmental change over distance."[38] Evenden examines the development of new geographies of production, military efforts to defend vital commodity chains, and environmental repercussions of these strategically important processes. As he shows, wartime expansion of aluminum production increased the character and the extent of environmental effects.[39] Commodity-chain analysis highlights the "unprecedented capacity of the Second World War to gather and scatter materials with untold human and environmental consequences, linking diverse locations with no necessary former connections."[40]

As a conceptual framework for investigating links between war and the environment, commodity-chain analysis also melds nicely with the mode of analysis employed in this study of World War II and its aftermath in Henan, which focuses on energy transfers to understand the ecological dimensions of war and militarization. Most significantly for our purposes, wartime expansion of aluminum commodity chains "required massive material and energy inputs" derived from multiple world regions, from extraction of tropical soils to the damming of rivers for hydroelectricity. What is more, as Evenden notes, "These critical links in the supply chain were bound together by a fossil-fueled, long-distance transportation system."[41] Commodity-chain analysis, like the concept of metabolism, directs our attention to how military systems acquire the inputs of energy and materials that they need to survive and function, as well as to the environmental consequences of these flows. Taking a cue from the approaches proposed by Evenden and Russell, we can explore the history of war through the lens of energy conversion to better understand its environmental consequences.

Because they also depended on constant energy inputs for their maintenance, intensively exploited agricultural landscapes and hydraulic systems like those that prevailed on the North China Plain proved especially vulnerable to war-induced disruption. Warfare upset finely honed human relationships with anthropogenic environments in Henan, triggering rapid and acute shocks. The military's insatiable appetite for energy drained the labor and resources needed to maintain environmental infrastructure, making it impossible to re-create a viable human-ecological order until conflict came to an end. Warfare, in other words, rendered postdisaster recovery even more difficult than usual by monopolizing the energy flows needed to pull things back together.

Multifaceted and multidirectional relationships played out between war, society, and the environment. Armies in eastern Henan intentionally disrupted hydraulic systems, after which they expended huge quantities of energy trying to manipulate rivers for strategic purposes. Warfare tore asunder agroecosystems and disrupted agricultural production, as Chinese and Japanese armies consumed increasing amounts of food energy to meet their metabolic demands. Military systems likewise extracted tremendous amounts of labor power from local societies in the form of soldiers and conscript labor, even as their actions caused population loss through death and displacement. Owing to precipitous war-induced population decline, local societies (or what remained of them) could not invest the labor and resources needed to maintain the agroecological systems from which they derived food and biomass. When hydraulic systems suffered war-induced disruption, flooding led to additional loss of labor and materials.

With military actors extracting even larger amounts of energy in their efforts to manipulate waterways, maintaining militarized hydraulic systems placed greater burdens on devastated localities. At the same time, Henan lost additional energy sources as refugees migrated west to Shaanxi, where their survival strategies generated additional environmental change. After World War II came to an end in the late 1940s, it took massive influxes of energy to draw human labor power back into eastern Henan's flooded area, making it possible to repair war-ravaged environmental infrastructure and restore agricultural landscapes to productivity.

Notes

1 For existing works that examine the flood from a military perspective, see Qu Changgen, *Gongzui qiåanqiu: Huayuankou shijian yanjiu* [Merits and wrongdoings for a thousand years: research on the Huayuankou incident] (Lanzhou: Lanzhou daxue chubanshe, 2003); Diana Lary, "The Waters Covered the Earth: China's War-Induced Natural Disasters," in *War and State Terrorism: The United States, Japan, and the Asia-Pacific in the Long Twentieth Century*, ed. Mark Selden and Alvin So (Lanham, MD: Rowman and Littlefield, 2004). Most of the empirical information in this chapter draws upon Micah S. Muscolino, *The Ecology of War in North China: Henan Province, the Yellow River, and Beyond, 1938–1950* (Cambridge: Cambridge University Press, 2015).

2 Han Qitong and Nan Zhongwan, *Huang fan qu de sunhai yu shanhou jiuji* [Damage and recovery and relief in the Yellow River flooded area] (Shanghai: Xingzhengyuan shanhou jiuji zongshu, 1948), 22–23.

3 The oldest and still the most influential account is Theodore H. White and Annalee Jacoby, *Thunder Out of China* (New York: W. Sloane, 1946). For more recent accounts, see Lary, "Waters Covered the Earth"; Odoric Wou, "Food Shortage and Japanese Grain Extraction in Henan," in *China at War: Regions of China, 1937–1945*, ed. Stephen R. MacKinnon, Diana Lary, and Ezra F. Vogel (Stanford, CA: Stanford University Press, 2007).

4 Largely because of flooding and famine, during the Sino-Japanese War, Henan had a larger refugee population than any other province in China. From 1937 to 1945, an estimated 14,533,200 people in Henan, an astounding 43 percent of the province's total prewar population, lived as refugees for a least a time. Stephen R. MacKinnon, "Refugee Flight at the Outset of the Anti-Japanese War," in *The Impact of War on Modern China*, ed. Diana Lary and Stephen MacKinnon (Vancouver: University of British Columbia Press, 2001), 122.

5 My conceptual framework draws considerable inspiration from the insights provided by Mark Fiege, "Gettysburg and the Organic Nature of the American Civil War," in *Natural Enemy, Natural Ally: Toward an Environmental History of War*, ed. Richard P. Tucker and Edmund Russell (Corvallis: Oregon State University Press, 2004); and Simo Laakkonen, "War, an Ecological Alternative to Peace? Indirect Impacts of World War II on the Finnish Environment," in Tucker and Russell, eds., *Natural Enemy, Natural Ally*.

6 Helga Weisz, "Combining Social Metabolism and Input-Output Analyses to Account for Ecologically Unequal Trade," in *Rethinking Environmental History: World-System History and Global Environmental Change*, ed. Alf Hornborg, John Robert McNeill, and Juan Martínez-Alier (Lanham, MD: Altamira Press, 2007), 291–92. This approach has also benefited greatly from Joan Martinez-Alier, *Ecological Economics: Energy, Environment and Society* (New York: Blackwell, 1987); Joan Martinez-Alier, "Marxism, Social Metabolism, and International Trade," in Hornborg, McNeill, and Martinez-Alier, eds., *Rethinking Environmental History*; Marina Fischer-Kowalski and Helmut Haberl, *Socioecological Transitions and Global Change: Trajectories of Social Metabolism and Land Use* (Northampton, UK: Edward Elgar, 2007).

7 Edmund Burke III, "The Big Story: Human History, Energy Regimes, and the Environment," in *The Environment and World History*, ed. Edmund Burke III and Kenneth Pomeranz (Berkeley: University of California Press, 2009), 35. See also Richard White, *Organic Machine: The Remaking of the Columbia River* (New York: Hill and Wang, 1995), 4–5.

8 David Pimentel and Marcia H. Pimentel, *Food, Energy, and Society*, 3rd ed. (Boca Raton, FL: CRC Press, 2007), 9; Gerald G. Marten, *Human Ecology: Basic Concepts for Sustainable Development* (London: Earthscan, 2001), 109; Vaclav Smil, *Energy in Nature and Society:*

General Energetics of Complex Systems (Cambridge: Massachusetts Institute of Technology Press, 2008), 4–5.

9 Smil, *Energy in Nature and Society*, 6–7; Marten, *Human Ecology*, 109–10; Pimentel and Pimentel, *Food, Energy, and Society*, 9–11; David Christian, *Maps of Time: An Introduction to Big History* (Berkeley: University of California Press, 2005), Appendix II; Burke, "The Big Story," 34.

10 Richard White, *Organic Machine: The Remaking of the Columbia* River (New York: Hill and Wang, 1995), 4–5.

11 Ibid., 14.

12 J. R. McNeill, "China's Environmental History in World Perspective," in *Sediments of Time: Environment and Society in Chinese History*, ed. Mark Elvin and Liu Ts'ui-jung (Cambridge: Cambridge University Press, 1998), 46. See also Ralph Sawyer, *Fire and Water: The Art of Incendiary and Aquatic Warfare in China* (Boulder, CO: Westview, 2004).

13 Mark Elvin, *The Retreat of the Elephants: An Environmental History of China* (New Haven, CT: Yale University Press, 2004), 164.

14 Robert B. Marks, *China: Its Environment and History* (Lanham, MD: Rowman and Littlefield, 2012), 336; Elvin, *Retreat of the Elephants*; idem, "Three Thousand Years of Unsustainable Growth: China's Environment from Archaic Times to the Present," *East Asian History* 6 (December 1993): 7–46.

15 Emmanuel Kreike, "Architects of Nature: Environmental Infrastructure and the Nature-Culture Dichotomy" (PhD thesis, Wageningen University, 2006), 18–19. These ideas are refined and elaborated in Emmanuel Kreike, *Environmental Infrastructure in African History: Examining the Myth of Natural Resource Management in Namibia* (Cambridge: Cambridge University Press, 2013).

16 Marks, *China*, 243.

17 A relatively abundant literature exists on the environmental crisis that prevailed in North China from the 1800s through the 1930s. See especially Kenneth Pomeranz, *The Making of a Hinterland: State, Society, and Economy in Inland North China, 1853–1937* (Berkeley: University of California Press, 1993); Randall A. Dodgen, *Controlling the Dragon: Confucian Engineers and the Yellow River in Late Imperial China* (Honolulu: University of Hawai'i Press, 2001); Elvin, *Retreat of the Elephants*; Lillian M. Li, *Fighting Famine in North China: State, Market, and Environmental Decline, 1690s–1990s* (Stanford, CA: Stanford University Press, 2007); Kenneth Pomeranz, "The Transformation of China's Environment, 1500–2000," in Burke and Kenneth, eds., *The Environment and World History*; Marks, *China*, 235–43.

18 Li, *Fighting Famine in North China*, 307. For a survey of flood and drought disasters in Henan during the Republican period, see Su Xinliu, *Minguo shiqi Henan shui han zaihai yu xiangcun shehui* (Zhengzhou: Huanghe shuili chubanshe, 2004). Scattered oral histories and memoir literature are usefully collected in Wen Fang, ed., *Tianzai renhuo—Tianhuo* (Beijing: Zhongguo wenshi chubanshe, 2004).

19 Micah S. Muscolino, "Violence against People and the Land: The Environment and Refugee Migration from China's Henan Province, 1938–1945," *Environment and History* 17, no. 2 (May 2011): 291–311.

20 The term "warscapes" is drawn from Carolyn Nordstrom, *A Different Kind of War Story* (Philadelphia: University of Pennsylvania Press, 1997).

21 S. Brönnimann et al., "Extreme Climate of the Global Troposphere and Stratosphere in 1940–1942 Related to El Niño," *Nature* 431 (October 2004).

22 Peter Perdue, "Is There a Chinese View of Technology and Nature?," in *The Illusory Boundary: Environment and Technology in History*, ed. Martin Reuss and Stephen H. Cutcliffe (Charlottesville: University of Virginia Press, 2010), 114.

23 Hans van de Ven, "The Sino-Japanese War in History," in Peattie, Drea, and van de Ven, eds., *Battle for China*, 458.

24 The concept of "advanced organic economy" originated with E. A. Wrigley, *Continuity, Chance, and Change: The Character of the Industrial Revolution in England* (Cambridge:

Cambridge University Press, 1990). Late imperial and early twentieth-century China has been characterized as an advanced organic economy by Marks, *China*, as well as idem, *The Origins of the Modern World: A Global and Ecological Narrative* (Lanham, MD: Rowman and Littlefield, 2002).

25 J. R. McNeill, *Something New under the Sun: An Environmental History of the Twentieth-Century World* (New York: Norton, 2001), 11–12.

26 Burke, "The Big Story," 36.

27 Micah S. Muscolino, "Refugees, Land Reclamation, and Militarized Landscapes in Wartime China: Huanglongshan, Shaanxi, 1937–45," *Journal of Asian Studies* 69, no. 2 (2010) 453–78; idem, "Violence against People and the Land."

28 Richard P. Tucker, "War and the Environment," *World History Connected* 8, no. 2 (May 2011): http://worldhistoryconnected.press.illinois.edu/8.2/forum_tucker.html.

29 William M. Tsutsui, "Landscapes in the Dark Valley: Toward an Environmental History of Wartime Japan," *Environmental History* 8, no. 2 (April 2003): 295.

30 McNeill, *Something New under the Sun.*

31 William H. McNeill, *Plagues and Peoples* (New York: Anchor, 1998), 72.

32 Christian, *Maps of Time*, 322.

33 Edmund Russell, "Afterword: Militarized Landscapes," in Coates and Cole, eds., *Militarized Landscapes*, 237.

34 Ibid., 236.

35 Ibid., 236. On energy transfer within food webs, see Smil, *Energy in Nature and Society*, 113–18.

36 Russell, "Afterword," 236–37.

37 Matthew Evenden, "Aluminum, Commodity Chains, and the Environmental History of the Second World War," *Environmental History* 16, no. 1 (January 2011): 70.

38 Ibid.

39 Ibid., 71.

40 Ibid., 88.

41 Ibid., 83.

Chapter 7
Environmental Scars in Northeastern India and Burma

RICHARD P. TUCKER

Along the great arc from China through Southeast Asia to the Indian subcontinent, the period of war and mass upheaval lasted far longer than 1945. In South and Southeast Asia, warfare and its ecological impacts were inextricably linked to the final stage of decolonization. Environmental disruption resulted from the combined stresses of war and freedom struggle. In these final years of British colonial administration, the scars of battle centered in the Brahmaputra watershed in northeastern India and the Irrawaddy watershed in Burma.[1] But wartime impacts continued to reverberate thereafter. Turbulent times led directly to Burma's independence in January 1948, at the cost of a fractured country, and India's independence in August 1947, at the cost of the creation of Pakistan out of northeastern and western India. The enormous population displacement and administrative reorganization of the late 1940s intensified ethnic and religious conflicts and extended the environmental consequences of the previous wartime conditions. The political and social dramas of the entire decade have commanded almost all historians' attention; the environmental consequences have never been seriously studied.

GEOGRAPHICAL, ETHNIC, AND POLITICAL BACKGROUND

The war between imperial Japan and the combined forces of Nationalist China, Britain, and the United States had two converging fronts in this region: one along the Bay of Bengal lowlands from Lower Burma into Bengal, and the other in the remote region where the eastern ranges of the Himalayas meet the north-south ridges of Upper Burma and the Brahmaputra River basin. The rugged uplands boast peaks as high as 20,000 feet that define the

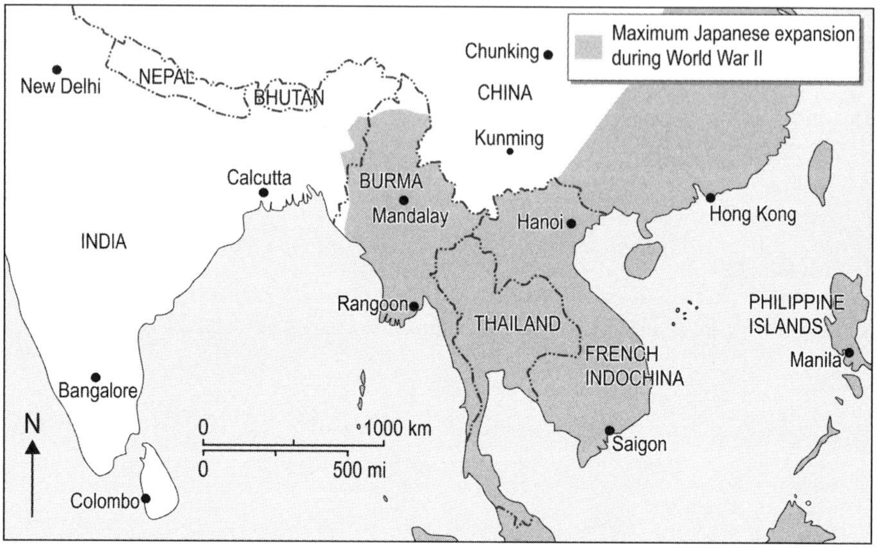

Figure 7.1. Burma and northeast India in the broader region of Japan's wartime expansion.

watersheds of the Irrawaddy River and its western branch, the Chindwin, and the Salween River to the east. The lowlands of the river basins have supported dense rice-growing populations for many centuries. But the mountain region is so remote from centers of people and power that the outside world hardly began to penetrate it before World War II.

Heavy monsoon rains, some of the heaviest on the planet, wash the entire region every June through September, which makes any sort of travel extremely difficult. Tropical monsoon forests cover most of the mountain region, except for heights above timberline. As a transition zone connecting several major biotic regions of Asia, this China-Burma-India border region is home to some of the world's greatest variety of flora and fauna species. In the lower hills to the south, hardwood species dominate the forest: sal (*Shorea robusta*) for construction, and teak (*Tectona grandis*) for many uses on international markets.[2] Mountain forests' soils are susceptible to severe erosion during heavy summer monsoon rains. In parts of this region the soils are vulnerable to laterization, whereby iron oxides turn them into degraded hardpan that is difficult to till for settled agriculture.

The highlands of the Brahmaputra above lowland Assam and eastward into Upper Burma have supported a wide ethnic variety of hill peoples. These tribal groups practice subsistence by hunting and gathering, and burning small patches of forest in the dry season to grow a variety of food crops.[3] Relations between indigenous tribal groups and ethnically different lowland Bengalis and Burmese were always complex and often conflicted. Those rivalries intensified under British colonial administration, as increasingly dense lowland populations began to encroach on hill tribes and their lands. Northeastern India fell under British rule in stages from the 1750s onward, beginning with the establishment of the previously insignificant Ganges delta town of Calcutta as the imperial capital. The Brahmaputra Valley northeastward into Assam came gradually under effective British administration in the course of the 1800s, driven by the availability of tropical timber from the rich forest resource, as well as wide acreage in the middle-elevation hills that became the Assam tea plantations.

Beyond the middle hills, the high mountain hinterland of Assam that linked to northern Burma experienced only a tenuous extension of British administration. Nagas, Garos, Khasis, and other ethnic minorities saw running conflict with both British and Bengalis over the decades. Naga headhunters in particular fought encroachment from below. After World War I, a British Army force invaded the Naga hills, subduing rebellious tribes by burning villages and crops.[4] The resulting uneasy peace allowed traders, displaced lowland peasants, military veterans, and missionaries to immigrate. The missionaries' impact was striking: by 1940, nearly one-third of the northeast tribal population was Christian.

The rice paddies and teak forests of Burma[5] also attracted aggressive interest from private British firms, and British India extended its reach into Rangoon and the Irawaddy River basin.[6] By 1855, the imperial British controlled Lower Burma, the flat plains dominated by ethnic Burmese (also called Burmans), administering it thereafter as a province of British India until it was granted internal autonomy in 1935.[7]

The British conquest of Upper Burma was completed in 1886, though its remote borders were only vaguely delineated on the ground. The British achieved limited administrative control thereafter, centering on systematic management of the teak forests.[8] To open the region to resource extraction, they constructed a railway north from Rangoon through Mandalay to the town of Lashio, near the frontier with Yunnan on the Chinese side of the mountains. That rail line was the first major transport artery cut through the teak forests of the region; it was to play a critical role in the war against Japan.

British Frontier Service officials consolidated their influence by organizing militias under Chin, Kachin, and Shan chiefs, thereby strengthening tribal power structures and intensifying the perennial stress between the ethnic groups of the northern hills and the ethnic Burmese of the lowlands. This strategy would be vital to shattering the Japanese occupation armies in 1944, but it had long-term consequences for both politics and the environment. The tension between ethnic Burmese and the minorities was magnified during the war, exacerbating the resource-use turmoil that the Japanese invasion caused.

MILITARY OPERATIONS IN THE WAR ZONE

The environmental disruptions of international war began in the mountain region as early as 1937, shortly after the Japanese invasion of the Chinese mainland, which forced the Chinese Nationalist (Kuomintang) command southwest into Yunnan.[9] At first the British were able to supply the Kuomintang through Canton and the southeast coast. But when the Japanese military closed Canton to the Europeans, the British had to look for a different supply route to the new Nationalist headquarters in Kunming.[10] One possible route lay through the previously unpenetrated mountain ranges of northeastern India and northern Burma. With maximum urgency they began building the Burma Road, which stretched westward for 717 tortuous miles from Kunming to Lashio, where it met the terminus of the north-south railway. The first motorable road across the eastern Himalayas, 200,000 Chinese and Burmese laborers completed the project in two years.

In the early months of 1942, Japanese forces swept through Southeast Asia into Rangoon and its hinterland, capturing Lower Burma with its vital rice production in March. Chaos descended on Rangoon; from then on, throughout the war, the Burmese capital underwent the massive human suffering and environmental disruption of many war-torn cities. Refugees struggled in from the countryside, hoping for safety and supplies. They faced hunger and disease as garbage and sewage accumulated, and public health and distribution networks were torn apart. Malaria in particular was rampant, compounded by dysentery and other diseases.

The British command retreated along the Arakan and Chittagong coasts to Calcutta, destroying as they left most oil and mining installations, along with tens of thousands of boats and motor vehicles.[11] As one writer described the scene, "Commanderless soldiers, civilian officials, armyless commanders, hordes of refugees, individual planes and plane crews, all falling back in confusion before the Japanese advance, were choking . . . highways, railways, airplanes . . . and communications."[12] Many Indians, perhaps as many as 100,000,

Figure 7.2. Serpentine section of the Burma Road, 1944. Source: US National Archives, photo # 292561.

also attempted to flee over the Taungup Pass to the Arakan coastal route. In the process, they disrupted villages and water supplies, spread cholera, and created massive waste and pollution along the way.[13]

Another flood of refugees fled northward from Rangoon and other lowland cities, including more than 100,000 Europeans and Indians. Japanese forces quickly took control of the southern portion of the Burma Road, preventing the British Fourteenth Army and American forces from supplying the Nationalist Chinese in Yunnan by land. Communication lines became extremely tenuous for the Allies along a seven-hundred-mile front, which had almost no all-weather roads. There was one narrow-gauge railway, reaching to within 110 miles of Imphal, but "it was washed away by floods, broken by bombing, swept down by landslides, closed by train wrecks and finally shattered by an earthquake. . . . A huge programme of all-weather road construction proceeded from 1942 to the end of the war."[14] The Japanese continued northward through the premonsoon months, occupying the key town of Myitkyina, and forcing the Chinese Army northward beyond the Upper Chindwin River Valley, to defend the only tenuous land connection between India and China.

Descriptions of the military operations and the refugee flood give vivid indications of the pressures on the land. In May, the British retreated up the Chindwin Valley. One major route led through the remote Hukawng Valley, where they left corpses, baggage, and muddy damage in the attempt to reach Imphal, Kohima, and other hill towns on the Indian side of the border.[15] Their chaotic retinue included over 100,000 Indian refugees, "accompanied . . . by many hundreds of terrified low-grade Indian troops . . . without officers, without discipline, without shame, who robbed and murdered the villagers, in some degree justifying the hatred felt by Burmans for all Indians. . . . Perhaps half the refugees died in the last stages of the retreat, of starvation, malaria, cholera and sheer loss of hope. The jungle on either side of the track was littered with their putrefying corpses, the stench overpowering and the disease-risk considerable."[16]

Teak forests in central Burma also felt immediate pressure. As Japanese troops moved up the Irrawaddy River into the hills, they began intensive logging for the military campaign. (This was consistent throughout the war: throughout the South Pacific and Southeast Asia, the Japanese intensively exploited timber resources.) The colonial forest management system had

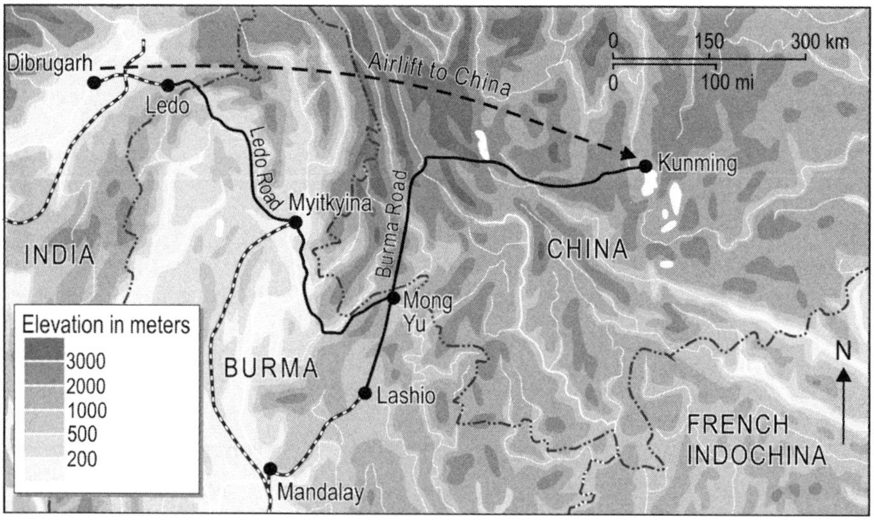

Figure 7.3. Connecting Allies. Assam, northern Burma, and the supply routes over the Himalayas to Chinese Nationalist headquarters in Kunming.

established sustainable rotational fellings, but Japanese timber requisitions paid no attention. The Japanese occupation also enabled lowland peasants and hill tribes to ignore former restrictions on their access to the reserved forests. Landless peasants, desperate under wartime conditions, took advantage of the collapse of administrative authority, competing with Japanese timber cutters to turn teak into money.[17]

On the Indian side of the battle lines, Assam underwent intensive environmental pressures during the war, on the rich soils of the Brahmaputra River Valley, the vast tea plantations of the middle hills and the high regions beyond in the tribal areas. Refugees from the coastal lowlands under Japanese invasion fled northward into the Chittagong hills of eastern Bengal, where improvised and friction-filled relations with local tribal villagers produced forests pockmarked by clearings. This migration had been a long-term process since the 1800s, but it accelerated chaotically during and immediately after the war, as many displaced lowlanders became long-term squatters on Forest Department lands.[18] Beyond, in the upper reaches of the Brahmaputra Valley, 40,000 laborers conscripted from the tea estates built roads and airfields as far northeast as Dibrugarh, the town where they built a major airfield for supply planes to fly "over the Hump" across the eastern Himalayas to Kunming. As the conflict wore on, a number of smaller airstrips were also constructed through the region.[19] The penetration of the forest had begun. The long-term postwar environmental impact in that area would also reflect the settlements that arose around the new transport system: workshops, housing, and bazaars.[20] In more distant support of these operations, throughout the four war years 1942–45, British and US forces from Delhi to Dimapur in eastern Assam placed heavy demand on transport, especially the railways, which led (among other consequences) to critical problems in food distribution throughout eastern India.

In northern Burma, two Chinese armies had been stationed by mid-1941. British officers were in charge of them, supported by the American Volunteer Group of about a hundred planes and pilots. At first the American pilots were outmaneuvered by the superior Japanese planes, but as the fighting wore on, the Americans gradually gained control of the air.

Both sides in the struggle faced desperate disease problems. "In 1942, for every [British] man evacuated with wounds, 120 were evacuated sick. The common ailments were dysentery, jaundice, jungle sores, and, far worse than all the rest, malaria."[21] The struggle to control malaria was an immemorial

Figure 7.4. Constructing the Ledo Road.

challenge for military operations. Malaria remained a scourge of all forces throughout tropical areas in the early stages of this war as well. By 1945, British medical personnel achieved considerable malaria control in Burma with the prophylactic pill mepacrine. The Japanese Army had none.

In a countermove in 1943–44, British and US troops built the legendary Ledo Road, three hundred miles from Ledo at the upper end of the Assam Valley across the mountains to Myitkyina, where it connected to the Burma Road north of the farthest Japanese advance. In contrast with the earlier Chinese effort, this road was built with heavy earth-moving equipment, and included two oil pipelines at its side. Even with that industrial power, the task mobilized 17,000 American engineers and 35,000 Chinese troops at the cost of $150 million. The first supply trucks drove to Kunming in January 1945; by the end of the war in August, 26,000 trucks arrived there and were left for the Chinese Nationalists in their subsequent futile struggle against Mao's Red Army.[22] The road was abandoned in November 1945, almost as soon as it was completed, and the forest began to heal along the gash.

Military roads constructed in fragile mountain environments under emergency conditions inevitably leave deep scars. They also open remote terrain to subsequent population movements, providing deforestation pressures for overflow populations from the lowlands. In the case of the Ledo Road in

Figure 7.5. American troop convoy on the Ledo Road, 1945. Source: US National Archives, photo #535540.

Assam, this effect was limited, as the postindependence Indian government closed the India-Burma frontier for some thirty years. But the war did create unprecedented access to the Burmese uplands, contributing to the pressures on the region's species-rich wildlife habitat.

In the lowlands to the south, Burmese Nationalists used the war to increase their leverage to wrest freedom from the British yoke. At first they tended to see the Japanese as liberators, but Japanese authority ranged from severe to brutal, soon alienating the Burmese population and weakening the Japanese Army's ability to extract food and other resources. In their turn, ethnic Burmese military units used the situation to exacerbate ethnic tensions by harshly repressing the lowland Karen minority.[23]

Many semiautonomous armed guerrilla groups arose in rural areas, as well as Burmese cadres operating in Rangoon by war's end.[24] Mary Callahan's analysis shows that

> the need for immediate, cost-effective access to resources and the Burma Road necessitated negotiation and compromise with indigenous social forces. . . . The imperial military dispensed to cultivators, workers, students, traders, bandits, and others a range of

> new skills, new outlooks, and new reasons to leave their homes. When Japan's Burma Area Army headquarters attempted to bring this unruly situation under control, they found that disbanding and sending home guerrilla fighters was a far more difficult task than mobilizing them. . . . Once armed, most of these guerrilla groups stayed armed [and] reasserted themselves in the anti-Japanese resistance, the postwar anti-British movement, and the postcolonial insurgency.
>
> The brief unification [1943–45] of more than a dozen different anti-Japanese groups was less significant for the future of Burma than the institutionalization of difference and intolerance among indigenous groups competing for political power over territory in Burma.[25]

The turmoil intensified ethnic violence in the hills as well. In the Kachin and Shan areas, Chinese troops retreated eastward into Yunnan in 1942, living off the land and looting villages as they went. By 1943, British and American troops intensified the arming of ethnic groups in the frontier region, motivating them by emphasizing the tribes' long-standing distrust of the lowland Burmese. They included American-led Chinese posted in the Hukawng Valley, and Chindit irregular forces and Kachins based in Fort Hertz, which had been built in 1914 as the most remote British outpost. Fort Hertz was a key hub in the defense of Hump flights and the Ledo Road.[26]

The first half of 1944 was the turning point in the war. Early that year, "a build-up of Japanese divisions in Burma, the pattern of road construction, logs being stock-piled along the Chindwin, herds of cattle being driven westwards" presaged the final massive Japanese offensive toward Imphal and Kohima, to capture Assam's airfields and cut the main railway between Lower Bengal and Assam.[27] Fighting intensified as the Allies cut many new tank tracks through the hill forests east of Imphal. American planes were diverted from flying the Hump to drop supplies to tribal guerrilla units in the jungle. Aided by tribal spotters on the ground, the planes were soon "air-dropping nearly a million and a half pounds of supplies each month to the units behind Japanese lines."[28] Air power was an important element as the Allies gradually gained the upper hand.

By early 1945, the Japanese Army began a southward retreat. By early June, the start of intensive monsoon rains was disastrous for Japanese supply lines. "Now all they could hope for was to get back to the Chindwin Valley without dying of starvation, malaria, typhus, or sheer exhaustion.

Some rearguards still fought, but every road and track was littered with the debris of defeat—abandoned arms and equipment, heaps of rotting rice and corpses."[29]

The British Fourteenth Army, which included Punjabi, Gurkha, and African units, slaughtered Japanese troops who were abandoned by their officers. From the north, 10,000 British, American, and Chinese forces, with their Kachin and Chindit guerrilla units, cut off other Japanese forces.[30] The Japanese Army had lived off the land, but by now there was no food left in the devastated villages and farms as they retreated southward toward Rangoon. With no vehicles, they organized animal power for the campaign, including 12,000 horses and mules, 30,000 oxen, and 1,000 elephants. "The scale of animal fatality was colossal. During the campaign Japanese horses survived only fifty-five days on average and mules seventy-three days. All the horses and mules had died by August and the cattle had also perished or been eaten. Only the elephants survived."[31]

By July, 80,000 Japanese troops were slaughtered, drowned, or starved, the largest number of losses in any Japanese battle of the entire war. In an ironic twist, this campaign of Allied bombing of transport and communications infrastructure reduced teak cutting and the rate of damage to the teak forests.[32] The war officially ended in August, when the Japanese emperor surrendered to the Allies shortly after two atomic bombs destroyed Hiroshima and Nagasaki. Over a period of months, Japanese fighters fled from Southeast Asia, leaving chaos in their wake.

PRESSURES ON LOWLAND FOOD AND AGRICULTURE

From early 1942 onward, another social tragedy had developed in the lowland regions from Burma to Bengal, far greater in human numbers than the nightmare in the hills, but ultimately much less environmentally damaging than on steeper terrain. In March 1942, Japanese warplanes had bombed the coastal towns on the eastern shores of the Bay of Bengal, along the Arakan and Chittagong coasts. Population displacement from the coastal rice paddies severely reduced the region's rice production. Refugees headed into the cities, placing severe stress on housing and infrastructure, including water and food supplies. In the densely populated lowlands of Bengal, there was urgent additional need for food, especially for the military, but also for workers in military industries, who required increased intake of calories beyond what their peacetime diet provided. A region-wide food crisis erupted after

Burmese rice imports were cut off in March; it intensified that October, when a cyclone flooded the Ganges-Brahmaputra Delta and destroyed the autumn rice harvest.

From early 1943 onward, the market distribution system for rice suffered severe inflation and then a massive breakdown. Widespread rural malnutrition was exacerbated by epidemics of smallpox, cholera, and malaria. Most migrants from the countryside went to cities, towns, and military cantonments, where they could hope for food distribution. But grain traders hoarded their stocks, forcing retail prices to spiral upward. The wartime administration, stretched to its limits, was unable to control the market. And with heavy military demands on the railway system, the administration gave priority in food deliveries to troops and their civilian support workers. The immense Bengal famine of 1943–44 resulted.

One authoritative study estimates that 3.5 million people died in this massive human tragedy.[33] In the urban areas in and around Dacca and Calcutta, the influx of refugees was so immense that no administration could coherently handle it. Water supplies were more inadequate and polluted than ever, and refuse collection was overwhelmed. Squatter settlements grew rapidly in and around marshy areas in these deltaic cities, their malnourished population growing more susceptible to malaria and other diseases.

FOREST RESOURCES DURING THE WAR

Timber was urgently needed for the war effort. The impact of the war on the forests of northeastern India was part of the larger picture of logging operations throughout the subcontinent. In the government's reserved forests throughout British India, the Forest Department managed to accelerate hardwood timber harvests for the Middle East battle zone despite serious manpower shortages. Two-man saws and axes were wielded throughout the subcontinent: in the timber reserves of northern Bengal and Assam, beyond in the deodar cedar forests of high elevations in the western Himalayas, in the pine forests of the lower hills, and in the sal forests of the Terai lowlands (some of which were soon cleared for settling Punjabi refugees in 1947–49), the wide forests of the Vindhya hills in central India, and the teak plantations along the southwest coast.[34] New logging roads were built, many of them cutting across steep slopes. There was unavoidable damage to the dominant timber species, as well as collateral damage to understory vegetation, soil, and downstream riverine environments.

After the war, the Indian Forest Service issued a remarkable assessment that no permanent damage had been done to the silvicultural cycle in the reserved forests. Some observers suspected that this conclusion embodied a strong element of self-justification, but there was little disagreement that in privately owned forests, in response to wartime inflation, there had been rapid and unsystematic fellings for urban markets. The administration had no control over that profiteering, and many smaller woodlands were largely denuded by Independence Day in August 1947.[35]

CONTINUATION, 1945–49: BURMESE INDEPENDENCE AND THE PARTITION OF INDIA AND PAKISTAN

The final days of the Allied reconquest of Lower Burma during the monsoon months of 1945 left the political situation, centering in Rangoon, totally unstable.[36] The Burmese provisional government under Aung San claimed full independence, but it had little capacity to function as an effective administration. Moreover, intensified wartime rivalries with the newly armed ethnic minorities in the north and east left the Rangoon government incapable of controlling the diverse country. Burma became officially independent in January 1948, but ethnic insurgencies erupted a year later, beginning with the Karens, initiating what some call the world's longest civil war in modern times. The long-standing tension between ethnic Burmese and the minorities had intensified during the war, exacerbating the resource-use turmoil caused by the Japanese invasion. Thereafter the Burmese Army attacked the minorities, many of whom fled across the eastern border into Thailand.[37] The junta's attempts to wear away its ethnic opponents included massive uncontrolled cutting of formerly managed teak forests, as the government's military leaders struck deals with Thai timber merchants, while some minority resistance groups adopted similar strategies.[38] By the 1980s, this became a classic instance of reckless extraction of valuable natural resources to finance civil wars. An unstable truce was declared in 1994, but orderly management of the country's forest resources remains an elusive goal.

In the natural forests of the mountain region of northern Burma, flora and fauna began to reclaim territory after 1945, as the international armies left and villagers began reestablishing their lives as best they could. The Ledo Road, which had bisected the forest from Ledo to Fort Hertz, began to crumble. Longer-term perspective on the damage and secondary succession of neonatural forest comes from the writings of Alan Rabinowitz, an

American wildlife ecologist who was admitted into Burma by the military junta to study wildlife in the north in the 1990s.[39] His reports showed that in the Hukawng Valley there were still just a few scattered tribal villages and only a rudimentary transport system. "Once the road fell into disrepair, anything and everything salvageable was used or hauled away by the local people. When the [Burmese government's army] controlled the area during the civil conflicts of the 1980s, many of the bridges that hadn't already collapsed were intentionally destroyed to hamper movement of the Myanmar army from Yangon." But there were still many "skeletons of bridges, along with miles of old rusting pipe, battered oil drums, and numerous unrecognizable pieces of military hardware, now taken over by the jungle." At one town that had been a staging area for American forces from India, there was "a large airstrip cut out of the jungle that still looked usable, and the remains of an American-built motor pool facility in the village."[40]

The tribal men were hunters, selling various wildlife trophies to Chinese buyers from Yunnan, but "there were many indicators of a relatively healthy, intact forest, including our almost daily sightings of great hornbills and gibbons, the pugmarks of leopards and dholes, and the sign of tigers, wild elephants, and gaur. Only the rhino was gone."[41]

In India it was becoming clear by 1945 that independence was near, but at the cost of creating a separate Pakistan in areas of Muslim majority. Mohandas Gandhi, Jawaharlal Nehru, and most leaders of the secular but Hindu-dominated Indian National Congress had refused to cooperate with the British without an unequivocal promise of full independence after the war, a vow that Churchill's government refused to take. So the Congress leaders were imprisoned from 1942 onward, while Muhammad Ali Jinnah and the leaders of the Muslim League were free to pursue their demand for an independent country. Communal tensions between Muslims and Hindus and Punjabi Sikhs during the wartime turbulence led to increasing communal panic in both Bengal and Punjab. In May 1947, Viceroy Mountbatten announced his decision to create Pakistan and leave the subcontinent in August. Between June and August, a British commission created the new national boundaries in a process defined primarily by the contiguity of socioreligious groups.

Wartime pressures had intensified administrative, economic, and political stress that led to independence and the partition of the subcontinent that August. The Partition upheavals, which witnessed some fifteen million refugees and one million deaths in Bengal and Punjab combined, produced the next chapter in the social and environmental consequences of mass violence in the subcontinent.[42] These events and their political, social, and communal dimensions have been endlessly recounted since then, but the published record still lacks an analysis of their environmental history.

The turmoil of refugee resettlement continued through the late 1940s, as documented in the massive literature on the social trauma of those years.[43] As environmental history, however, the ecological distortion resulting from malnutrition and population dislocation was only gradual.[44] From 1947 to 1950, a continuing stream of refugees crossed the new border between India and East Pakistan. Most were Hindus who moved into the cities of Bengal, especially Calcutta, and Lower Assam. A large percentage of displaced Hindus were townspeople, who would not choose to resettle in rural areas. In Calcutta, for example, many wetlands rapidly became squatter slums. This subject invites a full study of the stresses on urban land tenancy, water supplies, and waste disposal, the standard elements of urban environmental history in times of upheaval.

In the hill tracts of Assam's borderlands, thousands more displaced Bengali lowlanders brought tribal populations and hill forests under increasing pressure of competition for homesteads. From then on, the hill forests became steadily more pockmarked with survival clearings created by thousands of immigrants.[45]

CONCLUSION

On the governmental level, the tangled environmental legacy of these years was shaped through the varying administrative capacity of wartime and postwar regimes to cope with such complex dynamics as the delivery of food and energy supplies to both military and civilians. In the mountain region encompassing the India-Burma-China borderlands, the exigencies of World War II cut jagged scars through fragile regions that had previously never been penetrated by modern transport networks. The environmental damage underwent some degree of natural regeneration after 1945, but by then the region had been opened to longer-term penetration from the outside world, intensified by the refugee movements of the war years and their aftermath. The complex, poorly studied impact of refugees depended in part on the fit or misfit of their social status and land-use skills in the settings where they had to construct their new lives. Peasants uprooted from rice paddy lowlands could adjust fairly well to new surroundings that were environmentally familiar. But if fate took them into the hills, where survival depended on the tribal cultures' intimate knowledge of soil, flora, and fauna, the displaced people tended to cause more serious environmental degradation.[46]

More important in the long run was the intensification of ethnic and religious conflicts throughout the wider region, linking the mountains with lowland population pressures and the struggles of postindependence governments to assert control over the ethnic minorities of the hills. In India, fledgling East Pakistan (renamed Bangladesh after it split with West Pakistan in 1971), and Burma, long years of tribal insurgencies, governments' military counterinsurgency campaigns, and disruptions of both traditional and bureaucratic systems of resource management all contributed to the rising environmental stress that the war had thrust upon both the mountain region and the adjacent lowland regions of each country.

Notes

1 For context, see Srinath Raghavan, *War and Peace in Modern India* (London: Palgrave Macmillan, 2010); Kaushik Roy, ed., *War and Society in Colonial India, 1807–1945* (New Delhi: Oxford University Press, 2006); Johannes H. Voigt, *India in the Second World War* (New Delhi: Arnold-Heinemann, 1987).

2 T. C. Whitmore, *Tropical Rain Forests of the Far East*, 2nd ed. (Oxford: Clarendon, 1984), chap. 12.

3 Classic ethnographic studies on this region include Edmund R. Leach, *Political Systems of Highland Burma: A Study of Kachin Social Structure* (London: Athlone, 1964); Robbins Burling, *Hill Farms and Padi Fields: Life in Upland Southeast Asia* (Englewood Cliffs, NJ: Prentice-Hall, 1965); and Christoph von Fürer-Haimendorf, *The Naked Nagas: Head-Hunters of Assam in Peace and War* (New York: AMS Press, 1978).

4 Christopher Bayly and Tim Harper, *Forgotten Armies: The Fall of British Asia, 1941–1945* (Cambridge, MA: Harvard University Press, 2005), 83.

5 The colonial-era spellings of major locations have recently changed. The former Burma is now Myanmar; the former Rangoon is now Yangon; and the former Irrawaddy River is now officially the Ayeyarwady River. Calcutta is now Kolkata; Dacca has been renamed Dhaka. I use the colonial spellings for the years of this account, spellings that are still best known internationally.

6 For the environmental consequences, as rice cultivation steadily displaced the lowland rain forest, see Michael Adas, *The Burma Delta* (Madison: University of Wisconsin Press, 1974).

7 For a summary of Burma's political and administrative history, see D. G. E. Hall, *Burma* (New York: Hutchinson, 1950); Victor B. Lieberman, *Burmese Administrative Cycles* (Princeton, NJ: Princeton University Press, 1984); Thant Myint-U, *The River of Lost Footsteps: Histories of Burma* (New York: Farrar, Straus and Giroux, 2006).

8 Raymond Bryant, *The Political Ecology of Forestry in Burma, 1824–1994* (London: Hurst, 1997); Arupjyoti Saikya, *Forests and Ecological History of Assam, 1826–2000* (New Delhi: Oxford, 2011).

9 Throughout the war, Chinese Communist forces operated far to the north of this region. For context, see Rana Mitter, *Forgotten Ally: China's World War II, 1937–1945* (New York: Houghton Mifflin Harcourt, 2013).

10 Franco David Macri, *Clash of Empires in South China: The Allied Nations' Proxy War with Japan, 1935–1941* (Lawrence: University Press of Kansas, 2012).

11 Michael D. Leigh, *The Evacuation of Civilians from Burma: Analysing the 1942 Colonial Disaster* (London: Bloomsbury, 2014).

12 Bayly and Harper, *Forgotten Armies*, 180–81.

13 For the grisly details, see Leigh, *Evacuation of Civilians from Burma*, chap. 5.

14 Charles C. Trench, *The Indian Army and the King's Enemies, 1900–1947* (London: Thames and Hudson, 1988), 254.

15 Walter Baxter, *Look Down in Mercy* (New York: G. P. Putnam's Sons, 1951); Bayly and Harper, *Forgotten Armies*, 181–84.

16 Trench, *Indian Army and the King's Enemies*, 215.

17 For details of how limited Japanese capacity to exercise administrative control was, see Mary P. Callahan, *Making Enemies: War and State Building in Burma* (Ithaca, NY: Cornell University Press, 2003), 51–68.

18 Richard P. Tucker, "The Depletion of India's Forests under British Imperialism: Planters, Foresters and Peasants in Assam and Kerala," in *The Ends of the Earth*, ed. Donald Worster (Cambridge: Cambridge University Press, 1988), 118–40; Joya Chatterji, *Bengal Divided: Hindu Communalism and Partition, 1932–1947* (Cambridge: Cambridge University Press, 1994);Joya Chatterji, *Spoils of Partition: Bengal and India, 1947–1967* (Cambridge: Cambridge University Press, 2007); Anindita Dasgupta, "Denial and Resistance: Sylheti 'Refugees' in Assam," *Contemporary South Asia* 10, no. 3 (2001): 343–60; Girin Phukon, ed., *Inter-Ethnic Conflict in Northeast India* (New Delhi: South Asia Publishers, 2005); Willem van Schendel, *The Chittagong Hill Tracts: Living in a Borderland* (Bangkok: White Lotus, 2000); Willem van Schendel, *The Bengal Borderland: Beyond State and Nation in South Asia* (London: Anthem, 2005).

19 Joe G. Taylor, *Air Supply in the Burma Campaigns*, USAF Historical Studies 75 (Washington, DC: US Air Force Historical Division, n.d.); Chick Marrs Quinn, *How and Where They Died* (Typescript, 1989).

20 There is detailed material in other US military records of the China-Burma-India theater, much of it declassified now, as well as shelves of published military reminiscences of this "forgotten war," from which a patchwork quilt of environmental details could be laboriously extracted. I am grateful to Jean Mansavage for pointing to these materials.

21 Accounts of medical work in response to the crisis in public health and war-induced epidemic disease, for both military personnel and refugees, give vivid indications of the struggle between civilians and the natural world. See Gordon S. Seagrave, *Burma Surgeon* (New York: Norton, 1943); Gordon S. Seagrave, *Burma Surgeon Returns* (New York: Norton, 1946).

22 Charles F. Romanus and Riley Sunderland, *Time Runs Out in CBI*, 1st ed. (Washington, DC: Center of Military History, US Army, 1959); Joseph W. Stilwell, *The Stilwell Papers*, ed. Theodore H. White (New York: Shocken, 1948). A spate of novels and memoirs appeared in the postwar years. Some of them give vivid indications of the experience of soldiers and civilians in these settings, anecdotal evidence of the thousands of local impacts inflicted on landscapes. See, for example, Baxter, *Look Down in Mercy*.

23 The Karens constitute 10 percent of today's Myanmar population; most live in the Irrawaddy Delta and the mountain region bisected by the Thai border. Many are Buddhist, like the ethnic Burmese; others are various varieties of Christian and animist. In 1949, the predominantly Christian Karen National Army began a long-running insurgency along the eastern borderlands.

24 Callahan, *Making Enemies*, chap. 2.

25 Ibid., 66–67, 70.

26 Ibid., 255, 268–69.

27 Trench, *Indian Army and the King's Enemies*, 261–62.
28 William R. Peers and Dean Brelis, *Behind the Burma Road* (Boston: Little, Brown, 1963), 179.
29 Trench, *Indian Army and the King's Enemies*, 268.
30 Bayly and Harper, *Forgotten Armies*, 379–80, 388–90.
31 Ibid., 382.
32 Bryant, *Political Ecology of Forestry in Burma*, 133–34, 142–47.
33 Paul Greenough, *Prosperity and Misery in Modern Bengal: The Famine of 1943–44* (New York: Oxford University Press, 1982).
34 Richard P. Tucker, "The World Wars and the Globalization of Timber Cutting," in *Natural Enemy, Natural Ally: Historical Studies in War and the Environment*, ed. Richard P. Tucker and Edmund Russell (Corvallis: Oregon State University Press, 2004), 110–41.
35 Ministry of Agriculture, *India's Forests and the War* (New Delhi, 1948).
36 Christopher Bayly and Tim Harper, *Forgotten Wars: Freedom and Revolution in Southeast Asia* (Cambridge, MA: Harvard University Press, 2007).
37 A significant remnant of the Kuomintang forces stayed on in the Shan hills of eastern Burma, controlling the opium export trade to Bangkok and Saigon. For their disruptive role in the multiethnic tangle and continuing fighting in that region, see Robert H. Taylor, *Foreign and Domestic Consequences of the KMT Intervention in Burma*, Data Paper 93 (Ithaca, NY: Southeast Asia Program, Cornell University, 1973).
38 On timber smuggling, see Bryant, *Political Ecology of Forestry in Burma*, chap. 7; Martin Smith, *Burma: Insurgency and the Politics of Ethnicity*, 2nd ed. (London: Zed Books, 1999), chap. 20.
39 Alan Rabinowitz, *Beyond the Last Village* (Washington, DC: Island Press, 2001); idem, *Life in the Valley of Death* (Washington, DC: Island Press, 2008).
40 Rabinowitz, *Beyond the Last Village*, 239–40.
41 Ibid., 242.
42 Ian Talbot and Gurharpal Singh, eds., *Region and Partition: Bengal, Punjab and the Partition of the Subcontinent* (Oxford: Oxford University Press, 1999).
43 Ibid.; Madhusree Mukerjee, *Churchill's Secret War: The British Empire and the Ravaging of India during World War II* (New York: Perseus, 2010). Mukerjee argues that Churchill followed a "scorched-earth" policy toward rural Bengal. In an otherwise convincing revisionist study of his wartime policy, she uses this term metaphorically rather than in its usual environmental sense. For detailed analyses of the famine, see Greenough, *Prosperity and Misery in Modern Bengal*; Amartya Sen, *Poverty and Famines: An Essay on Entitlement and Deprivation* (Oxford: Clarendon Press, 1982), chap. 6; Mark B. Tauger, "Entitlement, Shortage and the 1943 Bengal Famine: Another Look," *Journal of Peasant Studies* 31, no. 1 (2003): 45–72.
44 Beyond Bengal, the wartime imperial British government's wartime Grow More Food campaign made great efforts to increase production of rice, wheat, and other grains by double cropping, expanding irrigation, switching from nonfood crops, and tilling "marginal" lands. This was a complex chapter in India's agricultural history, but it seems to have resulted in little net change in total arable area, and little destructive pressure on fragile marginal soils or grazing lands. Henry Knight, *Food Administration in India, 1939–1947* (Stanford, CA: Stanford University Press, 1947).
45 I am grateful to Anup Saikia for this analysis. Much more detailed analysis of vegetation changes, clearings for cropping and livestock, settlement patterns, and road construction may be possible by compiling a chronological series of aerial photos and Landsat images. For a recent perspective on forest depletion in the Garo, Khasia, and Jaintia hills of Assam, which were reorganized into Meghalaya State in 1972, see Bengt G. Karlsson, *Unruly Hills: Nature and Nation in India's Northeast* (New Delhi: Orient Black Swan, 2011).
46 For a finely tuned analysis of this sort, see Nigel Allan, "Impact of Afghan Refugees on the Vegetation Resources of Pakistan's Hindukush-Himalaya," *Mountain Research and Development* 7, no. 3 (August 1987): 200–204.

Chapter 8
Hawai'i
Before and After Pearl Harbor

CAROL MACLENNAN

America's militarization of the Oceanic Pacific began with Pearl Harbor. The Second World War was only a late installment in a long story, representing the second act in a tale of military consequences for Hawai'i's island environment. It began nearly one hundred years earlier when the United States became interested in one of the two Pacific Ocean harbors that provided ample protection for American naval vessels. Pearl Harbor's history starts in the 1850s with American designs to acquire its waters for a Pacific station. The period prior to the Japanese attack on Pearl Harbor in 1941 witnessed the gradual militarization of the harbor's land and water environment. The Second World War set in motion a second act of cascading consequences. The postwar decades brought a military base expansion with environmental changes across the island archipelago, tying Hawai'i's economy to US defense policy. Today, Pearl Harbor houses the US Pacific Command (USPACOM), one of nine (and the oldest) centralized commands, and as a result the Hawaiian Islands provide a home for all four branches of the military oriented toward security for the entire Pacific region. As the United States turns its attention toward the rise of China in the Pacific, it is likely that PACOM will become the most important command center in the United States.

Hawai'i's militarized environment is evidenced by the sheer magnitude of the military footprint for bases and training facilities, as well as its residual physical and cultural effects. The military occupies about 5.7 percent of Hawai'i's total land area, and about 22 percent of the most populated island, O'ahu. It owns about 231,000 acres and leases 177,000 acres, thus controlling the use of about 407,000 acres. There are a total of 118 military installations (seven of them very large), which employ about 95,000 military and civilian personnel (not including dependents).[1] About 54,000 square miles of nearby ocean are considered military operating areas, and about 58,000 square miles designated military special use airspace.[2] The consequences of this large defense presence are illustrated today by the existence of numerous toxic waste sites (several of which are part of a large Superfund Site at Pearl Harbor), and by threats to endangered species and to culturally sensitive areas.

Pearl Harbor and other US military bases in the Pacific find their origins in the early expansion of the American empire into the Pacific Ocean after 1870. The seventy years before the Second World War provided the infrastructure and institutional footprint for expansion after the war, and propelled the military into the dominant economic and environmental force that it is today. It began not with the threat of war, nor even with a strong desire of Congress and presidents to build naval defenses in the Pacific. Rather, it came from a campaign by American sugar industrialists to attract an American market for their Hawaiian-grown sugars beginning in the 1850s. Their persistent efforts to sell Hawai'i's deepwater port to the navy as a coaling station with unique defensive capabilities set in motion the events that ensured Hawai'i's role. The drive to acquire Pearl Harbor began in earnest in 1872–73 when the navy's energy needs for steam travel in the Pacific region inspired President Ulysses S. Grant to send Albert Steinberger to investigate the suitability of Pago Pago (Samoa) and General John Schofield to Hawai'i to examine Pearl Harbor. Rights to use each harbor were acquired in the 1870s (Samoa) and 1880s (Hawai'i). By 1900, the United States had secured access for its current large military bases in Guam, Hawai'i, and American Samoa. The environmental consequences of these actions at Pearl Harbor and eventually throughout Hawai'i are complex and tell a story important to our examination of the long shadows of wars both before and after the Second World War.

PU'ULOA TO PEARL HARBOR

Pearl Harbor, known to Hawaiians as Pu'uloa,[3] had a deep history of cultural and economic significance well before its American occupation. Polynesians migrated from the central Pacific in their voyaging canoes, settling the deep valleys throughout the Hawaiian Islands around 1000 AD. As population increased and voyaging ceased, they spread outward from their taro-growing valleys to occupy drier landscapes in the islands and to establish villages near

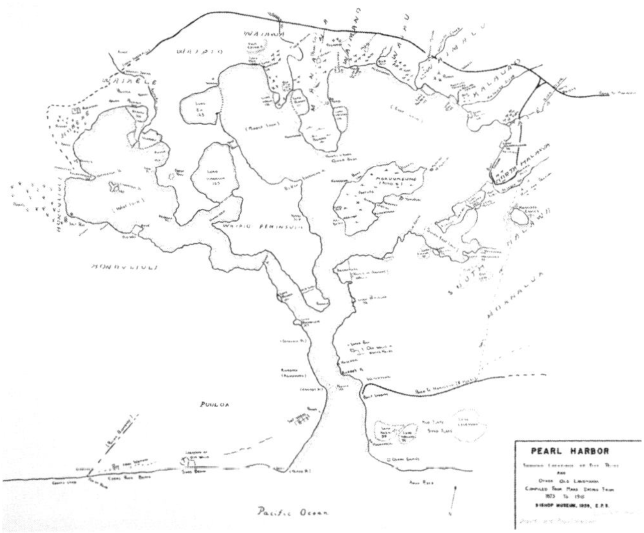

Figure 8.1. Historic fishponds in Pearl Harbor. Source: Pearl Harbor—Showing Locations of Fishponds and other Old Landmarks. Compiled from maps dating from 1873 to 1915. Bishop Museum, 1959, E.P.S.

shallow coastal waters to construct and manage fishponds. Archaeologists identify this time as one of agricultural intensification, the period when the early Hawaiian state began to form. Probably occupied around 1500 AD, Pu'uloa, which is located adjacent to the dry tablelands of west O'ahu, provided a rich and expansive water environment for Hawaiian fishponds. For nearly three hundred years, Pu'uloa developed into an active site of constructed ponds such that, by the time of European visitation in the late 1700s, foreign observers describe an elaborate system of lava walls underwater that segmented harbors and inlets from the sea where specific fish species were tended for consumption. A map showing ponds that remained visible in the late 1800s in the harbor illustrates their size and numbers. Over twenty managed fishponds in Pearl Harbor that fed Hawaiians existed well before the arrival of Europeans. Each pond has a name beginning with *loko* (the Hawaiian word for fishpond) and is identified by a dotted line. Illustrating the vibrancy of the pond fisheries on O'ahu, this map shows walled ponds in the harbor, and others are ponds built in low-lying sites on land near the harbor.[4]

The Hawaiian fishpond was primarily a grazing area in which the fishpond keeper cultivated algae for his fish, much in the way a cattle rancher cultivates grass for his cattle. Thus pond conditions were kept optimum for the cultivation of algae. This included designing the ponds with depths of only two or three feet, so that sunshine could penetrate the water, providing energy for the growth of algae.[5]

Pu'uloa has been a native Hawaiian sacred landscape for over a millennium. It is the home of the shark goddess Ka'ahupahau—who Hawaiians claimed caused the famed collapse of the large naval dry dock on the shores of the east loch in 1908. The numerous fishponds of O'ahu and other islands appear regularly in the *mo'olelos* (legends), and imbue fishing and pond construction with sacred meanings. For Hawaiians, eating is sacred, as the gods find form in the native foods of the islands. The *'aina* (land) is embodied in the persons of the chiefs and their *mō'ī* (king). Before Euro-American influences, the chiefs cared for the *maka'*āinana (commoners) by regulating irrigation, fishponds, and agriculture with *kapus* (taboos). The *maka'ainana* cared for the chiefs by caring for the land and following the *kapus*. The relationship between *mo'i*, *maka'ainana*, and *'aina* was (and still is) a sacred political ecology as represented at Pu'uloa.[6]

Europeans and Americans, who gave Pearl Harbor its contemporary name, recognized its strategic value early. Fishponds and their natural oyster beds and fisheries were of little concern to them, and the harbor often described as "idle and gone to waste." A coral reef and large sandbar prevented access to the three lochs of the harbor by large sailing vessels, thus protecting fishponds, fisheries, and shoreline taro and rice fields until the 1920s. It took five decades for the Americans to acquire the right to control the destiny of Pearl Harbor. The first attempts by the United States to gain exclusive access to the harbor in the 1850s were finally realized with annexation of Hawai'i. From the beginning, Hawaiians protested even the mention of any foreign use of their Pu'uloa. This harbor belonged to the crown and the nation, and giving away rights to its use violated the sacred relationship between *maka'ainana*, *mo'i*, and *'aina*.

After the US Civil War, the sugar industry in Hawai'i assumed a strategic role in financing the sovereign native kingdom that had developed a Western constitutional-style government and a literate Hawaiian population. Income from the Pacific whaling trade was in decline, and there was pressure on the king from the sugar planters to seek a reciprocity treaty with the United States so that Hawaiian sugars could enter American ports duty-free. This would be a huge boost to the island sugar economy and, in the eyes of the king, also to the security of Hawaiian sovereignty. In 1873, Congress and the Hawaiian Legislature openly discussed the possible cession of Pearl Harbor to the United States in trade for reciprocity, resulting in loud protest from all quarters of native society. Hawaiian Legislature representative Joseph Nāwahī called the proposed agreement to cede the harbor to the United States in return for duty-free sugar imports a "nation-snatching treaty."[7]

Not to be deterred, the sugar planters quietly invited General John Schofield to Hawai'i for a "vacation" to prepare a report for the War Department on the feasibility of a naval station at the harbor. But it went no further than a positive recommendation to locate a naval coaling station there because Hawaiian protests had effectively discouraged Congress from a trade of Pearl Harbor for open markets to Hawaiian goods. The sugar interests dropped the issue, and the treaty was signed in 1875 without the Pearl Harbor clause. But the idea did not die. The sugar industry grew rapidly after reciprocity, empowering the planter class to bend the Hawaiian government to its wishes. Schofield's report circulated through Congress and the White House. The idea never really disappeared from backroom negotiations.[8]

The political landscape changed in 1887 with Hawai'i's Bayonet Constitution. The economic power of the sugar planters was now substantial as a result of tremendous growth of sugar production and profits from reciprocity with the United States. Under a threat of violence from an armed militia, the sugar capitalists convinced King Kalākaua to rewrite the Hawaiian Constitution. The American minister in Honolulu backed their demands. The new Bayonet Constitution, as it was nicknamed, greatly reduced the voting rights of Hawaiians with a requirement of property ownership, and it limited the authority of the king over his cabinet. The sugar capitalists then forced the king to sign a renewal of the Reciprocity Treaty with a new clause that provided the United States with exclusive rights to utilize Pearl Harbor for the seven-year duration of the treaty. Hawaiians viewed Kalākaua's acquiescence to American planters as a fatal blow to national sovereignty and to the cultural right to protect their sacred landscapes.[9] Lili'uokalani, the subsequent heir to the monarchy, wrote in her diary, "It should not have been done."[10] The outpouring of grief in the Hawaiian nation over the cession of Pearl Harbor was palpable. Petitions to the Hawaiian Legislature, letters and columns in the Hawaiian-language newspapers, and appeals to Congress and the White House testified to the powerful tie between Hawaiian culture, the king, and sacred lands.

When Queen Lili'uokalani assumed the throne upon Kalākaua's death, she decided in 1893 to return to the original Hawaiian Constitution (thus revoking the Bayonet Constitution and reestablishing the powers of the monarchy). The American sugar capitalists utilized the threat of the US Marines present in Honolulu Harbor to dissuade her, proclaiming her action "illegal." They imprisoned the queen and declared Hawai'i a republic and reorganized the government. President Grover Cleveland, at the request of the queen, sent an investigator to sort out the events. The planters, however, ignored the finding from President Cleveland's Blount investigation that concluded their seizure of the government was itself illegal. President Cleveland did not enforce Blount's recommendation, and the sugar planters awaited a new administration in Washington, keeping their new republic intact for a few more years. The sugar capitalists eventually won support for annexation of Hawai'i to the United States from the newly elected President William McKinley and a new Congress in 1898. A decisive factor in annexation was the attraction of Pearl Harbor as a potential naval base, which coincided with the Spanish-American War and the expansion of the US Empire and its navy into the Pacific. The queen's pleas to Congress for redress and the formal Hawaiian petitions sent to Washington representing thousands of signatures were ignored.[11]

The actual use and occupation of Pearl Harbor by the United States took some time to occur. Yet the shadow cast by the actions of the sugar planters tying the harbor to their economic interests set in motion events that culminated nearly one hundred years later in the Japanese assault on Hawai'i. Dangling Pearl Harbor in front of Congress and the president during the last quarter of the nineteenth century proved to be the undoing of the Hawaiian cultural claim over the harbor's resources.

MILITARIZATION

The consequences for the gradual passing of Pearl Harbor from a food production system rich with fisheries and ponds into naval waters for warfare were substantial. During the first three decades of the twentieth century, the shadow of military threats in the Pacific enveloped the island of O'ahu and gradually militarized its natural resources. What began as a naval outpost for coaling and harboring the US naval fleet in the Pacific became a strategic outpost by the Second World War.

The growing militarization of selected Hawaiian environments on O'ahu before the Second World War occurred in three phases: acquisition, occupation, and enclosure. It culminated in 1947 with the National Security Act, which established Hawai'i as the headquarters for the Pacific Command and firmly secured the rules of military use of island lands and waters. The advancing control over Hawai'i's resources illustrates a process of increased military influence over the environment. Naval and army forces gradually pushed out others who utilized resources for food and cultural practices. Hawaiians and sometimes others resisted each significant acquisition and eventual enclosure through petitions, letters, and in contemporary times protests. This eventually resulted in something of a compromised military presence. Since the 1970s, Hawaiians have made some recent gains in moderating military activity in sensitive areas, with the aid of federal environmental and cultural legal protections. Yet military ownership with the rights of occupation and use created a long shadow over island environments. Taking in the scope of military activity in Hawai'i, today one is struck by the totality of the Pacific Command's presence on the landscape and adjacent waters, and by its authority over access and use of resources.[12]

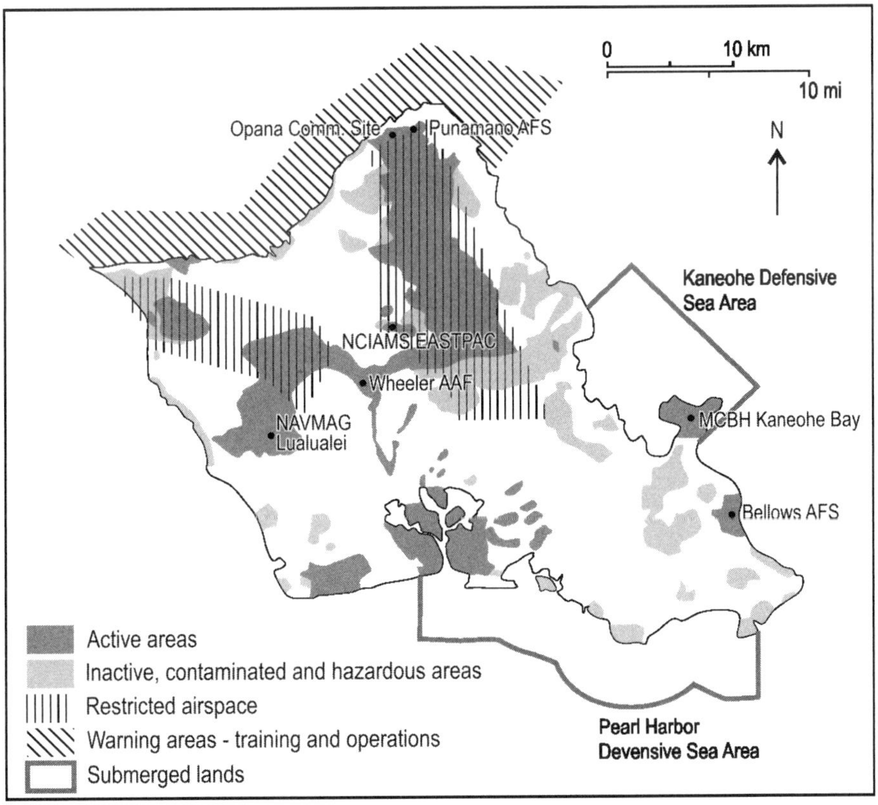

Figure 8.2. Current military areas on the island of O'ahu. Source: Hawai'i Peace and Justice, http://www.wp.hawaiipeaceandjustice.org/wp-content/uploads/2012/10/12.07.17Military-on-Oahu.jpg (accessed 15 June 2015).

The legal acquisition of land and water resources by the navy and army began with the annexation of Hawai'i in 1898 under the authority of a joint congressional resolution. The Newland Resolution bypassed the standard means of annexing new territory (a treaty passed by the Senate), thus casting a shadow over its legality. This was significant because of the resolution that called for the cession of all Hawaiian government lands (including crown lands) to the United States. Subsequently, the 1900 Organic Act, which defined the political structure of the new territory, contained provisions by which the United States could claim public lands for defense purposes.[13]

Condemnation by presidential executive order was the primary technical means by which the military acquired land and rights to harbor water and its fishing rights. The 1893 overthrow of Queen Lili'uokalani by the sugar planters and establishment of an independent republic set in motion the consolidation of native Hawaiian lands under the new government. Prior to the overthrow, the Hawaiian monarchy had managed a huge portion of island lands under two categories—crown (sacred) and government (secular) lands—in which two distinct policies applied for lease and sale. Crown lands belonged to the person of the monarch and were held in trust for the Hawaiian people, symbolizing the relationship between the *mō'ī* and *maka'āinana*. Under the new republic, the 1895 Land Act combined the crown and government lands under one set of policies, thus severing the sacred relationship of *'āina, mō'ī,* and *maka'āinana* and dealing a major blow to the future campaigns for regaining Hawaiian sovereignty. Before the Land Act, the crown lands totaled 971,463 acres, with a value in 1895 of $2,314,250. The combined crown and government lands, now called public lands and administered under a single policy, totaled approximately 1,750,000 acres. They included some of Hawai'i's more sacred landscapes, including Pu'uloa.[14]

With annexation, these public lands were ceded to the US government, and subject to condemnation for military use. Much of Pearl Harbor waters and a large portion of its watershed were condemned by executive order. In addition, President McKinley set aside over 15,000 acres for the army, which became Schofield Barracks and Fort Shafter, two installations established to protect Pearl Harbor. As the United States started in 1898 to plan for annexation, it identified government lands for future military use. Because they would become US public lands upon annexation in 1900, the military claimed acreage around Pearl Harbor and upslope in Wahiawā, which within a few years was developed into the naval station and an army base (Schofield

Barracks). Later, additional military sites such as Punchbowl and Hickam Field were acquired through condemnation proceedings or purchase. Similar acquisitions occurred immediately after the First World War when the United States expanded its military presence throughout O'ahu to include land on the windward side of the island.

When public lands would not suffice, the United States sought out private lands with the aid of the governor's office. To ease this process, a system of land exchanges between the territorial government and private parties was developed, after which the newly acquired private land (now public) was easily subject to federal condemnation. Willing partners to military expansion in the islands, the early territorial governors (appointed by the president) believed that military development was a form of economic development. When private owners balked (especially sugar plantations and ranching interests in the Pearl Harbor watershed), the governors and their legal staff (often members of the sugar planter class) helped resolve the differences by providing valuable public land in exchange.[15] This process continued apace such that by the early 1930s most of the major land acquisitions were completed at Diamond Head, Punchbowl, Fort Armstrong, Fort Kamehameha, Fort Ruger, Bellows Field, Ford Island, and Hickam Field. The acquisition of ceded Hawaiian lands established in the early 1900s a significant military presence on the island of O'ahu.

The occupation, or actual settlement and use of army- and navy-owned lands, occurred over a longer period of time, and thus began the toll on specific natural and cultural resources. The naval station at Pearl Harbor required the cutting of a massive coral reef at its entrance and the dredging of the harbor. Slowly, beginning in 1912, the navy relocated some of its operations from Honolulu to Pearl Harbor and began construction of the first dry dock for ship repairs. Eight years later, in 1920, the harbor opened to its first large ships.[16] The army established its reservation, Schofield Barracks, upslope from Pearl Harbor on the Wahiawā plains in 1908, and by the First World War it housed six thousand men. This war brought two additional installations to the windward side of the island: Waimanalo Reservation (now Bellows Air Force Base) and the Kaneohe Naval Air Station (now Marine Corps Base Hawaii). During these early years of base development, occupation did not immediately result in rigid restrictions of land use and fishing rights by native Hawaiians or others. In fact, Schofield Barracks was located in favorite hunting grounds, and frequently the army opened the reservation to local hunters. Similarly, although the navy had the right to forbid fishermen from utilizing Pearl Harbor, it did not impose a fishing ban except during the First World War. Native Hawaiian protest over loss of crown lands and annexation remained fresh during these first couple of decades. Easy relations between the territorial governors and the Department of War (US Army) and the naval station were reflected in the routine correspondence in which land exchanges were negotiated, services such as road building and water development were discussed, and complaints about military activities handed amicably.

Military security in Hawaiian waters required control over access to fishing as well as property for installations, but it was more difficult to acquire and enforce. Under customary Hawaiian law, nearshore water and fishery rights were attached to land rights. Because the naval station held title to a small portion of the vast coastal shore of the harbor, adjacent to the eastern side of the east loch of the harbor, it had no rights to control fishing and access to the entire harbor. Pearl Harbor's fisheries (which included fish ponds) were rich in species that fed a growing Honolulu fish market in the early 1900s, and the Hawaiian and Asian commercial fishermen had been leasing the rights from landowners to fish these waters for many years. With Hawai'i subject to US law, under ordinary circumstances these fisheries would be publicly owned and access and use automatically controlled by the government. However, Hawai'i became an exception. Private ownership of lands and fishing rights in Pearl Harbor by elite Hawaiian and business families complicated the situation. The large landowners within the harbor's watershed were among some of the more powerful landholders of the island: the James Campbell Estate, M. P. Robinson, the John 'Ī'ī Estate, Oahu Railway and Land Co., the Bishop Estate, and Samuel M. Damon. Most of these estates were reluctant to sever their fishing rights from landholdings because of their value, and proved to be difficult contestants in court battles with the US government after annexation.

In an effort to render void the traditional Hawaiian rights to fisheries, in 1902 the territorial government claimed that they were not "vested rights" but merely a license. In order to sustain traditional Hawaiian practice, the landowners (some of them wealthy native Hawaiian families) sued the government and were defeated in lower courts and the Territorial Supreme Court. In a 1904 appeal to the US Supreme Court, Samuel M. Damon (a prominent missionary descendant, banker, and owner of nine thousand acres of land and fishing rights in the Pearl Harbor watershed) won a decision that a fishery right was a vested right.[17] This was a major setback for the military. The contest over these rights dragged on into the 1950s, and over time the United States purchased the rights from landholding family estates and established its control over the access to Pearl Harbor and the right to exclude fishing.

The first step toward the enclosure of the harbor's waters (and other occupied military land) occurred in 1912 when President W. H. Taft issued an executive order banning foreign ships (commercial and military) from entering Pearl Harbor. Then, in 1917, when the United States entered World War I, the commandant of the naval station in Hawai'i secured a presidential executive order to temporarily prohibit commercial fishing vessels from working the harbor's channels for the duration of the war.[18] After the 1919 opening of the first naval dry dock in the eastern loch, the secretary of the navy and Congress pursued a strategy of negotiation for the purchase of fishing rights from private landowners in the harbor's watershed. As noted in a congressional report, "The base is so vitally necessary to our naval establishment and has cost the Government so much money and will cost so much more in the future that it ought to be given the utmost possible protection from approach by unauthorized persons, both in time of peace and in time of war."[19] One issue of particular concern to military interests was the increasing control over commercial fishing by Asian (primarily Japanese) fishermen.

It would take congressional action and funding to finally enclose the harbor and prohibit entry. With an appropriation of federal funds to purchase fishing rights, in 1930 the navy began negotiations with landowners with the promise to appropriately value fishing rights as monetized property rights. Several of the parties finally settled in 1940 and 1944. The John 'Ī'ī Estate settled in the early 1950s. Suffice it to say that after two decades, the navy board's recommendation of $61,170 for the fishery rights in 1930 eventually totaled up to $208,133 by the final purchase in 1950.[20] Although the court battles were primarily over valuation of the fisheries, the process of gradual enclosure as these rights were acquired shut down much of the commercial fishing in the harbor. Fishpond use also declined rapidly during this era as the naval station and shipyards expanded.

Enclosure of Pearl Harbor and surrounding lands had economic and cultural consequences. In the early 1900s, the harbor supplied the fish market in Honolulu with ten to fifteen tons of fish per month.[21] A federal agent from the US Bureau of Fisheries investigated Hawai'i fisheries between 1900 and 1903, describing the types of fish, market conditions, and commercial fishing economy in great detail. He called Pearl Harbor one of the most important fisheries in the islands that supported over 39,000 residents in Honolulu. In 1903, Hawaiians were only a portion of the fishermen and women on Oahu. Of the approximately 1,400 fishermen on the island, over half were Chinese and Japanese.[22] The gradual closure of Pearl Harbor to commercial fishing after the First World War marked an economic shift in the lives of independent Hawaiian and Asian families living nearby. Naval shipyards, ammunition supply warehouses, airfields, and military housing eventually covered Hawaiian villages, salt ponds, fishponds, and cultural properties, such as Queen Emma's Estate at the mouth of Pearl Harbor (now under Hickam Air Force Base). Further, with the growing use of the harbor for dumping naval waste and sewage and the prohibition of fishing, the deterioration of water quality and resulting decline in the fish population went little noticed until the 1960s. Beach closings at ʻEwa and other popular swimming spots woke up citizens in the islands to the consequences of pollution. Caused primarily by disposal of raw sewerage into the harbor and nearby ocean areas, the beach closings brought attention to other types of contamination caused by heavy military use, such as oil spills and nuclear waste. The parallel enclosure of military-owned lands prohibiting access to hunters and other users after the 1930s created large zones off-limits to civilians. Military bases were gated, and the growing number of training installations established after 1945 created dangerous landscapes with their live ordnance. Eventually, military-acquired and -leased land and designated coastal waters on Hawaiʻi Island, Maui, Kahoʻolawe, and Kauaʻi expanded the reach of no-go zones after the war.

One can witness the expansion of the army and navy in their increased personnel numbers as the United States shifted its policy from utilization of Hawaiʻi as a shipyard and protected harbor to a training and base outpost for its Pacific maneuvers. Military installations in 1920 housed 5,942 individuals. By 1940, that number was 29,831—nearly a fivefold increase.[23] By the Second World War, the navy and army occupied 24,556 acres in Hawaiʻi, mostly on Oʻahu. With a nervous eye on Japanese expansion, by 1941 the United States had built a sizable garrison in preparation for war in the Pacific. The National Industrial Recovery Act and other relief programs financed the new facilities and dredging of the harbor, especially after 1936, when Congress significantly increased funding each year. In 1940 alone, a total of $32 million of public works appropriations were made. During this period, the army also expanded its territorial reach with purchases of additional lands on Oʻahu for training in areas adjacent to Schofield Barracks in the remote reaches of Wahiawā (1924) and shortly thereafter in Mākua Valley (1929). The Second World War brought martial law to Hawai'i for three years (1941–44), thus completely enclosing the entire territory in militarized environment. Finally, significant additions to military bases and training installations occurred with the close of the war. Today, nearly 109,000 acres on Hawai'i Island are devoted to the Pōhakuloa Training Area stretching across Saddle Road and up the flanks of Mauna Loa and Mauna Kea, acquired through purchases between 1955 and 2006 and now the largest military installation in the Pacific. The Pacific Missile Range Facility on Kaua'i (acquired by the navy in 1954) tests missiles and controls over two thousand acres of the Barking Sands Beach, plus one thousand square miles of underwater range and 42,000 square miles of airspace. The military use of land, water, and airspace is not without consequence for Hawai'i's natural resources and communities.

THE POSTWAR SHADOW

After the war, the environmental effects of the militarized environment were notable. Increased pollution of Pearl Harbor, O'ahu's aquifers, and its fragile forests and valleys resulted from the rapid expansion of military activity on the major bases and training areas. Ordnance from bombers, infantry maneuvers with live fire, and naval vessel and submarine exercises damaged cultural and natural resources, expanding the military presence in more remote locations on the other islands and surrounding shores. In the 1950s, the introduction of nuclear armaments and submarines and the use of chemicals and heavy metals at base facilities for ship repairs and maintenance brought new forms of pollution. In an oceanic island environment, this leaves little room for error. Hawai'i and its people also became increasingly dependent on the military for a dependable economic base. By the time sugar and pineapple fields were disappearing in the 1980s and 1990s, military civilian employment and business in the local economy had become second only to tourism in economic importance. Economically wedded to defense employment and expenditures, Hawai'i and its residents accepted the burden of military dependence.

The earliest warnings of environmental damage from growing military activity in the islands came from the first beach closures around Pearl Harbor in the 1960s, owing to the large naval housing complex that for decades had emptied its sewage into the estuary. During an era when environmental awareness of pollution was on the rise, Pearl Harbor's contamination became a notable public concern. Congressional representative Patsy Mink voiced citizen concern over naval dumping of radioactive waste into Pearl Harbor.[24] The city and county of Honolulu sewerage plants and naval facilities continued to release untreated waste into the harbor. Bacterial levels in the harbor were high, and unchecked sediments were smothering the oyster beds. Oil slicks were also common.[25]

The 1970s were also a period of federal action to clean up Hawai'i's degraded environment. Federal grants enabled Honolulu to build a new sewage treatment plant. Local newspapers reported on a navy plan to establish Pearl Harbor as a model for restoring water quality and the environment. Long known as a sink for industrial facilities at the shipyard and repair facilities along the harbor where drains washed oily waters laden with cleaning solvents and polychlorinated biphenyls (PCBs) directly into the water, the harbor's pollution came under scrutiny. A 1973 navy study of harbor sediments and soil samples found cadmium, chromium, copper, lead, mercury, nickel, silver, and zinc from man-made sources in the harbor sediments. It also concluded that the biology of the harbor "appears to be affected by the metals found in the sediments."[26]

As agricultural employment declined in the 1970s and 1980s along with a wave of plantation closings, the military economy in Hawai'i increased in importance. Ironically, local awareness of pollution (military and civilian) at Pearl Harbor coincided with the state government's recognition that military spending along with tourism would become the foundation of Hawai'i's future economy.

Multiple pollution issues surfaced in the 1980s, gaining the attention of the state government and the Environmental Protection Agency (EPA). The state, now enforcing federal clean air and water, waste disposal, and toxic chemical standards, found itself in a difficult position. The once-dominant sugar- and pineapple-based agricultural economy had ceded its role to federal (military) employment and procurement and to tourism at the same moment that dangerous environmental hazards at Pearl Harbor began to surface. Small leaks of radioactive water from nuclear submarines at the naval station were the first signal to the public that nuclear materials were present in the harbor and also stored in arsenals nearby. Until that point, the navy neither confirmed nor denied claims by protest organizations of storage and transportation of nuclear materials.[27] Declining water levels in O'ahu's aquifers led the state Department of Land and Natural Resources to claim jurisdiction over the Pearl Harbor aquifer.[28] A rapidly growing military personnel and their families added a significant burden, increasing withdrawals from the aquifer. Sewerage spills from naval facilities and from city treatment plants were a regular occurrence, frequently noted in the newspapers and leading to the conclusion that sewage treatment plants were obsolete and hazardous. Oil spills reported in the daily papers also led to the overall impression of extreme pollution in Pearl Harbor. In 1985, Hawai'i's Department of Health informed the army that TCE (trichloroethylene), an industrial solvent, was present in the drinking water at Schofield Barracks, and by 1990 this base was placed on the National Priorities List (Superfund).[29]

The Pearl Harbor Naval Complex (6,300 acres) was declared a Superfund Site in 1992, adding to a growing list of over 130 military sites on EPA's National Priority List (NPL) of the most contaminated areas in the United States. Placement on the NPL took nearly ten years because the navy had originally identified thirty-one potential sources of hazardous waste that included "unlined landfills, pesticide disposal pits, chromic acid disposal areas, PCB disposal areas, mercury-contaminated harbor sediments, leaking underground solvent tanks, [and] waste oil facilities."[30] Placement on the NPL occurred nearly twenty years after the navy and the city of Honolulu, utilizing federal funds, had already begun to tackle pollution and environmental degradation in the harbor, and is a testament to the complexity of the problems and the solutions necessary for protection of public health and ecological restoration. In 1998, the Hawai'i Department of Health issued an advisory to the public to not eat fish and crabs from Pearl Harbor because of elevated levels of PCBs in the fish tissue reported by the ATSDR (Agency for Toxic Substances and Disease Registry) of the US Centers for Disease Control.[31] Today, the Pearl Harbor Complex Superfund Site is divided into eighteen geographical units, of which only one has been remediated.[32] Within this area, there exist seven hundred discrete sites of contamination. The process for site characterization, feasibility studies, and final remediation is painfully slow. EPA did address the most hazardous sites through early removal action of contaminants. Only the medium- and low-risk sites remain for cleanup. While Pearl Harbor may be a complex and expansive site to remediate, it is not atypical of other military Superfund Sites plagued by oil and fuel spills, threatened aquifers, PCB contamination, and nuclear waste carelessly disposed. EPA lists over 130 military sites in the Superfund program, including at the large Pacific bases in Guam and American Samoa. Hawai'i is not unique.

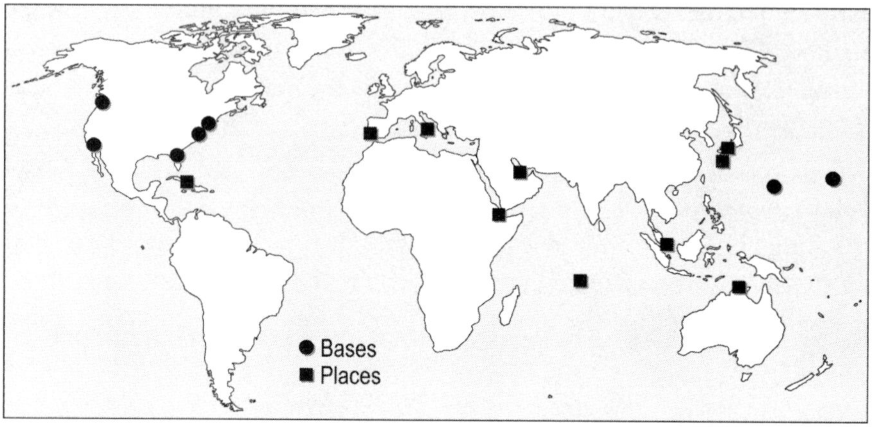

Figure 8.3. Power over the blue planet. Main US Navy bases and places today. Source: US Navy.

During and after the Second World War, the military expanded its presence on Hawai'i Island, Kaho'olawe (in Maui County), and Kaua'i. The intensified use of training installations through use of live ordnance and bombardment rendered whole regions off-limits because of the hazards, even during periods of nonoccupation. Three of the multiple training locations have become important flashpoints for native Hawaiians, who have begun to seek reclamation of culturally and biologically sensitive military lands and, in some cases, complete cessation of military use. Since the 1970s, beginning with the contemporary Hawaiian renaissance, live-fire training at Kaho'olawe Island, Mākua Valley (Oahu), and Pōhakuloa (Hawai'i Island) has been contested on these militarized landscapes. In each of these cases, military training interferes with cultural practices and the protection of endemic Hawaiian species. The alliance between indigenous and environmental organizations has proven a powerful challenge to military land-use policies.

The most long-lived and successful struggle has been at Kaho'olawe, a small and uninhabited island off the western coast of Maui. Starting in the late 1930s, navy bombers used the island for target practice, which continued and escalated through the Vietnam War. Protests by native Hawaiian organizations and direct action that drew national attention eventually resulted in victory.[33] In 1994, Congress ended the use of Kaho'olawe as a naval target and authorized $400 million for cleanup, including $44 million to the state for archeological and educational work. Over twenty years later, after the navy had removed ten million pounds of surface ordnance from the island, recovery and restoration remain unfinished. Although most of the funds have been expended, significant hazards remain. With subsurface ordnance still to be removed, visits to Kaho'olawe for cultural purposes continue to be restricted.[34] Although it has been returned to the state, which is now in charge of continued funding of cleanup and restoration, it is still a dangerous landscape.

CONCLUSION

Since the Second World War, Hawai'i's military presence has engulfed a large portion of island territory and economy and has evolved into a major political force. Civilian employment at military installations and defense contracts replaced sugar and pineapple as an economic mainstay. Hawai'i's congressional representatives continue to fight to keep or increase military spending in the islands when reduced troop levels or base expenditures appear in proposed federal budgets. When he was alive, Senator Daniel Inouye held powerful committee positions in which he could protect and expand military spending in Hawai'i. But along with economic dominance came problems. The expansion of PACOM activity in the Pacific increased the presence of live-fire training in the islands, with environmental and cultural repercussions. Increased security around military installations quickly resulted in restricted access to cultural and publicly utilized landscapes. An economy dependent on military spending put Hawai'i at the mercy of volatile trends in defense appropriations.

But the long shadow of the Second World War is also the continuation of a longer experience of this oceanic island environment with US military interests and war preparation. Dating to the 1850s, American designs on Pu'uloa for its Pacific adventures set the stage for the rapid increase of defense immediately before and after 1945. The acquisition and enclosure of Pearl Harbor after 1900, a direct result of the Spanish-American War, foreshadowed the encroaching militarization of the island lands, designating once-viable natural resources that served economy and culture as productive solely for military benefit. The loss of a vital estuary and control over fisheries brought by the First World War, and the eventual pollution of the harbor's sediments and water, culminated in the removal of a sacred native Hawaiian and food-producing landscape from beneficial use.

World War II extended the military footprint beyond Pearl Harbor and intensified an environmental and social process well under way by the mid-twentieth century. The Japanese bombing of the harbor on December 7, 1941, marked a significant transition in Hawai'i's militarization. The evolution of PACOM as a feature of island landscape and economy, with its four-service naval, army, air, and marine bases dotting O'ahu, arose in the shadow of that fateful day. It set the stage for significant environmental damage in its wake, symbolized by the Superfund status of Pearl Harbor and the USS *Arizona,* which to this day leaks oil from its sunken position at the harbor's bottom.

Just as important as that shadow is the fluorescence of native Hawaiian protests to reclaim crown lands illegally ceded to the US government. The work to protect cultural sites from military expansion and continued use for live-fire training has altered the dynamic between the US Armed Services and Hawai'i's people. Allied with environmental organizations and equipped with post-1970 environmental laws, Hawaiians have had some success in altering military land-use policies. The return of Kaho'olawe to public (state) control was a pivotal event. Temporary cessation of live-fire exercises in Mākua Valley, a more recent victory, testifies to the persistence of efforts to limit military authority in matters of land use. The power of environmental laws to protect water and endangered species, and to clean up toxic sites, plays a critical role in curtailment of military designs. The long shadow of the Second World War cast over Hawai'i, while demonstrating the consequences of militarization, also contains the seeds of mitigation of some of its worst effects. It is in the contest between native Hawaiian and their environmental allies on one side and the US military command in Hawai'i on the other where the final details of the long shadow will become apparent.

Notes

1 US Department of Defense, *Base Structure Report, 2012* (Washington, DC, 2012), accessed June 15, 2015, http://www.acq.osd.mil/ie/download/bsr/BSR2012Baseline.pdf.

2 "Militarized Sites in Hawai'i," *DMZ Hawaii,* accessed June 15, 2015, http://www.dmzhawaii.org/?page_id=244.

3 More accurately, Pearl Harbor is known as Ke Awalau o Pu'ulou, which translates "many-harbored sea at Pu'uloa." Pu'uloa is a site in the Ewa District on the western shore of the harbor, one of the first settled areas around the lochs. I frequently use the shortened version—Pu'uloa—to signify the whole of Pearl Harbor.

4 Edith P. Sterling at the Bishop Museum compiled this map in 1959 from numerous sources identifying all the ponds located in or near Pearl Harbor. Authoritative sources on Hawaiian fishponds include Patrick Kirch, *On the Road of the Winds: An Archaeological History of the Pacific Islands before European Contact* (Berkeley: University of California Press, 2000); and idem, *A Shark Going Inland Is My Chief: The Island Civilization of Ancient Hawai'i* (Berkeley: University of California Press, 2012); Marion Kelly, "Dynamics of Production in Precontact

Hawaii," in *What's New: A Closer Look at the Process of Innovation*, ed. S. E. van der Leeuw and R. Torrence (London: Unwin Hyman, 1989), 82–106; and C. Summers, *Hawaiian Archaeology: Fishponds* (Honolulu: Bernice P. Bishop Museum, 1964).

5 Kelly, "Dynamics of Production in Precontact Hawaii," 86.

6 See Kyle Kajihiro, "Becoming Pearl Harbor: A Lost Geography of Empire" (master's thesis in geography, University of Hawai'i, 2014).

7 Quoted in Jonathon K. K. Osorio, *Dismembering Lāhui: A History of the Hawaiian Nation to 1887* (Honolulu: University of Hawai'i Press, 2002), 168. See also Carol A. MacLennan, *Sovereign Sugar: Industry and Environment in Hawai'i* (Honolulu: University of Hawai'i Press, 2014).

8 General Schofield's 1873 report was published in 1925 as "Pearl Harbor, 1873," *American Historical Review* 30, no. 3 (April 1925): 560–65. See also Merze Tate, *Hawaii: Reciprocity or Annexation?* (East Lansing: Michigan State University Press, 1968).

9 See Osorio, *Dismembering Lāhui*. The agreement to renew the treaty was concluded between the Hawaiian government and the United States in December 1884 without a Pearl Harbor clause. However, the US Senate did not ratify it until January 1887, with an amendment granting the United States exclusive right to use the harbor and to establish a coaling station. Kalākaua signed it in September (after being forced to sign the Bayonet Constitution), and President Cleveland signed it in November. Without the revolt by the sugar planters and the diminished power of the king, it is likely that the Hawaiian monarch would not have agreed to cession of the harbor.

10 Queen Lili'uokalani, Diaries, September 26, 1887, Bishop Museum Library, Honolulu. Quoted in Sylvester K. Stevens, *American Expansion in Hawaii 1842–1948* (Harrisburg: Archives Publishing Company of Pennsylvania, 1945), 189.

11 See Noenoe K. Silva, *Aloha Betrayed: Native Hawaiian Resistance to American Colonialism* (Durham, NC: Duke University Press, 2004).

12 The US Pacific Command was established in 1947 and is the oldest and largest unified command of the US military. Under this structure the army, navy, air force, and marines operate large bases on O'ahu and control the use of land on other islands for training and bombardment. The US military also has the authority to limit access to the airspace and waters surrounding the islands.

13 The resolution also declared that US public lands policies would not be applied to Hawai'i, therefore recognizing the "continuing obligation of the United States toward Native Hawaiians," which was acknowledged in subsequent legislation in 1900 (Organic Act), 1920 (Hawaiian Homes Commission Act), and 1959 (Admission Act). See Jon M. Van Dyke, *Who Owns the Crown Lands of Hawai'i?* (Honolulu: University of Hawai'i Press, 2008), 212–14.

14 Ibid., 175, 216.

15 See the governor's correspondence under the US departments of war, navy, and army for this early period, available at the Archives of Hawai'i. See also Robert H. Horowitz, *Public Land Policy in Hawaii: Land Exchanges* (Honolulu: University of Hawai'i Legislative Reference Bureau, 1964), for examples of the lack of accountability and negligence within the governor's office in managing the valuation and procedures for land exchanges.

16 H. L. Stickney, "Early History of Pearl Harbor" (unpublished manuscript, University of Hawai'i Library, 1946).

17 John N. Cobb, *The Commercial Fisheries of the Hawaiian Islands in 1903* (Washington, DC: Department of Commerce and Labor, US Government Printing Office, 1904), 436.

18 Estate of John Ii, 81st Cong., 2nd sess., Report No. 2159, July 25, 1950, 5.

19 Rights of Fishery, Pearl Harbor, Hawai'i, H.R. HR10873, 66th Cong., 2nd sess., Report No. 508, December 17, 1919, 2.

20 Estate of John Ii, 3.

21 Ibid., 2.

22 Cobb, *Commercial Fisheries of the Hawaiian Islands*, 507.

23 Robert C. Schmitt, *Historical Statistics of Hawaii* (Honolulu: University of Hawaiʻi Press, 1977), 660. This includes military personnel, dependents, and civilians housed on military and naval installations.

24 *Honolulu Advertiser*, September 15, 1970.

25 US Department of the Interior, Federal Water Pollution Control Administration, Pacific Southwest Region, *Report on Pollution of the Navigable Waters of Pearl Harbor* (Washington, DC, October 1969).

26 Naval Civil Engineering Laboratory, *A Study of Sediments and Soil Samples from Pearl Harbor Area, March 1973* (Honolulu: Hawaiian Collection, Hamilton Library, University of Hawaiʻi, 1973), iii.

27 *Honolulu Advertiser*, April 7, 1981.

28 *Honolulu Advertiser*, May 11, 1982.

29 *Honolulu Advertiser*, August 20, 1990.

30 US Environmental Protection Agency, "NPL Site Narrative for Pearl Harbor Naval Complex," accessed June 15, 2015, http://www.epa.gov/superfund/sites/npl/nar1302.htm.

31 Agency for Toxic Substances and Disease Registry, US Department of Health and Human Services, *Public Health Assessment for Pearl Harbor Complex* (Washington, DC, December 2005), http://www.atsdr.cdc.gov/hac/pha/PearlHarborNavalComplex/PearlHarborNavalComplexPHA122805.pdf.

32 See EPA's webpage on the Pearl Harbor Complex for a summary of the current status, accessed June 15, 2015, http://yosemite.epa.gov/r9/sfund/r9sfdocw.nsf/ViewByEPAID/hi4170090076?OpenDocument.

33 See Mansel G. Blackford, "Environmental Justice, Native Rights, Tourism, and Opposition to Military Control: The Case of Kahoʻolawe," *Journal of American History* 91, no. 2 (September 2004): 544–71.

34 *Honolulu Star Advertiser*, July 12, 2013.

Chapter 9
The Great Louse War
Control of Typhus Fever

HELENE LAURENT

Until World War I, the main threat to survival during wartime was not enemy fire but the microbes that often killed in numbers many times greater than weapons.[1] Infectious diseases frequently spread from soldiers to civilians, causing the biggest wartime losses.[2] New diseases traveled efficiently with armies, and epidemics functioned as biological weapons: America was conquered with a relatively small army as the European diseases smallpox and measles decimated the indigenous population.[3] On the other hand, tropical diseases foreign to Europeans, foremost among them malaria, functioned as a barrier to imperialist expansion in Asia and Africa, complicating battles even after World War II.[4]

It was not until the end of the nineteenth century, which saw such advances in microbiology, that improvements in hygiene and the development of vaccines made prevention of epidemics efficacious. Furthermore, the control of disease-spreading insects and pests had also evolved by this point. Western armies implemented this new knowledge in their everyday practices, making possible large-scale, permanent concentration of combat troops during World War I without fear of epidemics. As a result of this progress in microbiology, for the first time in the history of war, more soldiers died on the battlefield than in the sick bed.[5]

The impact of contagious diseases on war has interested historians for decades. Friedrich Prinzing discussed war-induced epidemics depopulating vast civilian areas, as took place in seventeenth-century Germany during the Thirty Years' War. William McNeill applied the notion of macro- and microparasitism in describing the conquests attempted by humans on one hand

and by microbes on the other. McNeill describes convincingly how microbes, the "microparasites," were inadvertently carried along by the conquerors, the "macroparasites," and how the effect of the microbes could be much more far-reaching than the effect of a conqueror's armed invasion.[6]

In recent decades, a totally new aspect of the essential connection between wars and infectious diseases has been investigated. Historical research has brought under scrutiny the connection between chemical warfare and pest control. During World War I, many chemical substances lethal to humans—such as the sodium cyanide that was eventually used against humans by the Nazis during World War II—were used against vermin. Modern military technology has been involved in developing chemical poisons to protect people against disease-spreading insects as well as in developing agents of chemical warfare against humans. Many chemical substances widely applied during the war, DDT (or dichlorodiphenyltrichloroethane) being perhaps the best known, have later had far-reaching environmental effects.[7] Edmund Russell, above all, has opened up a new perspective in the ecological connection between war and peace.[8]

The subject of this chapter is louse-borne typhus fever, which historically spread among soldiers like wildfire; this much-feared fever threatened Finnish troops as well as others during World War II. During the Continuation War against the Soviet Union, from 1941 to 1944, typhus epidemics flared up among the Soviet prisoners of war, evacuees, and refugees. In total there were 1,407 typhus cases, but only forty-eight of them were Finns.[9] Why was the control of typhus fever so successful in Finland? How did the Finns manage to conquer this enemy, which had demanded such high casualties among both soldiers and civilians during earlier wars? I argue that the explanation for Finland's success may be found in the fact that the war against typhus in Finland was mainly fought with a natural, ecological weapon, practically without any chemicals. This essay describes the prevention methods used and organizations established to combat typhus fever in Finland, and compares them to those of other countries, mainly the Soviet Union, Germany, and the United States.[10]

LOUSE-BORNE TYPHUS, THE GREAT WARTIME KILLER

Typhus fever was a dreaded disease that spread particularly during famine and wars. It is caused by an intracellular bacterium called *Rickettsia prowazekii* that is transmitted from person to person by human body lice. The body louse attaches itself and lays its eggs, or "nits," in the inner seams of clothing. It feeds on the blood of its host four to six times a day and can lay up to four thousand nits during its lifetime. The louse becomes infected with *Rickettsia prowazekii* by ingesting the blood of a host suffering from typhus fever. The infection is always lethal for the louse, but before it dies it can transmit the infection to a new human host. In principle, just one louse bite is sufficient. In crowded and unhygienic quarters, lice move easily from person to person. If the lice are infected with rickettsiae, typhus spreads in a flash and is extremely difficult to contain.

The disease's long incubation time, approximately two weeks, makes it difficult to pinpoint the initial source of infection. Lice feces often remain infective in dust for months, making it possible for typhus to spread to the lungs via inhalation.[11] Symptoms typical of typhus include high fever, muscle pains, headaches, and mental confusion. Because the illness can be mild and only 50 percent of patients develop a rash, the symptoms can easily be misinterpreted and a positive diagnosis missed. Mortality rates vary between 5 and 40 percent. Patients recover slowly and are not capable of working for six to eight weeks.[12]

The names given to typhus, such as "war fever" and "camp fever," reflect its links to warfare. In Europe, typhus fever was for the first time referred to by the name of *tabardillo,* in connection with the Spanish *reconquista* in 1492. After that, it spread all over Europe with mercenaries. It played an important role in the devastation of the Thirty Years' War (1618–48), but the most famous of the catastrophes caused by typhus occurred in the Napoleonic Army during 1812–13. Of the 500,000 to 700,000 men who set off to conquer Russia, only 3,000 returned home. A major part of the Napoleonic Grande Armeé succumbed to "typhus," a fever that in many cases was louse-born typhus fever.[13] Typhus was also a significant killer in World War II, where it was used as a biological weapon in German prisons and concentration camps.[14]

Although never endemic in Finland, typhus has occasionally made an appearance during times of crisis. For instance, during the great famine years of 1866–68, almost 100,000 people in Finland died of "typhus," which probably included both typhoid and typhus fever. The famished people who left their homes looking for food infected each other in cramped and dirty poorhouses, which became death traps for the already-weakened populace.[15] During the Finnish Civil War, fought between the Whites and the Reds in 1918, typhus did not appear in prison camps, even if the fever was rampant in the vicinity of the eastern and southern borders.[16] On the threshold of the

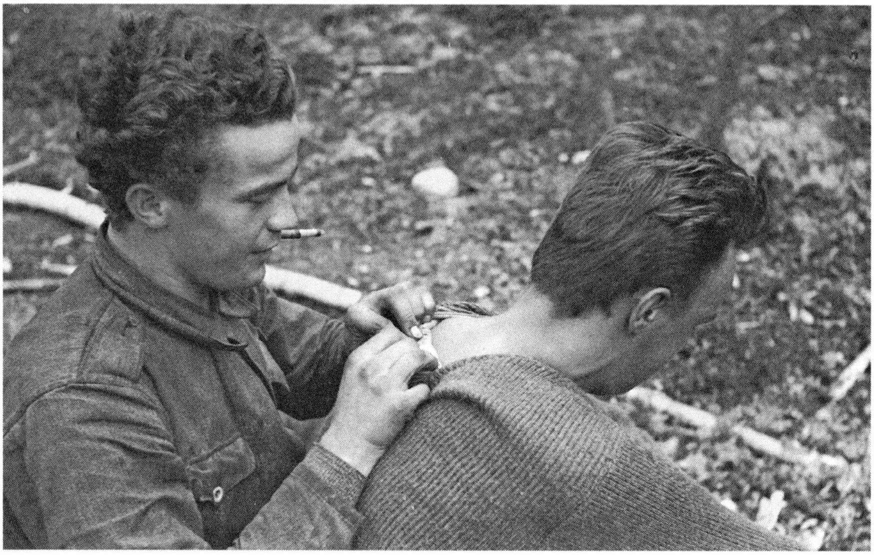

Figure 9.1. A war without end? A soldier looking for body lice. Source: Finnish Defence Forces Photo Archive.

Winter War in 1939, body lice were quite common in Finland, evident by a small typhus epidemic in central Finland, where a defector from the Soviet Union had joined a wandering Gypsy group.[17] Finland was thus still susceptible to typhus on the threshold of World War II.

THE PREVENTION OF TYPHUS FEVER

The best way to avoid a typhus epidemic, besides strict cleanliness, is the elimination of body lice, the only vector transmitting the disease. Low temperatures do not affect the lice, which are known to survive at temperatures of −12°C. Heat, however, destroys both the lice and the disease-causing organisms, the rickettsiae. When fumigating dwellings, chemical compounds, disinfection agents, and gases must be used. When sulfur is burned, sulfurous acid is released into the air, killing the lice in approximately six hours. Gaseous hydrogen cyanide (HCN), known to kill all vermin, including rats, was used first in the 1880s in California as a pesticide in the orange groves. It was extremely efficient in disinfecting factories and ships as well, but HCN proved to be lethal to humans. Disinfection with cyanide has to be conducted at a temperature of more than 18°C, and afterward the building must be aired for several hours.[18] In Germany, a new product called Zyklon was developed from hydrogen cyanide. The gaseous cyanide had been transformed into crystalline form; it was stored in metal containers and enhanced with a warning substance that irritated the mucous membranes, which made it easy to transport and safe to use.[19] It was not until the mass production of DDT commenced that efficient eradication of lice became possible, preventing the outbreak of a typhus epidemic in post–World War II Europe. DDT was effective and easy to administer, and at the time considered safe for humans.[20] The typhus bacteria, the rickettsiae, had to be destroyed by heating or by using disinfectants, such as Lysol.[21]

With typhus, the causative organism and method of transmission remained long unknown. In 1909, the French scientist Charles Nicolle determined the role of lice in the transmission of typhus.[22] Typhus was subjected to intensive research during World War I, when the causative bacterium was found.[23] At that time, in contrast to World War II, international cooperation and exchange of new scientific advances across the trenches through neutral countries were lively. Research results on the role of body lice in spreading the contagion could be applied in practice. On the western front, typhus could be avoided. "Lice-killing institutions" or showers and heating devices for clothes became routine. Delousing was mandatory when arriving at or leaving the front.[24]

The disinfection systems used on the eastern front collapsed after the Russian Revolution in 1917 and the ensuing civil war. Typhus started to spread rapidly. The epidemic raged in Russia on both the Red and White sides during the entire civil war and did not recede until after social unrest diminished in the mid-1920s. The seriousness of the epidemic and the role of lice were illustrated by V. I. Lenin's oft-cited 1919 remark, "All attention to this problem, comrades. Either lice will conquer socialism, or socialism will conquer lice." Between 1917 and 1925, twenty-five million people are thought to have contracted typhus, three million of which succumbed to the disease.[25]

In light of the tremendous losses claimed by typhus, the development of efficient anti-rickettsial drugs took a long time. Until antimicrobials chloramphenicol and tetracycline entered the market in 1948, other typhus control methods had to be applied.

PREVENTION OF TYPHUS FEVER IN THE FINNISH ARMY

During the spring of 1915, Germany sprayed chlorine gas for the first time on the battlefield of Ypres, thus starting a totally new, sinister phase in the history of warfare. Germany's gas attack led to an armament race for new poison gases. As a result of gas warfare, approximately 100,000 soldiers were asphyxiated in the battlefield, and more than a million men were disabled. Many animals too, particularly horses, suffered from gas warfare. Several battle gases were discovered to be pesticidal.[26]

After the war, the chemical industry in Germany continued to research poison gases. Fritz Haber, the developer of chemical warfare agents, received the Nobel Prize in chemistry in 1919; in his acceptance speech, he described gas warfare as "the highest form of killing." Chemical industry laboratories developed both pesticides and poison gases. The I. G. Farben Company synthesized compounds in the organophosphate group, some of which, such as parathion, were efficient pesticides, while some, like tabun or sarin, were neurotoxins that acted like poison gases. As a result of this research, the likelihood of future gas warfare became significant. In 1935, the Italians used mustard gas when attacking Ethiopia.[27]

In Finland, gas warfare was a particularly distinct possibility. The Society for Gas Protection, which was founded in 1927 and later renamed the Finnish Organization for Civil Defense, took over instruction and propaganda related to gas protection. It grew into a wide civic organization with close ties to the armed forces.[28] In 1927, the military founded a gas protection office headed by Major, later Colonel, Uolevi Poppius.[29] The gas protection troops made up approximately 0.5 percent of the regular combat troops, which were also equipped with gas masks.

But there was no proof that the Soviet Union utilized poison gases against Finland. Consequently, the profile of the gas protection units was modified, and more stress was placed on disinfection. The army had no practical experience in disinfection when the Winter War began. Controlling rats and insects, particularly body lice, became essential within a few months of the start of war, however, both to prevent the spread of diseases and to boost the fighting morale of the troops. The short duration of the Winter War, less than four months, impeded the building of efficient disinfection systems.[30] But the experiences gained during the Winter War brought about changes in military sanitation systems. Disinfection became the responsibility of the gas protection units, whose basic duties consisted of putting out fires and analyzing chemical samples in field laboratories. Now they were responsible for all disinfection measures, including fumigation of buildings and delousing of troops. The methods of gas protection and disinfection were similar; for example, the same chemicals were used both to neutralize poison gases and to kill microbes.[31]

In the Continuation War, the main assault of the Finnish troops on the Soviet Union began on July 10, 1941. The Finnish troops advanced rapidly, crossing the former national border by the end of July. The front line stopped deep in Soviet East Karelia in December 1941, where it remained until the summer of 1944. During the attack phase, the soldiers were constantly on the move, sleeping in open air or in tents, and maintaining their hygiene by swimming in lakes.[32] The onset of stationary warfare meant digging trenches and sleeping in dugouts that were hastily constructed, cold, cramped, and dark during the first winter of the war. In 1941, winter came early. Troop maintenance faltered, clean underwear was unattainable, and washing possibilities were almost nonexistent. Considering all these adversities, body lice naturally made an appearance.[33] At the beginning of 1942, lice infection among troops grew to alarming proportions: in some places, up to 50 percent of the soldiers were lice infested.[34] The army high command was too busy elsewhere to take notice of the pest situation. Its attitude changed abruptly when the first typhus fever cases appeared in prisoner of war (POW) camps in March 1942.

Finland took about 65,000 prisoners of war between 1941 and 1944, most of them within the first months. About 18,000, or 29 percent, died, mainly during winter 1941–42 as a result of malnutrition and infections.[35] The number of captives being unexpectedly high, the prisoners had to be placed in tents and even in the open air without shelter in the summer of 1941 while waiting for the construction of new barracks. Even during the following winter, a considerable number of the prisoners were still housed in cardboard tents, and the hygiene in the camps was abominable. The food was insufficient and of low quality, which resulted in mortality rates of up to 5 percent per month in February and March 1942.[36]

Army epidemiologists had already for some time dreaded the outbreak of an epidemic of typhus fever among the POWs, as the louse situation in the camps was uncontrollable. They knew that there was typhus in the Soviet Union as well as at the eastern front, from Germany up to Estonia.[37] It was only a question of time before typhus would make an appearance in Finland. In March 1942, the first typhus cases were diagnosed in a large transit camp in central Finland. Typhus spread with the prisoners to several POW camps around Finland. The epidemic did not recede until June 1942. In all, 741 prisoners fell ill and 105 died, the mortality rate being 14.2 percent, which is typical for typhus fever. Typhus was contained within the barbed wire of the camps, even if serious potential for wider spreading existed. The prisoners were in close contact with both soldiers and civilians, as they were used as a workforce in factories, forestry, and farms and, near the front line, in fortification and cleaning tasks.[38]

THE NATURAL WEAPON

The army high command reacted immediately to the typhus epidemic when the fitness of the combat troops was endangered. Now the army decided to resort to an old method to get rid of the lice and the disease-causing rickettsiae as well. Sauna bathing was an old, time-honored tradition in Finland. Practically all farms had saunas, situated preferably near a lake to have access to water. In towns the saunas were often communal. As the sauna was the warmest and cleanest place on the farm and fresh water was readily available, childbirth frequently took place there. Sauna bathing was also supposed to cure ailments of all sorts. There was an old saying, "If sauna, tar and spirits don't cure you, the disease is fatal." The traditional Finnish sauna is of the steam bath type, where water is thrown onto hot stones on the sauna stove, producing hot steam, or *löyly*. When taking the *löyly*, the bathers often slap themselves gently with a birch whisk, which cleans and relaxes the body. The temperature of the sauna is normally in the range of 70°C to 80°C.[39]

Although men built saunas at the front when possible, insects could not be controlled owing to the continuous movement of troops. Therefore the army decided that the wartime control of lice in Finland was to be based on a new invention: portable disinfection saunas. The planning and construction of portable disinfection saunas began in the fall of 1941. In January 1942, the first saunas were delivered to the troops. Over the spring of 1942, production increased to up to a hundred saunas a month, after extra funding for their construction was received. The gas protection office was responsible for the construction, delivery, and maintenance of the saunas. It published the *Manual for the Disinfection Saunas*, which contained detailed instructions for both the sauna crew and the bathers.[40] To function, the portable saunas needed only a trained crew, firewood, and water. They were inexpensive, easy to build, and could be transported quickly from one place to another.

The dimensions of the small disinfection sauna were 4.5 meters × 6.5 meters. It held six people, which meant that the sauna could process one hundred soldiers every twenty-four hours. The barrack-like sauna was simple,

Figure 9.2. A small mobile disinfection sauna. Source: *Terveydenhoitolehti*, 1942.

constructed partially of cardboard and partially of wood, and it could be assembled or dismantled within a few hours. It was planned for continuous use close to the front. The sauna parts could be transported in trucks or even in sleighs. The furnace heated the sauna bathing room and, with the help of a fan, also heated the adjacent space, where uniforms were heated to approximately 80°C. Half an hour was enough to kill lice, nits, and other insects. The capacity of the large disinfection sauna was 150 soldiers per twenty-four hours, and it was placed farther from the front line.

Discipline was vital for the bathing to produce the expected effect. The important principles of the disinfection sauna were as follows:[41]

1. Simultaneity: all men living in the same tent or dugout had to bathe at the same time.
2. Completeness: everybody had to bathe.
3. Separation: the unclean and the clean had to be kept apart.
4. Discipline: the bathers had to follow strict hygiene.

The bathing was regulated in order to reach the daily bathing quota:

> The leader of the disinfection team is the gas protection officer of the detachment, taking care of the bookkeeping, making

> observations on the effects of bathing, and attending to the eventual repairs of the sauna.
>
> The sauna crew, consisting of 1 + 4 men, is formed from the gas protection unit and is responsible for the practical operation of the sauna. The detail is responsible for the building, dismantling, and heating of the sauna.
>
> The inspector for the disinfection saunas is responsible for the disinfection measures in his division and for the assembly, review, and reporting of the experience gained.

Thirty minutes were reserved for undressing, forty minutes for bathing, and thirty minutes for dressing, and the uniforms were placed in dry heat cabinets for forty minutes. Bathing had to be performed in the following order:

1. Sweating, with no *löyly* (throwing water on the hot stones on the sauna stove) allowed.
2. Washing with a piece of soap, size 1 centimeter × 2 centimeters × 3 centimeters. Washing performed on benches.
3. Final sweating, with *löyly* allowed.
4. Rinsing with cold water.

The detachments were responsible for their bathing, even if inspections were made and detailed books were kept on the lousiness of the troops. The gas protection units were a part of the combat troops, and managing the disinfection saunas was not their only task. Every soldier had to be able to bathe in the sauna and change his underwear weekly. With the help of disinfection saunas and dry heat cabinets, the combat troops were quickly freed from lice infestation. The uniforms had to be disinfected every two to three weeks.[42] The motivation of the bathers was high, not only for hygienic reasons but also because of furloughs: soldiers were not allowed to go on leave with lice. If lice were found, the leave could be postponed by weeks, even months.[43] Sauna bathing was popular among the troops because, besides getting rid of the lice, it brought variety to the often-monotonous life on the front. Hence soldiers tended to follow carefully all bathing instructions.

When delousing of the front was solved, a new problem arose. The officers did not necessarily think that the rules applied to them. They could get a louse certificate without bathing and, consequently, officers were seen on

Figure 9.3. Horse in a delousing box. Source: Finnish Defence Forces Photo Archive.

leave with both lice and a louse certificate.[44] Men also often carried lice from their homes to the front, which is why officers and soldiers returning to the front had to be deloused. So, in the end, soldiers were required to bathe in the disinfection sauna any time they left or returned to the front. In summer 1942, the National Board of Health initiated a louse war among the civilian population, which showed that both head and body lice were quite common, particularly in eastern and northern Finland.

At the front, disinfection grew to huge proportions. In June 1942, for example, 130,000 men were deloused. Horses, too, were deloused in special mobile cabins. In August 1944, the gas protection office of military headquarters assessed the success of the army delousing campaign between July 1943 and June 1944. Disinfection had been thorough, and through diligent sauna bathing, the troops had become virtually louse free. In May 1944, the degree of lice infestation in field troops was 0.5 to 3 percent. In the whole military organization, 2.5 million men had bathed in the disinfection saunas during the year. In June 1944, owing to a rapid retreat, two hundred disinfection saunas and cabinets had to be left in the hands of the enemy. After withdrawal, the army still possessed a total of 157 saunas and 209 disinfection cabinets. Within the military, officers and soldiers alike were eventually motivated to use the disinfection saunas because the epidemics had been contained with their help.[45]

The construction of the saunas was constantly being tweaked, as their builders made use of reports received from frontline gas protection units. Even during the retreat during the summer of 1944, the troops were able to organize disinfection baths. By following strict bathing principles, the troops remained practically louse-free until the end of war, and the typhus fever never spread to the front.[46] It was a remarkable achievement.

Nevertheless, chemical poisons were necessary. Hydrogen cyanide was imported from Denmark and Zyklon from Germany, which was mainly utilized for disinfection of larger buildings such as schools and hospitals. Bunkers and dugouts were fumigated with foul-smelling sulfur and hydrogen cyanide, which killed insects such as fleas and bedbugs that often bothered the men more than lice, even if they were essentially harmless.[47] Sometimes they were fumigated with Zyklon, which, after opening its container, caused the bugs to fall down dead in heaps from chinks in the wall.[48] Because of the difficulties in importing hydrogen cyanide and Zyklon at the end of 1943, new experiments in insect control were conducted. Under the supervision of professor of biochemistry A. I. Virtanen, the 1945 Nobel Prize laureate, the army biochemical laboratory developed the so-called V-method of releasing hydrogen sulfide. The method was effective and fast, but it blackened metal objects and, like hydrogen cyanide, was dangerous to humans. The use of disinfectants was substantial: every year, 400,000 kilograms of sulfur, 34,000 kilograms of hydrogen cyanide, and 4,500 kilograms of Zyklon were used, and in 1944, 10,000 kilograms of substance V was consumed.[49]

PREVENTION STRATEGIES FOR TYPHUS IN GERMANY

The prevention strategies for typhus fever varied from country to country. Germany had a strong chemical industry that since World War I had synthesized new pesticides in addition to neurotoxins applicable in chemical warfare. Zyklon was commonly used for the disinfection of ships, buildings, and railroad cars. After World War I, when there was a westward flood of refugees from Russia troubled by typhus, the eastern border of Germany could be crossed only through certain control stations, where all refugees

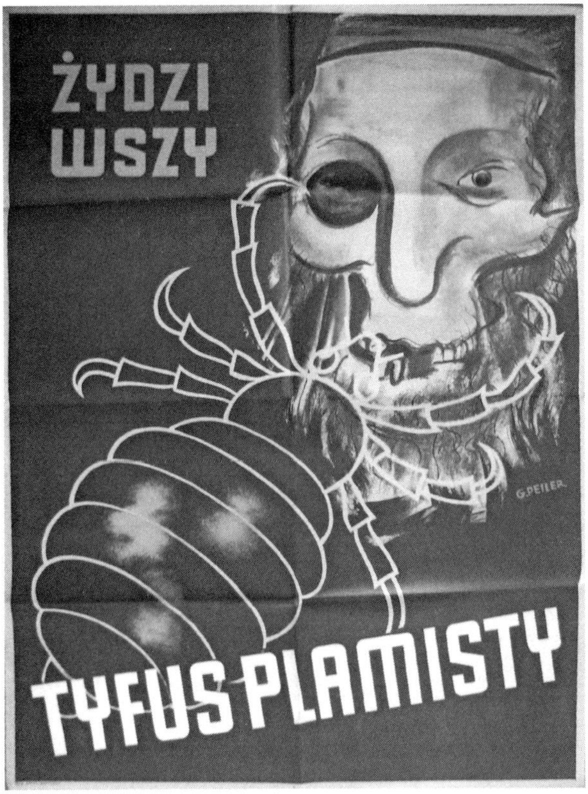

Figure 9.4. *Jews, lice, typhus fever.* A Nazi poster published in occupied Poland. Source: Collections of the Regional Museum in Rzeszów.

were disinfected. There they were showered and their belongings disinfected with Zyklon. The civilian population became used to the mass disinfections that continued during World War II, particularly in areas conquered by Germany.[50]

After the Germans had invaded Poland in the fall of 1939, the Jewish population was rounded up in ghettoes, where body lice started to spread in the cramped conditions, causing typhus epidemics. One of the ideological arguments was that Jews, often sellers of used clothing, spread vermin and diseases and were therefore to be isolated from the "clean" and healthy population. In Nazi propaganda, Jews were likened to vermin to be destroyed. The louse problem in the ghettoes was not attended to, so it can be said that typhus was actually used as a biological weapon. The same indifference repeated itself in the concentration camps and the prison camps at the eastern front, where typhus was rampant.[51] Zyklon was standard equipment in the camps, and the bulk of the poison was used to fumigate edifices and clothes. Some of the Zyklon was also used in an unprecedented manner as a weapon of mass destruction in the gas chambers, where millions of Jews and other oppressed peoples were killed.[52]

The development of the typhus vaccine was cumbersome because the intracellular *Rickettsia* bacteria could not be cultivated like other bacteria. At the Pasteur Institute in France, the vaccine was prepared from the lungs of infected mice or rabbits. In Poland, a tedious method requiring a lot of manpower was developed in 1930 by biologist Dr. Rudolf Weigl, in which the vaccine was prepared from the intestines of the infected body louse. For one dose of vaccine, fifty lice were required.[53] The Germans were not focused on vaccine development; they instead concentrated their efforts on producing disinfection equipment and lice-killing poison gases. They purchased vaccines for their own need, mostly from the occupied countries of France and Poland. There was not enough vaccine for the combat troops, however, and consequently only selected groups could be immunized against typhus. In the German-occupied Soviet Union, typhus was quite common, which resulted in transmission of the fever to the German combat troops. Germany could never totally eradicate typhus from their troops at the eastern front. Estimates of typhus cases in the German Army from different sources vary between 70,000 and 180,000.[54]

TYPHUS CONTROL AMONG ALLIED COUNTRIES

The United States and the United Kingdom collaborated in the development of typhus vaccines and insecticides that could be sprayed on skin or clothes. Between 1939 and 1940, the American scientist H. R. Cox succeeded in growing *Rickettsia* bacteria in the yolk sacks of chick embryos, which finally made possible mass production of the vaccine.[55] In 1942, President Roosevelt appointed a multidisciplinary Typhus Commission that combined the resources of the public sector and private industry to develop methods for typhus prevention. The Swiss chemical company Geigy had found in 1939 that DDT was an efficient insecticide and patented the compound in 1940. When testing it, the Americans observed that DDT had adverse effects on animals. Nevertheless, they bought its production rights in 1942 and immediately started mass production of DDT. The particular problem for US troops was waging war in the tropical zone of the Pacific Ocean, where scrub typhus transmitted by mites,[56] malaria, and hot and humid climate were much more intractable adversaries than the Japanese. Malaria caused on average five times more casualties than combat. After

DDT dustings on some islands, birds and mammals vanished, and even the vegetation suffered.[57]

In the part of North Africa conquered by the Allies, both body lice and typhus were widespread, which offered an opportunity to test the new vaccines and pesticides in practice. The Cox vaccine was tried out in Egypt and found effective. DDT powder was tested among the people in Arab villages. Treating with DDT posed a problem because it had to reach the inner seams of people's clothing while being worn. The solution was simple: the powder was dusted into the clothes from the sleeves, neckline, and waistline with a hand pump. The DDT spread out evenly inside the clothes and on the skin, giving excellent protection against lice.[58]

As a result of the efforts of the Typhus Commission, DDT was acquired for mass distribution in 1943. In Europe, DDT was applied for the first time in Naples in December 1943, when a typhus epidemic broke out among refugees in the overcrowded city. Within a few weeks, more than a million of its inhabitants were dusted with DDT, causing the epidemic to subside quickly. The US troops sent to central Europe were immunized with the Cox vaccine, with good results. Only 104 American soldiers fell ill, and all of them recovered.[59]

The Soviet Union did not publish its typhus statistics from World War II. According to the official figures sent to the League of Nations in the 1930s, the yearly incidence varied between 20,000 and 78,000 cases. Morbidity

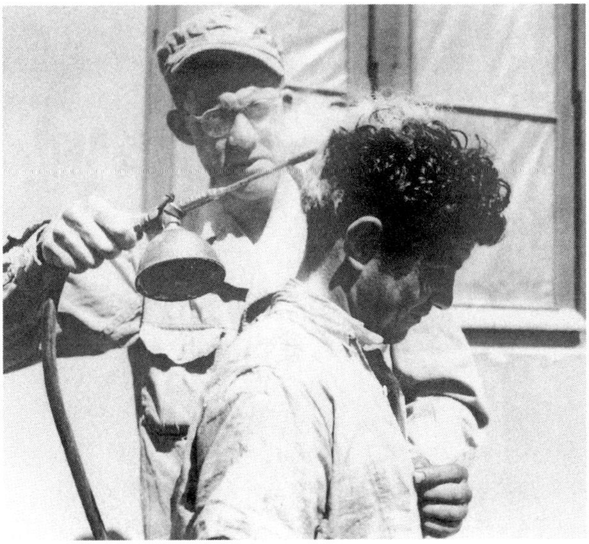

Figure 9.5. DDT dusting in Allied zone. The use of DDT increased enormously worldwide during and after World War II because of its effectiveness against mosquitos and lice. Source: Public Health Image Library, ID# 2621.

numbers of tenfold from that time have been proposed owing to the typhus epidemics raging in the gulags of Stalin.[60] According to unofficial information, the Soviet front was afflicted with typhus epidemics.[61] It must be assumed that typhus was common during World War II in the Soviet Union.

During the postwar period the use of DDT increased tremendously alongside the rising numbers of displaced people in the whole of Europe. It has been estimated that in 1945 over twenty-five million refugees and former soldiers were on the move in chaotic central Europe. Refugees were attended to by the United Nations Relief and Rehabilitation Administration (UNRRA) of the newly established United Nations, which distributed DDT and typhus vaccines in the camps to prevent a typhus epidemic. DDT rained over millions of people in Europe between 1945 and 1946 prevented a massive typhus epidemic.[62] By 1947, UNRRA had dispatched to Europe more than 3,000 tons of 10 percent DDT powder for typhus control as well as 1,500 tons of 100 percent DDT powder for malaria control.[63] The first application of DDT was thus the effective and rapid control of typhus in the aftermath of World War II.

CONCLUSION

Typhus fever was still a killer to be reckoned with during World War II; the precise number of victims is not known, but according to even conservative estimates, hundreds of thousands of people were afflicted. Strategies of typhus prevention differed greatly from country to country. Germany and the Soviet Union fared the worst, while Great Britain and the United States fared the best, having been able to develop modern prevention methods during the war. Considering the resources available, Finland succeeded outstandingly well. Finland is said to be the only country whose army remained louse-free without the use of DDT.[64]

In Finland, lice and therefore also typhus were conquered with the sauna, a method seen from today's viewpoint as environmentally friendly and effective, since the use of chemical poisons remained low compared with other warring countries. An important reason for the success of the campaign was cultural: in Finland, sauna bathing was a familiar and enjoyable way of getting rid of nasty body lice. In memories collected on wartime saunas, bathing is remembered with warmth and humor. Another essential factor was the professional and effective organization that existed before the war.

In both the First and the Second World Wars, battles were fought on two fronts. The first front was ideological and directed at people, and the other was microbiological and directed at germs and their carriers. This two-front war was a structural precondition for the commencement of both world wars and the ecological changes brought on by them, because wartime success against microbes enabled the building of vast armies. Even today, the war against microbes should not be forgotten. A concrete testimony to this was seen at the end of the 1990s, when a massive epidemic broke out in a refugee camp in Burundi and more than 100,000 people fell ill with typhus.[65]

The two world wars contributed organizationally to the emergence of environmental damage, too, by bringing together the interests of the state and the chemical industry. In consequence, the industry developed lethal chemical compounds that were effective against microbes and pests as well as humans, when needed. The application of these products in malaria and agricultural pest control led to the accumulation of pesticides in the food chain and the wide-ranging postwar environmental destruction described by the marine biologist Rachel Carson in her famous book from the 1960s. Let us not forget that the danger that *Silent Spring* described was born on the battlefield of Ypres in Belgium in April 1915.

Notes

An earlier version of this article was published as "Control of Typhus Fever in Finland during World War II," *Vesalius* 15, no. 2 (December 2009): 71–79. Published with permission from *Vesalius*.

1 Matthew Smallman-Raynor and Andrew D. Cliff, "Impact of Infectious Diseases on War," *Infectious Disease Clinics of North America* 18, no. 2 (June 2004): 341–68.

2 Edmund C. Tramont, "The Impact of Syphilis on Humankind," *Infectious Disease Clinics of North America* 18, no. 1 (March 2004): 101–10.

3 William McNeill, *Plagues and Peoples* (New York: Anchor Books Doubleday, 1998 [1976]), 208-18; Alfred W. Crosby, *Ecological Imperialism: The Biological Expansion of Europe, 900–1900* (New York: Cambridge University Press, 2004); Kirsten I. Bos et al., "Pre-Columbian Mycobacterial Genomes Reveal Seals as a Source of New World Human Tuberculosis," *Nature* 514 (October 23, 2014): 494–97.

4 Mark Harrison, "Medicine and the Culture of Command: The Case of Malaria Control in the British Army during the Two World Wars," *Medical History* 40, no. 4 (October 1996): 437–45; J. R. McNeill, *Mosquito Empires: Ecology and War in the Greater Caribbean, 1620–1914* (Cambridge: Cambridge University Press, 2010).

5 Smallman-Raynor and Cliff, "Impact of Infectious Diseases on War."

6 Friedrich Prinzing, *Epidemics Resulting from Wars* (Oxford: Clarendon Press, 1916); McNeill, *Mosquito Empires.*

7 John H. Perkins, "Reshaping Technology in Wartime: The Effect of Military Goals on Entomological Research and Insect-Control Practices," *Technology and Culture* 19, no. 2 (April 1978): 169–86; Edmund Russell, "Speaking of Annihilation: Mobilizing for War against Human and Insect Enemies 1914–1945," in *Natural Enemy, Natural Ally: Toward an Environmental History of Warfare*, ed. Richard P. Tucker and Edmund Russell (Corvallis: Oregon State University Press, 2004), 142–75.

8 Edmund Russell, *War and Nature: Fighting Humans and Insects with Chemicals from World War I to Silent Spring* (New York: Cambridge University Press, 2001).

9 Erkki Klemola, "Pilkkukuume Suomessa ja Itä-Karjalassa vuosina 1942–1944," *Duodecim* 1 (1948): 1–23; Helene Laurent, "Pilkkukuumeen vastainen taistelu jatkosodassa," in *Sodan ekologia: Nykyaikaisen sodankäynnin ympäristöhistoriaa*, ed. Simo Laakkonen and Timo Vuorisalo (Helsinki: Suomalaisen Kirjallisuuden Seura, 2007), 271–303.

10 The Finnish sources used in this article are compiled for the most part from the National War Archive (NWA), from the previously secret files of the gas protection and medical offices of army headquarters. Educational leaflets produced by the military were used as well. These sources give a reasonably clear picture of the prevention measures used in the campaign against typhus fever and their results.

11 Viktoria A. Harden, "Typhus, Epidemic," in *The Cambridge World History of Human Disease*, ed. Kenneth F. Kiple (New York: Cambridge University Press, 1993), 1080–84.

12 Ibid.; Didier Raoult, Theodore Woodward, and Stephen Dumler, "The History of Epidemic Typhus," *Infectious Disease Clinics of North America* 18, no. 1 (March 2004): 127–40.

13 Raoult et al., "History of Epidemic Typhus"; Jeffrey A. Lockwood, *Six-Legged Soldiers: Using Insects as Weapons of War* (Oxford: Oxford University Press, 2009), 63–64.

14 Hans Zinsser, *Rats, Lice and History* (London: Penguin, 2000 [1935]); Paul Weindling, *Epidemics and Genocide in Eastern Europe 1890–1945* (Oxford: Oxford University Press, 2000).

15 Kari J. Pitkänen, *Deprivation and Disease: Mortality during the Great Finnish Famine of the 1860s* (Helsinki: Publications of the Finnish Demographic Society, 1993).

16 Lars Westerlund, ed., *Sotaoloissa vuosina 1914–1922 surmansa saaneet*, Statistical Report 10/2004 (Helsinki: Publications of the Prime Minister's Office, 2004), 127.

17 Pauli Pylkkänen, "Pilkkukuumeesta," *Duodecim* 7 (1937): 728–33.

18 Weindling, *Epidemics and Genocide in Eastern Europe*, 46; Puhdistusosasto Kokko [Disinfection Division Kokko], September 24, 1942, no. 286, T7025, NWA.

19 Weindling, *Epidemics and Genocide in Eastern Europe*, 124.

20 Edmund P. Russell, "The Strange Career of DDT: Experts, Federal Capacity, and Environmentalism in World War II," *Technology and Culture* 40, no. 4 (November 1999): 770–96.

21 Military Headquarters, Medical Division 1, *Ohjeita pilkkukuumeen vastustamiseksi* [Instructions for typhus control] (Helsinki, 1943).

22 Nicolle was awarded the Nobel Prize in Medicine in 1928 for his discovery of the transmission method of typhus fever.

23 Harden, "Typhus, Epidemic"; Raoult et al., "History of Epidemic Typhus."

24 Weindling, *Epidemics and Genocide in Eastern Europe*, 73–76; Lockwood, *Six-Legged Soldiers*, 78–84.

25 See Victoria A. Harden, "Typhus, Scrub (Tsutsugamushi)," 1086–88; K. David Patterson, "Typhus and Its Control in Russia 1870–1940," *Medical History* 37, no. 4 (October 1993): 361-81; Lockwood, *Six-Legged Soldiers*, 79.

26 *Wikipedia*, s.v. "Chemical Warfare," accessed May 6, 2012, http://en.wikipedia.org/wiki/Chemical_warfare; Russell, *War and Nature*, 49.

27 For an eye-witness description, see N. Ayele, "The Horn of Africa and Eastern Africa in the World War Decade (1935-1945)," in *Africa and the Second World War* (Paris: United Nations, 1985), 87; David Day, *The Ecowars: A Layman's Guide to the Ecology Movement* (London: Paladin, 1989), 234.

28 Sakari Haahti, ed., *50 vuotta väestönsuojelutyötä: Suomen väestönsuojelujärjestö 1927–77* [50 years of civil defence: the Finnish Organisation for Civil Defense 1927–1977] (Helsinki: Väestönsuojelusäätiö, 1977), 1–6.

29 Ibid., 18.

30 Woldemar Lojander, "Nuorimman lääkärikuntamme omakohtaisista terveydenhoidollisista ja lääkintähuollollisista näkemyksistä Talvisotamme ajalta vv. 1939–1940," *Sotilaslääketieteellinen Aikakauslehti* 1 (1960): 3–78.

31 Yhteenveto kss.komppanioiden desinfektiotoiminnasta 31.8.41 saakka [Summary of disinfection activity of gas protection troops (GPT) until 31 Aug 41], Appendix to PM letter no. 2367/Kss.1/sal, October 7, 1941, T15803 Folder 17, NWA.

32 Yhteenveto kss.toiminnasta 1.9.–31.10.41 [Summary of GPT activity 1 Sep–31 Oct 41], T15803 Folder 17, NWA.

33 Aarre Nyman, Esko Salminen, and Urpo Vento, eds., *Sotasavotta: Korsuelämää 1939–1944* (Helsinki: Otava, 1975), 34–36.

34 Yhteenveto kss.komppanioiden sekä joukkojen kss.elimien desinfektion toiminnasta 1.1.–30.6.42 [Summary of GPT companies' and troops' GPT disinfection activity 1 Jan–30 Jun 42], T15806 Folder 16, NWA.

35 Lars Westerlund, "The Mortality Rate of Prisoners of War in Finnish Custody between 1939 and 1944," in *POW Deaths and People Handed Over to Germany and the Soviet Union in 1939–55: A Research Report by the Finnish National Archive*, ed. Lars Westerlund (Helsinki: Kansallisarkisto, 2008), 25–50; Gunnar Rosén, *Sata sodan ja rauhan vuotta: Suomen Punainen Risti 1877–1977* (Helsinki: Suomen Punainen Risti, 2002), 509.

36 Eversti S. Malm, Huomioita Suomen ja Neuvostoliiton välillä vv.1941–44 käydyn sodan aikana otettujen sotavankien kohtelusta ja hoidosta [Notes on the care and treatment of POWs taken during the war between Finland and the Soviet Union 1941–44], November 16, 1944, T20950 Folder F50, NWA; Rosén, *Sata sodan ja rauhan vuotta*, 523–26; Westerlund, "Mortality Rate of Prisoners of War," 71–74.

37 Weindling, *Epidemics and Genocide in Eastern Europe*, 287–90.

38 As a consequence of the typhus epidemic, improvements to the dwellings, hygiene, and healthcare of prisoners of war were made. Typhus did not recur in the camps after 1942. Klemola, "Pilkkukuume Suomessa ja Itä-Karjalassa."

39 Sauna bathing, an eastern tradition, has been common also in Estonia and in coastal Latvia. In Russia it is called *banya*. For the history and current use of sauna, see, e.g., *Wikipedia*, s.v. "Sauna," accessed November 8, 2009, http://en.wikipedia.org/wiki/Sauna.

40 GPT commander no 3706/Kss.17sal, December 18, 1941, T15803 Folder 17, NWA; Military Headquarters medical division 1 no 1776/lääk./7/28 sal, April 28, 1942, 8:63 Folder Fa 35, NWA; Military Headquarters, Gas Protection Office, *Desinfektiosaunan käyttöopas* [Guide to disinfection sauna use] (Helsinki, 1942).

41 *Desinfektiosaunan käyttöopas.*

42 Käsky puhtaudenpidon järjestelystä erikoisesti pilkkukuumevaaraa silmälläpitäen [Command on organization of hygiene, with a particular emphasis on the risk of typhus], Command section no. 1434/Kom. 2/21/sal, April 22, 1942, T20948 Folder F16, NWA.

43 GPT commander no 3923/Kss.4/sal, Military Headquarters, Gas Protection Office command no. 9/42, T 15806 Folder 15, NWA; Nyman, Salminen and Vento, *Sotasavotta*, 72.

44 Military Headquarters, Gas Protection Office, December 19, 1942, Mäkelä, Sortavala, T15806, Folder 16, NWA.

45 Report on army disinfection activity during the third war year 1 Jul 43–30 Jun 44, Military Headquarters, Gas Protection Office, June 17, 1944, T20950 Folder 49, NWA.

46 Ibid.

47 Report on the activities of the gas protection units 1 Jan–30 Jun 42, T15806 Folder 16, NWA.

48 Corporal Sakari Laakkonen (b. 1919). Interviewed by Simo Laakkonen, March 1, 2006. Interview notes are in the possession of the interviewer. Puhdistusosasto Kokko, September 24, 1942, no. 286, T7025, NWA.

49 Military Headquarters, Gas Protection Office, August 17, 1944, T20950 Folder F49, NWA.
50 Russell, "Speaking of Annihilation"; Weindling, *Epidemics and Genocide in Eastern Europe*, 112-14, 289.
51 Weindling, *Epidemics and Genocide in Eastern Europe*, 273–74; idem, " 'Belsenitis': Liberating Belsen, Its Hospitals, UNRRA, and Selection for Re-Emigration, 1945–48," *Science in Context* 19, no. 3 (Autumn 2006): 401–18.
52 Weindling, *Epidemics and Genocide in Eastern Europe*, 298–306.
53 Paul Weindling, "Between Bacteriology and Virology: The Development of Typhus Vaccines between the First and Second World Wars," *History and Philosophy of the Life Sciences* 17, no. 1 (1995): 81–90; Arthur Allen, *The Fantastic Laboratory of Dr. Weigl* (New York: W. W. Norton, 2014).
54 Weindling, *Epidemics and Genocide in Eastern Europe*, 284–88, 336.
55 Weindling, "Between Bacteriology and Virology."
56 Harden, "Typhus, Scrub (Tsutsugamushi)."
57 Weindling, *Epidemics and Genocide in Eastern Europe*, 374; Russell, *War and Nature*, 113, 124–26, 135–36; For a general picture, see Leon J. Warshaw, *Malaria: The Biography of a Killer* (New York: Rhinehart, 1949); Lockwood, *Six-Legged Soldiers*, 154–55.
58 The Fred L. Soper Papers, "World War II: Typhus Fever and Malaria in the Mediterranean," accessed May 4, 2006, http://profiles.nlm.nih.gov/VV/Views/Exhibit/narrative/typhus.html.
59 The Fred L. Soper Papers; Charles M. Wheeler, "Control of Typhus in Italy 1943–1944 by Use of DDT," *American Journal of Public Health and the Nation's Health* 36, no. 2 (February 1946): 119–29; Chris J. D. Zarafonetis, "The Typhus Fevers," in *Internal Medicine in World War II*, vol. 2, *Infectious Diseases* (Washington, DC: Medical Department, US Army, 2009), http://history.amedd.army.mil/booksdocs/wwii/infectiousdisvolii/chapter7.htm; Lockwood, *Six-Legged Soldiers*, 151–53.
60 Patterson, "Typhus and Its Control in Russia."
61 E. H. Beardsley, "No Help Wanted: Medical Research Exchange between Russia and the West during the Second World War," *Medical History* 22, no. 4 (October 1978): 365–77.
62 Weindling, *Epidemics and Genocide in Eastern Europe*, 398–400.
63 Wilbur A. Sawyer, "Achievements of UNRRA as an International Health Organization," *American Journal of Public Health* 37, no. 1 (January 1947): 41–58.
64 Ole Enkvist, "Hygienian valvonta Karjalan kannaksella 1941–44," *Sotilaslääketieteellinen Aikakauslehti* 1 (1947).
65 Didier Raoult et al., "Outbreak of Epidemic Typhus Associated with Trench Fever in Burundi," *Lancet* 352, no. 9125 (August 1998): 353–58.

Chapter 10
Perspectives on the Acoustic Ecology of War

OUTI AMPUJA

The environmental impacts of industrial warfare include intensive noise pollution, a dimension of conflict that environmental historians have generally overlooked. Preindustrial battlefields were also flooded with chaotic noise, but the explosive power of industrial armaments greatly intensified it, producing a wholly unique type of acoustic environment that had great significance for soldiers in terms of their well-being and even survival. The life of a soldier or horse on the front lines depended to a great extent on acoustic literacy, or the way in which and speed with which one reacted to certain sounds. In both world wars, noise pollution caused physical and mental damage to people, which often prevented them from enjoying the sounds of nature long after the conflicts were over. In World War II, traumatic wartime sonic experiences on the front lines and at the home front increased the sensitivity of numerous civilians and soldiers to peacetime noise and unwanted sounds. For those who lost their hearing during the war, a silent spring meant something different than it did for most people.

This essay explores the role that the recognition of the sounds of war plays in survival, the effects of noise on soldiers' hearing and mental health, and the use of noise as a psychological weapon. These are the aural elements present in the soundscape of war. These themes belong to the sphere of acoustic ecology, which examines the relationship between all living organisms and the acoustic environment. Acoustic ecologists are interested in what kinds of acoustic environments can be considered balanced, and the significance of the acoustic environment for the well-being of humans and other living organisms.

These topics have seldom been dealt with in literature. As a general rule, military histories and wartime memoirs touching on the physical and psychological health of veterans include at best brief mention of the sounds of war and the effects of noise. An exception is Charles D. Ross's article on the American Civil War, in which he relates how, relying solely on sound, commanders who had no visual or other method of communication were able to attack the enemy.[1] But Ross's article does not address the use of sound and noise as a psychological weapon. Various studies on "shell shock" of World War I have focused on the psychological impacts of noise but neglected the other dimensions of the wartime soundscape.[2] It is time to explore wartime soundscape as a holistic phenomenon comprising both negative and positive sounds and their impact on the body and the mind.

For this article I interviewed five Finnish veterans who had served on the front lines of World War II, asking them about their memories and impressions of the sounds of war as well as the significance of sounds in terms of survival. For these men, issues of sounds and noise were not the first topics to come to mind when they talked about war. After their initial confusion, however, most of the interviewees were able to call to mind numerous sounds, both pleasant and unpleasant, and events from the front that were linked to them. In addition to these interviews with veterans, the war diaries kept by Finnish military companies, the memoirs of soldiers who fought in various armies on various fronts, and published research that touches on the subject served as source data. The data deal primarily with World War II, but some examples from other wars are also included.

This essay focuses on the soundscape of World War II in Finland but presents examples from Germany and Russia as well. Questions related to the sounds and noise of war may be a way of opening up new perspectives that diverge from those offered by operation-focused military histories teeming with regiment numbers and battlefield names. In the best cases, the veteran's aural memories opened up a channel to his emotions: the interviewee, instead of recounting events, started talking about what it felt like being in the war. Inevitably, fear rose as a topic of discussion.

I use the term "noise" to refer to sounds that are experienced as unpleasant, disturbing, or unnecessary. "Sound" is a more neutral concept that covers all sounds in the acoustic environment—all sounds that can be detected by the human ear. "Soundscape" refers to the field of sound surrounding us at any given time; it can include noise, the sounds of nature, and sounds produced by people and machines.[3] Soundscape is not necessarily a synonym for acoustic environment; the concept of soundscape can also include how the individual or community experiences the acoustic environment.[4]

AN ACOUSTIC SURVIVAL GUIDE

Sound has the same significance for life during wartime as during times of peace: sounds are signals about events in the environment; they contain clues that we have learned to interpret and that guide our actions. The majority of sounds in our everyday soundscapes do not inspire reactions of any practical significance for us. Only exceptional sounds—in particular, loud sounds or sounds that in our experience denote danger—get us to react.

The clues embedded in the soundscape of war are interpreted according to the same logic: some sounds mean danger, others don't. In war, the importance of the sense of hearing, like that of other senses, grows in comparison to its significance during times of peace. One's life may depend on the correct interpretation of sounds, and rapid reactions and momentary inattentiveness may have immediate negative consequences. This is noted, for instance, in the memoirs of Erkki Huokuna, who didn't notice the shot of the launcher firing behind him because he was concentrating too intently on trying to repair his machine gun. As a result, the memoirist and his comrades were wounded.[5]

The sense of hearing also had great significance during guard duties. The Finns had a custom of keeping "listening posts" at the forefront of the

Figure 10.1. Sentry in a snowy wilderness. Source: Finnish Defence Forces Photo Archive.

line, sometimes even stationing them beyond it. By listening to the enemy's movements, the sentry was able to get an idea of the enemy's intentions and prevent surprise attacks, which were generally attempted in the dark. For a listening sentry, any abnormal sound could signify a surprise attack. Pentti Sipiläinen, who fought on Finland's eastern front, remembers of his time as listening sentry that if "a bird took off or a hare jumped, we had to steel ourselves, since that meant that something was going on."[6]

In war, gear that interfered with one's sense of hearing could prove life threatening. According to the experiences of Soviet soldiers, the tapered caps that were part of the original Soviet Army uniform prevented soldiers from hearing well. When on guard duty, the soldier would have to open it so he could hear with one ear. The woolen helmet covers used during heavy freezes also interfered with the Red soldiers' ability to hear and, according to their experiences, "simply helped the Finnish ski troops—the White Death," as they were called by the Soviet soldiers.[7] Nor did it aid the sentry to move around, as the sound that was emitted by the snow crunching under his feet could reveal him to the enemy. For this reason, soldiers laid branches of spruce underneath their feet as they moved through the snow.[8]

Soldiers on guard may have attempted to be particularly quiet, but the soundscape of war was frequently full of sounds and noise, especially during the day: explosions, shooting, and cries. But one had to learn how to distinguish sounds signifying imminent danger from the overall aural hubbub. New recruits had to be able to decide rapidly, based on the sounds of the shelling, whether taking cover was necessary. Lauri Anttolainen tells of the initial experiences of new arrivals at the front line, explaining that when hearing a suspicious sound, "at first, we always hit the ground."[9] Reino Joukamo has a similar recollection. By following the example of a more experienced soldier who acted as a guide for the new arrivals, Joukamo gradually learned when it was unnecessary to take cover.[10] Some of the veterans I interviewed guessed that they learned to distinguish the sound of an incoming shot from one flying overhead within a couple of weeks, or at most a month, of arriving at the front line.[11]

How did the soldiers distinguish the sounds "requiring" cover from other sounds? Based on the information gathered from interviews and memoirs,[12] we can compile the following "sound map" of interpretations of the sounds produced by weapons:

- A whistling projectile will fly over; one that's headed your way doesn't whistle.
- A shell that hisses will fly over; a sharp sound means it will strike nearby.
- The sound of a rifle bullet that will hit nearby is short, high, and sharp.
- You don't hear the shot that hits you.

The sounds of war also formed the basis of broader situational evaluations.[13] German field surgeon Peter Bamm, who served on the eastern front during World War II, writes about interpreting the sounds of war. Observing and correctly interpreting the sounds of war was vital for the surgeon in cases, especially when the Germans were retreating, when the front line would move extremely rapidly. The entire field hospital, including patients and staff, was the responsibility of the surgeon. A correct analysis of the situation made it possible to retrieve and treat the wounded until the very last moment, and could save the entire surgical staff from ending up as prisoners of war. In his memoirs, Bamm tells of having made decisions regarding retreat based specifically on the soundscape: "As long as there was still movement on the road, there was no imminent danger. . . . But sometimes the line was cut. Silence on the road was a warning signal that the front line was approaching. Behind the fighting infantry, the terrain was empty. From then on, the situation had to be evaluated according to the noise of battle. . . . We learned to distinguish down to the calibre our own shells from those of the enemy as they struck the ground. The final signal was the clanking of the tanks' tracks and the artillery fire."[14]

Other commanders and their men also listened attentively to the sounds coming from the opposing side. Observations regarding sound were recorded abundantly in the war diaries of Finnish companies. Below are a few excerpts of observations recorded in the war diary of the 44th Infantry Regiment, Company 3, which was stationed at the eastern front of Finland in spring of 1944:

> 25 Apr, 0200 hours: lots of commotion and the sounds of men working
> 2 May, 1900 hours: heard marching song from the Russki side, a large group singing.
> 10 May, 2200–2400 hours: motor humming from the left, maybe a tractor

19 May, 2000–2100 hours: at grid 58/34 celebrating, vodka—women
26 May, 0010 hours: something driven to line 5–15–6, probably lumber, after that the sounds of men working
27 May, 2200–2400 hours: in line 5, chopping and sawing
28 May, 1800 hours: at grid 64/50, two powerful detonations of earth[15]

The 44th Infantry Regiment, Company 3, moved to the reserves on May 30, and responsibility for the front was taken by another infantry regiment, which continued gathering "acoustic intelligence." In the sector of the front in question, the Soviet offensive began on June 21, 1944; the sounds recorded in the war diary above most likely portended this event. Troops were sometimes forced to rely on sound data when all other ways of acquiring information failed.[16]

SOUND AND NOISE AS A WEAPON

The significance of different kinds of sounds and noise as weapons and in generating fear has been recognized since prehistoric wars. Noise is perceived in all cultures as a weapon, a scourge.[17] Also, the walls of Jericho are described as having crumbled under the force of sound.[18] A war cry is perhaps the simplest example of the use of sound as a weapon. Its intention is to fan fear in the enemy while at the same time boosting one's own courage and that of one's comrades.[19]

Reino Joukamo relates an example of the significance of one's own voice in sustaining courage. In the bomb shelters, civilians sang "just about anything" so that they wouldn't hear the horrifying sounds of the bombing. The singing also functioned as a morale builder, as a countersound, and as inspiration for people to have confidence in their survival. Such singing, too, was a generator of transitional experience. Transitional experience refers to our ability to transfer our feelings and attachments to substitute objects, for instance, to another person or inanimate object. The transitional experience can be seen with small children, for whom a teddy bear can function as a substitute for the security offered by the mother, concretizing the connection of trust between child and mother. When the mother is absent, the teddy bear acts as a psychological link to her for the child.[20] For adults, music can function as a transitional object. Adults also frequently have a habit of humming, whistling, or drumming their fingers in an attempt to keep themselves calm if they are stressed, afraid,

Figure 10.2. A heavy grenade launcher firing at German positions in Lapland, 1944. The gunners are protecting their ears with their hands. Hearing protectors were not in use at the time. Source: Finnish Defence Forces Photo Archive.

or confused. The aural transitional object shows us that there is no need to fear.[21] A literary example is provided in *The Unknown Soldier,* a war novel that is deeply revered in Finland. The book depicts a private who cold-bloodedly shoots fifty enemies with a submachine gun. While waiting for the enemy to approach, the private counsels a comrade who is acting as the ammo man to keep mentally calm: "If you know how to sing, then hum quietly. It'll keep your mind easy. That's the mental strategy you gotta apply."[22]

The battle cry hasn't been the loudest wartime sound for ages, having been muffled by the mass of sound produced by more and more technologically advanced weapons. As the significance of firearms in warfare and especially artillery has increased, the roar of battles has grown deafening. During the Battle of Waterloo, the battle cry *Vive l'Empereur!* was still heard, as were war cries on the fronts during World War II,[23] but the rumble of artillery already dominated the soundscape on the battlefield at Waterloo, literally deafening the men.[24] The most extreme decibel figures were presumably reached during the artillery bombardments of World War II. According to Anthony Beevor, the Red Army's attack on Berlin began with the greatest artillery bombardment that the world had seen to that point. Of those that survived "the inferno," as that bombing was called, many "lost all sense of hearing."[25]

Arvo Seppinen, who served in the Finnish field artillery, describes how it felt to be the target of naval artillery: "Explosions that ripped at our ears slashed through the air, knocked over men, and made the earth roll. . . . Some men gave in and covered their ears, which felt like they were being jabbed with hot needles."[26] Lauri Anttolainen, who served as signalman in a forward-observer detachment, remembers some of the gunners trying to protect their ears by shoving cotton into them, as hearing protectors were not available.[27] Such attempts were made because extremely loud noise causes physical pain to the ears. The pain or discomfort caused by a loud sound or a sound at an unpleasant frequency can be so powerful that the broadcast of loud or unpleasant sounds has been used as a torture method.

An enemy's incessantly listening ears could be used against it as well. Military history recounts numerous cases where one side has produced misleading sounds. The sounds of bridge building or tank engines can be reproduced with amplifiers, leading the opposing side to concentrate its forces in the wrong places. Such diversionary sounds work particularly well at night.[28] The enemy can also be diverted with a wall of sound produced by heavy artillery. Behind the wall of sound it is possible to execute operations in secret.

Producing sounds that signify a threat is an effective way of fanning uncertainty and fear among the enemy. During World War II, simply hearing the clanking of the tanks' tracks could inspire fear in the infantry ranks.[29] After dark or when visibility was otherwise lacking, ominous sounds had an even more terrifying effect. According to Peter Watson, World War II soldiers feared nighttime bombing more than daytime bombing, even though the accuracy of the hits was greater during the day.[30]

The research results based on the experiences of these soldiers indicate that simply hearing a sound without a line-of-sight connection was "nerve wracking," as stated in the memoirs of the Finn Evald Laamanen, who ended up the target of a Soviet close air support aircraft attack. A few times, he had to sit through an attack in a tent that prevented him from seeing it.[31] Unable to act but hearing all kinds of sounds that signified danger, a soldier can fall victim to his own imagination. The sense of sight gives a better opportunity to prepare for what's coming than simply the sense of hearing, which is why Pentti Sipiläinen relates having always lain on his back during bombing so that he could "see when [the bombs] are released."[32] Seeing the fall of a bomb that predicted destruction wasn't as frightening as hearing its sound. According to a study conducted in 1944, the sound of falling and exploding

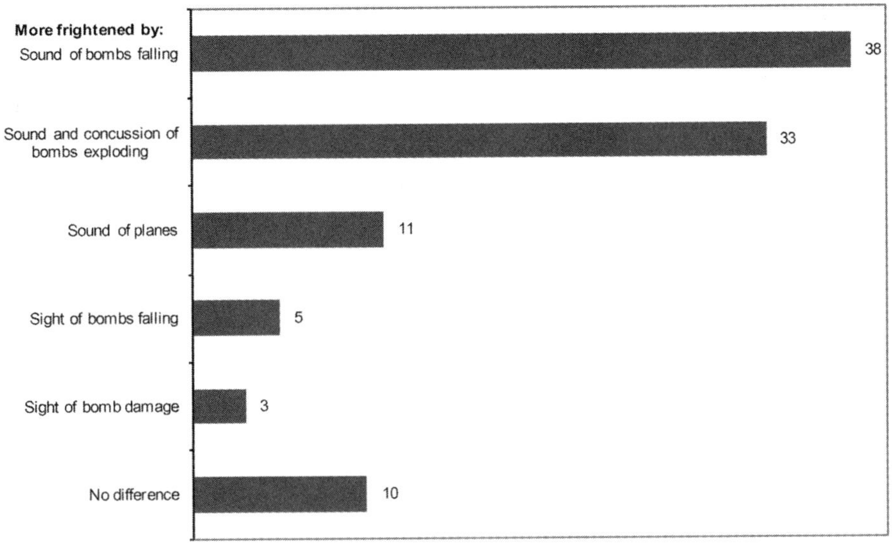

Table 10.1. Rate of Fear of Different Aspects of Bombing

Note: In a study conducted of US soldiers who volunteered to fight in the Spanish Civil War, participants were asked to rate their fear of different aspects of bombing. This table shows the percentage of men answering.

Source: John Dollard, *Fear in Battle* (Washington, DC: Infantry Journal, 1944), 17.

bombs aroused greater fear in soldiers than the sight of falling and exploding bombs.[33] According to the study, the difference in reactions aroused by visual versus aural observations was significant, as shown in table 10.1.

NOISE AS A PSYCHOLOGICAL WEAPON

There's a strong psychological aspect related to the use of noise as a weapon, recognized in a variety of ways in weapon design. Two particularly illustrative examples come from World War II. One was the Junkers Ju 87–type dive-bomber, also known as the Stuka, used by Germany at the time. The plane was fitted with a small, simple propeller, the sole function of which was to produce a loud racket when the plane dove. In terms of psychological warfare, this invention, in all its simplicity, was exceedingly effective. According to a survey, soldiers feared the German dive-bombers specifically because of their horrible sound. This finding is interesting, because normally the fear inspired by a weapon correlates with its danger and effectiveness.[34] The noise produced by the dive-bombers—and also by the heavy bombers that dropped their loads from a horizontal position—had a significant effect on how badly

Figure 10.3. Bombs being attached to a German Junkers Ju 87 ("Stuka") dive-bomber. The noise-producing sirens are located in front of the aircraft's landing gear. Source: Finnish Aviation Museum.

these weapons were feared by the opposing forces (see table 10.2). Another point of interest in the findings is that noise is the only factor noted as frightening, even though its consequences are not directly life threatening.

The Soviets also realized the possibilities offered by noise in the sphere of psychological warfare. The "noise weapon" of the Red Army was the Katyusha

Table 10.2. Men's Reasons for Fearing Different Weapon Types, by Percentage

	Mortar	Anti-Aircraft Gun	High Bomber	Machine Gun	Dive Bomber
Accuracy	37	31	23	16	11
Lack of warning	19	11	11	15	5
Rapidity of fire	8	7	1	42	0
Noise	**11**	**19**	**21**	**6**	**48**
No defense	1	2	14	0	4
Other	24	3	30	21	32

Source: Peter Watson, *War on the Mind: The Military Uses and Abuses of Psychology* (New York: Penguin Books, 1980), 153. Watson refers to J. C. Naylor, "Proposed Method for Determining the Psychological Effects of Weapons" (paper presented at the first Symposium of the Psychological Effects of Non-Nuclear Weapons, Eglin Air Force Base, April 29, 1964).

rocket launcher, which the Germans called Stalinorgel and the Finns knew as "Stalin's Organ." The Katyusha was an effective weapon in terms of speed of fire, and the rocket was most typically fired from a launcher erected in the bed of a truck.[35] Introduced by the Soviets against Germany during 1941,[36] the weapon was used against the Finns toward the end of World War II. It inspired terror, at least initially, among the soldiers who ended up targets of its fire. Pentti Sipiläinen remembers how the arrival of the "organ gun" "affected a lot of people pretty badly." He relates: "One morning they began firing the organ gun, and the reinforcements from the older age groups weren't able to do anything, they just lay there trembling at the bottom of the trenches."[37]

Sipiläinen describes how he and the other members of the younger age group received information in advance that the Katyusha wasn't actually that dangerous because its ammunition didn't shrapnel. The weapon only killed if it was a direct hit. Still, Sipiläinen remembers, "the first time I heard it I thought, this is the end of the world." But after seeing the exploded shells, which contained no shrapnel, he realized that the weapon wasn't particularly dangerous.[38] Lauri Anttolainen also remembers the arrival of this new Russian weapon at

Figure 10.4. A loudspeaker intended for broadcasting propaganda is transported into the terrain on the Eastern Front. Source: Finnish Defence Forces Photo Archive.

his sector of the front and recalls the sound as having been terrible. At first he and his comrades were baffled by the weapon, but based on the unexploded shells, they decided that its effect "was mostly psychological." According to him, "the moment of surprise was scary, because we didn't know what it was."[39]

Why are loud sounds frightening, even though noise alone isn't lethal? There seems to be no simple answer to this question. When preparing this essay, I did not come across a single study that would have presented the results of psychological research on or theories of the mechanisms of the human mind or brain that form the basis for the effectiveness of noise as a psychological weapon. But we can infer something from studies of environmental noise that deal with noise as a psychological stress factor and find that it causes a wide variety of stress-related symptoms. Noise stimulates negative feelings and stresses organisms' autonomic regulatory systems, such as hormone levels, heart and circulatory activity, and the immune system. Individual factors, such as sensitivity to noise, further shape the relationship between noise and reaction. An individual's interpretation of noise—the cognitive process of evaluation—plays a critical role in determining the risk of threat. One's individual interpretation of noise and ability to deal with a situation define whether one experiences the noise as stressful. Situations in which the relationship between a person and the environment is characterized by a state of imbalance, threat, harm, or demand beyond the person's control are perceived as stressful.[40] The capacity of a person who is subjected to and stressed by noise to manage other potential stress factors is weakened, because dealing with the situation at hand demands effort and causes exhaustion.

On this basis we can make the tentative inference that in war, the effectiveness of noise as a way of fanning fear is linked to the stress reactions it creates: noise signifies a threat beyond the soldier's control. This generates stress, which in turn hinders action. But if the source of the threat-causing noise—for instance, the weapon producing it—is familiar and perceived as being less dangerous than was initially thought, such stress diminishes, as does the fear related to the noise. This seems to have been the case with the Katyusha rocket: once the soldiers observed that its shells were relatively nondangerous, the noise produced by the weapon was no longer considered frightening.

NOISE AND PSYCHOLOGICAL COLLAPSE RESULTING FROM WAR

War abounds with stressful situations, and for many who ended up at the front, the fear of death, continuous exhaustion, thirst, and hunger, as well as the critical injuries sustained by and deaths of their comrades were overly devastating experiences. The phrase "Every man has his breaking point" came into use during World War I: any individual could develop psychological disorders during the demanding conditions of war. The term "shell shock" was used to describe these psychological disturbances during World War I; during World War II, the Americans used the terms "combat exhaustion" and "operational fatigue" to describe a state of psychological collapse.[41] In contemporary usage, the term "combat stress reaction," which is considered a type of post-traumatic stress disorder (PTSD), is used to describe acute psychic injury incurred during war.[42] Combat stress reaction is most typically preceded by anxiety, fear of death, crying spells, psychological numbing, and a feeling of vulnerability. During war, feelings of vulnerability, fear, and anxiety are experienced by almost everyone, but for soldiers suffering from PTSD, the fear grows so intense that it hampers their ability to evaluate situations, paralyses the limbs, causes unconsciousness, and generates a countereffect that leads to taking unnecessary risks. A psychologically numb soldier doesn't feel pain, horror, or sorrow. In addition, exhaustion, depression, and introversion prevent soldiers from taking effective action.[43]

There are indications that, for some individuals, the sounds and noise of war contribute to making the burden of war too great, and thus foster the emergence of psychological symptoms. Military psychiatry has seen cases of sudden deafness. Prior to the end of World War I, sudden deafness was thought to have been the result of powerful explosions and damage to the ears and brains resulting from physical injuries. During World War I, hundreds of autopsies were performed on soldiers who had gone deaf and later died, in an attempt to locate the damage in the brain and eardrums. Oftentimes these autopsies revealed nothing. Gradually it was understood that sudden deafness could result from psychological factors.[44] For a relatively large proportion of those who suddenly went deaf during World War II, the cause of deafness was proven to have been purely psychological. Sudden deafness is considered an unconscious method of shutting out an external reality that has grown too painful to bear.[45]

THE HUMAN CRY

Although war is full of loud and frightening noises, the horror of one sound rose above the rest in the stories of the veterans I interviewed: the screams of the wounded. Human cries of pain appeared to arouse horror and distress more effectively than any weapon of technological warfare psychologically tuned to portend destruction. Lauri Anttolainen recalls one dark autumn night when the enemy made a violent reconnaissance without artillery preparations in order to be able to surprise the Finns at their stations. After the fierce combat following the surprise attack, the Finns heard terrible screaming from the Russian side that lasted until morning. Anttolainen recalls considering with his comrades whether the screaming was an intentional attempt to lure the Finns into helping the wounded and thus being taken prisoner. The "horrible screaming" echoed in Anttolainen's ears for days afterward.[46]

Pentti Sipiläinen remembered one unpleasant instance in particular. During one Finnish attack, the purpose of which was to reclaim a certain village from the enemy, a soldier in one of Sipiläinen's groups was severely injured. But the situation was such that no one was able to reach the wounded man to help him. He cried, "God, God, mother, help me!" The voice grew fainter and finally ended. Kalevi Kaltto recounted the day that his friend was left wounded behind enemy lines, out of help's reach. It was December 22, 1939. The friend yelled, "Don't leave me here!"

Finnish soldiers were also captured as prisoners. One of these was Reino Joukamo. Of his time in the camp, Joukamo recalled one aural memory that was perhaps for him the most painful. In 1944, a Belgian prisoner who had fought with the German troops was lying near him in the camp's hospital barrack. The prisoner was emaciated and in very bad shape. The Belgian got worse and worse, and before his death, he cried out ceaselessly in pain for three days: "Mama, mama!" Reino Joukamo was forced to watch the suffering of another human in close proximity, helplessly, hoping that the Russians would have at the least released the prisoner from his suffering by shooting him. But the man was allowed to slowly waste away.

THE SOUNDS OF WAR AND PSYCHOLOGICAL SCARS

The sounds of war continued to echo in the memories and dreams of veterans long after the war came to an end. All of the veterans I interviewed described thinking about wartime events daily. Of wartime sounds, the clamor of the bombs and artillery shells as well as the cries of the wounded seemed especially to infiltrate their dreams. They saw most of their nightmares about the war in the years directly following it, but for some of them they continue to this day, if less frequently. In a nightmare experienced by Pentti Sipiläinen, whose friend remained wounded behind enemy lines, the same events always

Figure 10.5. Musical performances brought welcome variety to the soldiers' lives. Source: Finnish Defence Forces Photo Archive.

recur: someone is wounded, but Sipiläinen is unable to help. Urho Sarvikivi relates that his brother, who had been at the front, saw such terrible nightmares after the war that he had a hard time sleeping at night. He would cry out: "The Russkis are coming, got to get to our stations!"[47] Lauri Anttolainen and Reino Joukamo had nightmares too.

For some of those who lived through the war, the associations related to the sounds of war can have an effect during waking times as well. When Joukamo is alone in the forest, a sound of a branch breaking can mentally take him back to the front. This kind of sound makes him feel the same kind of uncertainty and fear that he felt during battle. Unbearable memories arise when he hears Russian being spoken. Because of his time as a prisoner of war, Joukamo feared Russian "yelling, and angry snapping." He also identifies as sensitive to sound and can't stand any sort of background noise; he believes that his sensitivity to sound results from the war.

The sounds of severe thunderstorms remind Lauri Anttolainen of war, and then he "gets a weird feeling, even now." Pentti Sipiläinen's thoughts also almost undetectably shift to war, battles, and artillery fire during thunderstorms. My grandfather, who also lived through the war, was sensitive to the sound of thunder after returning from the front. He would order the entire family into the potato cellar during thunderstorms.

It's unlikely that anyone who lived through the war returned home untouched. For some, the effects of the war on the psyche were relatively mild, and generally these people attempted to continue living as best they could without unnecessarily remembering the past. Many of the veterans I interviewed didn't speak about the war at all for years or decades afterward, even with their former comrades-at-arms. But each of them eventually found a home and place in postwar Finland, despite their harsh experiences. Not everyone who was in the war fared as well. In addition to the fallen and the physically wounded, of the Finnish soldiers who fought in World War II, over 18,000 were hospitalized for psychological trauma.[48] By the late 1980s, 11.3 percent of them, approximately two thousand men, had applied for compensation for psychological disorders caused or worsened by the war. We can assume that tens of thousands of Finnish soldiers suffered war-induced psychological damage that would have made them eligible for care and compensation. This assumption is supported by two facts: first, only a portion of the traumatized soldiers sought care after the war,[49] and second, of US soldiers who fought in World War II, up to a third of those in combat were estimated to have been psychologically traumatized by the war. For those who ended up as prisoners of war, the lifetime incidence of trauma-induced symptoms can rise to 70–80 percent.[50]

Those who suffer psychological injury during war may exhibit symptoms for decades afterward. In these cases, the most typical symptoms include recurring nightmares; anxiety, particularly during situations reminiscent of traumatic experience; depression; disassociation; antisocial or unstable behavior; irritation; inexplicable bursts of rage; excessive cautiousness or fear; difficulty concentrating; or the feeling of having inadequate future opportunities. As a result of these and numerous other symptoms, the traumatized person may experience problems with friendships, spouses, and families, and in the workplace.[51] Abuse of alcohol and other intoxicants is also frequent.[52] It appears that those suffering from the long-term effects of psychological trauma also react to noise more sensitively than others.[53] A decreased capacity for tolerating loud sounds for up to fifteen years after the war has been found in those who suffer psychological symptoms.[54]

Being subjected to noise during war caused physical damage too. Over 10,000 (4.7 percent) Finnish soldiers suffered a hearing injury during the war.[55] It is estimated that these statistics do not include many combatants whose hearing was weakened in the war. According to an investigation conducted in 1992, in Finland the number of veterans who are hard of hearing is greater than among the general population.[56]

CONCLUSION

Aware of war's horrors and the resulting large-scale suffering and death, it at times felt trivial asking these veterans about the sounds of war. On the other hand, the perspective that opened up through sounds and aural memories seemed, perhaps surprisingly, to strike at the root of war: the experiences and sensations of individual people. It was as if the aural memories offered the interviewees a safe channel for speaking about these issues, and about the fear that is so fundamentally related to war. In that sense, then, this study serves a purpose, because individual feelings and traumatization among soldiers have until this point been dealt with quite minimally in military-historical research.

It was also surprising to learn how effective the psychological weapon of noise can be during wartime. Even though by World War II armies were already using modern weapons in terms of firepower and accuracy, still such a simple thing as loud noise could get soldiers to consider weapons producing the noise to be more frightening than more effective and accurate weapons. It was also new to me how alone each soldier was with his experiences and fears, both on the front and after the war. I had expected that the soldiers would have discussed the enemy's weapons amongst themselves, and perhaps exchanged points of view on the sounds they emitted, for instance. Yet all the soldiers I interviewed seemed to have painstakingly avoided speaking of anything that could be related to death, fear, or uncertainty over one's own fate. The sounds of war seemed to have played precisely those strings that the horror and stress of war had tensed to the extreme in the soldiers' minds, which is why they were not discussed. Veterans didn't seem to speak of these things after the war, either. They wanted to forget the war as quickly as possible.

Perhaps speaking about fear and feelings didn't fit into the masculine culture of the front lines. And demonstrating weakness didn't fit into the image of Finnish soldiers offered by "official Finland": a soldier from the backwoods who, jaw slightly clenched, could take just about anything. Some were so badly traumatized by the war that talking about it later was anything but appealing. Generally speaking, wartime sounds and noise were replaced in peacetime by a culture of silence.

War is an extreme experience, and those who return from it are no longer the same people they were before they left. Noise doesn't linger "in nature" the same way that poisonous chemicals do. But it can leave permanent imprints on humans. War leaves physical and psychological scars, and the sounds and noise of war appear to play a part in this process. The hearing of many soldiers worsened, they suffered sensitivity to loud noises, nightmares, and some veterans still return to the front in their memories when they hear sounds that remind them of war. Former soldiers must live with the memories and perhaps the traumas war generates for the rest of their lives. Each remembers the war in his own way. Some of these traumas are passed on, affecting the minds and lives of the following generations as well. War can truly cast long, invisible, and somber shadows.

Notes

The author wishes to extend sincerest thanks to those veterans who agreed to be interviewed and open-mindedly answered questions, some of which touched on painful topics. All the veterans' interview notes are in the author's possession.

1 Charles D. Ross, "Sight, Sound, and Tactics in the American Civil War," in *Hearing History: A Reader*, ed. Mark M. Smith (Athens: University of Georgia Press, 2004); also S. Goodman, *Sonic Warfare: Sound, Affect, and the Ecology of Fear* (Cambridge: Massachusetts Institute of Technology Press, 2010); Mike Goldsmith, *Discord: The Story of Noise* (Oxford: Oxford University Press, 2012), 181–87.

2 For example, see Richard Gabriel, *The Painful Field: The Psychiatric Dimension of Modern War* (New York: Greenwood Press, 1988); and Peter Barham, *Forgotten Lunatics of the Great War* (New Haven, CT: Yale University Press, 2004).

3 R. Murray Schafer, *A Sound Education* (Indian River, ON: Arcana Editions, 1992).

4 Barry Truax, *Acoustic Communication* (Westport, CT: Ablex, 1984 [2001]), xii.

5 Erkki Huokuna, "Tunturijääkärinä," in *Minuun sattui . . . Mikkelin seudun sotainvalidit kertovat* (Mikkeli: Mikkelin seudun sotainvalidit ry, 1994), 43.

6 Corporal Pentti Sipiläinen (b. 1922), farmer, currently retired. Interviewed by the author in Porvoo, Finland, February 17, 2004.

7 Finnish ski troops were long scout patrols. During the Continuation War (1941–44), Finnish tactics relied partially on sending patrols of skiers deep into enemy territory to carry out intelligence and sabotage operations.

8 Information on the Winter War experiences of Red Army soldiers is taken from Viktor Stepakov and Dimitri Orehov, "Paraatimarssi Suomeen," in *Talvisota venäläisten silmin*, ed. Matti Hyvärinen (Porvoo: WSOY, 1992), 245–46.

9 Private Lauri Anttolainen (b. 1920), farmer, currently retired. Interviewed by the author in Artjärvi, Finland, March 5, 2004

10 Sergeant Reino Joukamo (b. 1923), architect, currently retired. Interviewed by the author in Järvenpää, Finland, December 26, 2003.

11 E. B. Sledge, who served as a young marine in the Pacific and fought at Peleliu and Okinawa during World War II, recalls in his memoirs that he was trained to identify the weapon type of approaching enemy fire based on sound. He and his fellow marines were urged to quickly pick up this skill, as it was critical to survival. E. B. Sledge, *With the Old Breed: At Peleliu and Okinawa* (New York: Ballantine Books, 2007), 30.

12 Cf. Anatoli Gordijenko, *Kuoleman divisioona* (Jyväskylä: Gummerus, 2003), 172. The work is based on the diary of war correspondent and political commissar N. I. Klimov.

13 Such as in the US Civil War. Ross, "Sight, Sound, and Tactics in the American Civil War."

14 Peter Bamm, *Näkymätön lippu* (Jyväskylä: J. K. Gummerus, 1959), 212; orig. Peter Bamm, *Die unsichtbare Flagge* (Munich: Kösel-Verlag, 1952).

15 44th Infantry Regiment, Company 3 (3./JR 44) war diary from December 3, 1943, to August 24, 1944; war diary (SPK) 11878, Military Archives of Finland (SA).

16 For example, Lasse Laaksonen, *Todellisuus ja harhat: Kannaksen taistelut ja suomalaisten joukkojen tila talvisodan lopussa 1940* (Helsinki: Yliopistopaino, 1999), 259.

17 Jacques Attali, *Noise: The Political Economy of Music* (Minneapolis: University of Minnesota Press, 2002 [1985]), 27.

18 Jeremiah 19:3; Joshua 6:20 (King James Bible).

19 Peter Bailey, "Breaking the Sound Barrier: A Historian Listens to Noise," *Body & Society* 2, no. 2 (June 1996): 52. See also R. Murray Schafer, *The Soundscape: Our Sonic Environment and the Tuning of the World* (Rochester: Destiny Books, 1994 [1977]), 50.

20 Donald W. Winnicott, *Playing and Reality* (Harmondsworth, UK: Penguin, 1971).

21 Kari Kurkela, *Mielen maisemat ja musiikki: Musiikin esittämisen ja luovan asenteen psykodynamiikkaa* (Helsinki: Sibelius Academy, 1997), 459–60.

22 Väinö Linna, *Tuntematon sotilas* (Porvoo: WSOY, 1956), 265–66; a similar interpretation was provided by Susanna Välimäki in *Johdatus musiikintutkimukseen*, ed. Tuomas Eerola, Jukka Louhivuori, and Pirkko Moisala (Vaasa: Suomen musiikkitieteellinen seura, 2003), 305–6.

23 Cf. Antony Beevor, *Stalingrad* (London: Penguin, 1999), 42.

24 John Keegan, *The Face of Battle: A Study of Agincourt, Waterloo and the Somme* (Suffolk, UK: Penguin, 1976), 141–43.

25 Antony Beevor, *Berlin: The Downfall 1945* (London: Penguin, 2002), 218.

26 Arvo Seppinen, "Kun taistelulaiva Marat ja Saarenpään linnake 'haastelivat' keskenään," *Kansa Taisteli* 10 (1959): 281.

27 Interview with Anttolainen, 2004.

28 Jon Latimer, *Deception in War* (London: John Murray, 2003), 74.

29 Cf. Vilho Karhula, "Ensimmäinen haavoittumiseni," in *Minuun sattui*, 65.

30 Peter Watson, *War on the Mind: The Military Uses and Abuses of Psychology* (New York: Penguin, 1980), 153.

31 Evald Laamanen, "Miinat yllättivät," in *Minuun sattui*, 115.

32 Interview with Sipiläinen, 2004.

33 Study conducted with American soldiers who volunteered to fight in the Spanish Civil War. John Dollard, *Fear in Battle* (Washington, DC: Institute of Human Relations, Yale University, 1944), 17.

34 Watson, *War on the Mind*, 151–53. Watson refers to a study by J. C. Naylor, "A Proposed Method for Determining the Psychological Effects of Weapons" (paper presented at the Symposium of the Psychological Effects of Non-Nuclear Weapons, Eglin Air Force Base, May 29, 1964); about Stuka sound, see Goldsmith, *Discord*, 181–82.

35 Michael Foedrowitz, "Die Stalinorgel: Sowjetische Mehrfachraketenwerfer," *Waffen-Arsenal*, Sonderband S-30 (1993); David C. Isby, *Weapons and Tactics of the Soviet Army* (London: Jane's, 1981), 203.

36 Foedrowitz, "Die Stalinorgel," 7.

37 Interview with Sipiläinen, 2004.

38 Ibid. In terms of firepower, the Katyusha was rapid but imprecise, for which reason it was also known as the "hectare gun."

39 Interview with Anttolainen, 2004.

40 Marjut Wallenius, "Ympäristömelun kokeminen ja terveys," in *Melukylä vai mansikkapaikka? Asukkaiden ja asiantuntijoiden näkemyksiä asuinalueiden terveellisyydestä* (Helsinki: Ympäristöministeriö, 2001), 88–89; J. Campbell, "Ambient Stressors," *Environment and Behavior* 15, no. 3 (1983); G. Evans and S. Cohen, "Environmental Stress," in *Handbook of Environmental Psychology*, ed. D. Stokols and I. Altman (New York: Wiley, 1987); P. Lercher, "Environmental Noise and Health: An Integrative Research Perspective," *Environmental International* 22, no. 1 (1996): 117–29.

41 See Matti Ponteva, "Psykiatriset sairaudet Suomen puolustusvoimissa 1941–1944," *Sotilaslääketieteellinen Aikakauslehti* 52 (1977): 40–43, in which he refers to various studies in terms of military psychiatry during World War I.

42 Zahava Solomon, Nathaniel Laor, and Alexander C. McFarlane, "Acute Posttraumatic Reaction in Soldiers and Civilians," in *Traumatic Stress: Effects of Overwhelming Experience in Mind, Body, and Society*, ed. Bessel A. van der Kolk, Alexander C. McFarlane, and Lars Weisaeth (New York: Guilford Press, 1996).

43 Ibid., 105–7.

44 Gabriel, *Painful Field*, 13.

45 E. S. Levine, *The Psychology* of *Deafness* (New York: Columbia University Press, 1960), 69–70, which cites studies by Miller et al., *A Bibliography in Audition*, vol. 2. (Cambridge, MA: Harvard University Press, 1959); E. A. Strecker, *Fundamentals of Psychiatry*, 2nd ed. (Philadelphia: J. B. Lippincott, 1944).

46 Interview with Anttolainen, 2004.

47 Private Urho Sarvikivi (b. 1924), tinsmith. Interviewed in Järvenpää, Finland, on December 16, 2003.

48 Ponteva, "Psykiatriset sairaudet Suomen puolustusvoimissa 1941–1944," 51.

49 Lars Weisaeth, "The Psychosocial Effects of War and Other Disasters" (paper presented at the 20th General Assembly of the World Veterans Federation, Helsinki, Finland, October 20–24, 1991).

50 Seppo Hietanen et al., "Sotakokemusten jälkeiset traumaperäiset stressihäiriöt ja niiden aiheuttaman haitan korvaaminen Suomessa," *Sotilaslääketieteellinen Aikakauslehti* 2 (1995): 38; on the psychic wounds of US soldiers, see also Jonathan Shay, *Achilles in Vietnam: Combat Trauma and the Undoing of Character* (New York: Touchstone, 1994), 203.

51 Kalle Achtè, "Psyykkisistä sotavammoista: Posttraumaattisen stressireaktion käsite tuli Suomeen," *Sotainvalidi* 1, no. 55 (1995): 4–5; Hietanen et al., "Sotakokemusten jälkeiset traumaperäiset stressihäiriöt"; Kirby Farrell, *Post-Traumatic Culture: Injury and Interpretation in the Nineties* (Baltimore: Johns Hopkins University Press, 1998), 6–5.

52 Bessel A. van der Kolk, "The Complexity of Adaptation to Trauma: Self-Regulation, and Characterological Development," in Bessel et al., *Traumatic Stress*, 191.

53 According to a small-sample study conducted by Dobbs and Wilson in 1960, the systems of psychologically symptomatic soldiers reacted more powerfully to the sounds of battle than those of control groups. Dobbs and Wilson (1960) cited in Ponteva, "Psykiatriset sairaudet Suomen puolustusvoimissa 1941–1944," 56.

54 Archibald et al. (1962) cited in Ponteva, "Psykiatriset sairaudet Suomen puolustusvoimissa 1941–1944," 56; cf. also the prisoner-of-war experiences of Reino Joukamo presented in this essay.

55 Markku Honkasalo, *Suomalainen sotainvalidi: Tilasto-osa* (Helsinki: Sotainvalidien veljesliitto, 2001), 25.

56 Leena Ryynänen et al., *Suomalaisen rintamaveteraanin muotokuva: Veteraaniprojekti 1992:n loppuraportti*, Reports of the Ministry of Social Affairs and Health 4 (Helsinki: Ministry of Social Affairs and Health, 1994), 29.

Part III

RESOURCE EXTRACTION AND THE WAR

Chapter 11
Aluminum's Permanent Revolution

MATTHEW EVENDEN

In August 1942, Siegfried Moos, a Jewish expatriate economist from Berlin who had taken a position at the Oxford Institute of Statistics, declared an "aluminum revolution." This revolution, he noted, was the product of modern war, "in which the extent and character of airforces are determined by the quantities of aluminum available."[1] Aluminum production had first risen in World War I, he observed, and then increased more than tenfold in the interwar period to reach a global output by 1939 of almost one million tons. In the first two and half years of the war, output had almost doubled and would likely top two million tons by the year's end. This massive expansion had set in motion a train of consequences for bauxite producers, shipping companies, aluminum smelters, and aircraft manufacturers.[2] Although Moos did not conceive the problem this way, the aluminum revolution had also triggered profound changes in the social and environmental relationships that were necessary for, and produced by, aluminum production. Moos considered output and aircraft manufacture in England, but the Allied aluminum commodity chain's backward linkages reached aluminum smelters and dammed rivers in Quebec, a shipment station in Trinidad, and bauxite mines, and razed forests along the Demerara River in British Guiana. These were changes in what we might conceive of as aluminum's commodity chain.[3] If this was a revolution, then it was a revolution not simply of industry, but also of an expanded commodity trade, and of nature and society at an increasingly global scale. Although the war instigated this revolution, after 1945 its impacts would continue to redefine materials use, energy, and society. In many respects, what Moos identified as the aluminum revolution became a permanent feature of modernity.

ORIGINS OF THE ALUMINUM COMMODITY CHAIN

Although the element aluminum is found widely in the earth's crust, its use as a refined metal is relatively recent. Formal experimentation with aluminum reduction dates from the early nineteenth century. But before the development of the Hall-Héroult process in the 1880s, which smelted alumina at high temperatures using a cryolite bath, aluminum could not be produced in quantity. After the 1880s, the metal grew slowly in commercial importance, and capital was invested in hydroelectric plants to meet the energy-intensive smelting process. In the final stages of the First World War, aluminum became widely used in alloys for aircraft. As a lightweight, flexible, and durable material, aluminum had some considerable advantages. In the interwar period, as aircraft design underwent important developments, aluminum only increased in military importance. In alloys used to build aircraft bodies, aluminum often accounted for 90 percent of the whole. It also gained new uses in naval construction, wiring, munitions, and various domestic products.

As aircraft became increasingly deployed in military roles during the 1930s, aluminum's strategic importance rose. Nazi Germany's abrogation of the Treaty of Versailles's constraints on the size of the German Air Force, for example, created a burgeoning aluminum industry, which by 1939 was the leading world producer. The coordinated attacks of German air and ground forces through Belgium and northern France in the spring of 1940 only underlined the point that the rules of warfare had changed. Facing what was increasingly viewed as an air war, with the blitz unleashed on London and industrial Britain in the late summer of 1940, the British government moved toward a massive aircraft production program. In May 1940, the US government announced its own program to build fifty thousand planes a year in the name of national defense.[4] Apart from the military implications of these decisions, there were profound material consequences.

Aluminum production first started in Quebec in the interwar period when James Duke, an American tobacco magnate and power developer, exercised rights to develop waterpower at Lac Saint-Jean in the upper Saguenay basin. This development was subsequently assumed by the Northern Limited subsidiary of Alcoa in the 1920s through a corporate merger. It then became the core property of Alcan, after a corporate reorganization in 1928 split Alcoa into two separate companies.[5] Although Alcan struggled to survive the trying markets of the 1930s, it became a major global player during the war. Over six war years, Alcan's production of primary aluminum rose from

75.2 thousand metric tons in 1939 to 1,545.4 thousand metric tons in 1945, a more than twentyfold increase.[6] As a result of this expansion, by war's end, Canada ranked second only to the United States in total aluminum output. Through six years of war, Canada had supplied 90 percent of British and Commonwealth aluminum and 35 percent of total Allied demands.[7] Why this happened and what implications it bore for the aluminum commodity chain need to be explained.

The aluminum commodity chain that tied Quebec's rivers in forward linkages to the British war economy presented the most reliable and productive aluminum supply for the British during the war. Domestic production by the British Aluminum Company was badly curtailed after the loss of bauxite supplies from southwestern France in 1940, leaving it entirely dependent on bauxite from Alcan's properties in British Guiana.[8] Although British production rose somewhat over the war years, it never accounted for more than one-twentieth of Canadian production.[9] Following the German invasion of the Nordic countries, the trade in Norwegian aluminum also ended.[10] Fears about the exposure of Scottish smelters to potential air attack finally suggested the importance of placing aluminum expansion offshore. The constraints in the United Kingdom did not mean that risks did not exist in the Americas, as we shall see. But as early as 1937, British economic planners had begun to cultivate the Aluminum Company of Canada as a primary aluminum supplier, and with the advent of war, that cultivation turned into a basic position of dependence. In 1940, the British government paid Alcan $40 million to expand capacity.[11]

Despite its own massive aluminum industry, the United States would also come to rely on Canadian aluminum. This is because the ambitious aircraft construction program planned by the Roosevelt administration in the late 1930s ran into supply bottlenecks caused by the lack of electric power for aluminum smelters. As war production rose in major industrial regions like New York State, aluminum production had to compete with other demands. As generating demand approached capacity in a host of regional systems, major new power loads could not be met. Increasingly, new aluminum production would have to seek out regions with surplus power capacity. While Alcoa moved apace to expand production, the federal government broke the company's former monopoly by signing contracts with new firms and building federal dams and smelters.[12] To help alleviate power and aluminum shortages, the United States also placed pressure on Canada to export electric power, and entered forward supply contracts with Alcan for hundreds of thousands of tons of aluminum.[13] The seriousness with which the US president viewed the aluminum problem became apparent in 1941 when he inserted specific mention of the commodity in the diplomatic text outlining the Hyde Park declaration, signed with Canadian prime minister Mackenzie King, to integrate the continental economy for defense purposes.[14]

THE SITES AND SPACES OF THE COMMODITY CHAIN

One of the reasons the aluminum commodity chain helps reveal aspects of the global environmental history of war is because it unfolded over vast distances and imposed environmental burdens on several different locations. At the first point of production, Alcan operated under permissive colonial conditions in British Guiana to extract wealth from tropical soils and low-wage workers (see fig. 11.1). In the early twentieth century, silicate soils with high concentrations of bauxite were known to exist in commercially viable concentrations in parts of southern France, the Balkans, and Arkansas. Major new discoveries occurred shortly after the turn of the century in South America along the Demerara River in British Guiana and in Dutch Surinam.[15] During the First World War, when bauxite demand rose internationally, Alcan developed bauxite properties in British Guiana, owing in part to imperial preferences for commonwealth firms, while Alcoa built mines in Dutch Surinam. Through the Second World War, these South American mines would provide the bulk of North American bauxite supply. Despite the German invasion of the Netherlands, Dutch Surinam remained outside the Nazi orbit, and US troops were stationed in the colony, partly to protect the US bauxite supply.

The development of bauxite mining in British Guiana reads like a caricature of dependency theory. Alcan's leases ran for seventy-five years and imposed a light tax burden of 3 percent on export values using cheap wage labor, readily available in the context of a declining plantation economy. Later estimates suggest that British Guiana's resource rent before 1950 amounted to about 1 percent of aluminum market values.[16] Over the course of the war, Alcan's subsidiary, the Demerara Bauxite Company (or Demba), operated its original Three Friends and Trewern mines near Mackenzie; opened and exhausted the Hope mine on the western shore of the Demerara; and at the wartime production peak in 1943 opened two more mines, the Maria Elizabeth near Three Friends and the Topira, located near Itune.[17] To make way for these mines, tropical forests were felled, the sand and clay overburden removed, and a laborious

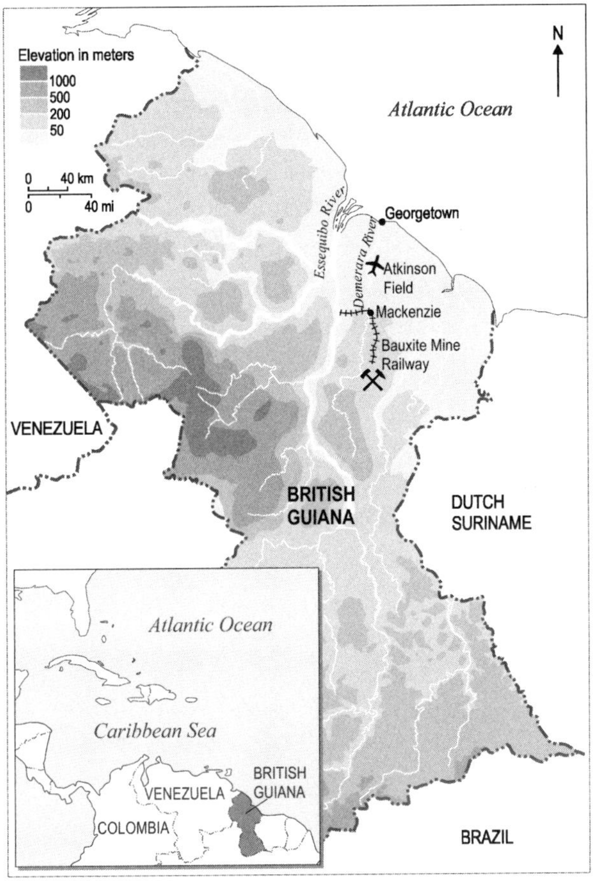

Figure 11.1. British Guiana and the Demerara River

extraction process engaged. Workers surface-mined the bauxite using steam-powered shovels and backhoes, and others filled rail cars with a mass of silicate soils colored red by iron oxidization. At a small plant on the river at Mackenzie, crews sorted, washed, and crushed the material, and loaded shallow-draught boats for export. The company town divided along racial lines, with a white outpost of Canadians, Americans, and Britons in a well-housed and serviced community (complete with a golf course) on the southern part of town, known as Watooka, and a separate and less-well-housed workers' settlement of primarily Afro-Guyanese stretching to the north, called Cockatara.[18] The town displayed its corporate origins in the name of its main street, Arvida, after Arthur Vining Davis, the founder of Alcoa and an important Alcan shareholder. The same name was applied to the company town in Quebec where Alcan's principal smelters were located, and where the bauxite was headed.

Bauxite moved north on steam-powered vessels from river to ocean to fjord. Shallow-draught boats worked up the Demerara River in small loads because the river's silty bottom, carrying suspended sediments from the interior, blocked the entry of larger vessels, particularly at the delta, where the Georgetown bar posed a significant barrier. In the early stages of the war, transshipment occurred at a "topping-off" location in the Virgin Islands at St. Thomas. However, located eight hundred miles from Georgetown, this site proved both too distant and potentially dangerous in the context of U-boat warfare. In 1942, Alcan established a new transshipment station at Chaguaramas Bay in Trinidad, half the distance to the Virgin Islands. Substantial oceangoing freighters, operated by the Alcan shipping subsidiary, Saguenay terminals, collected the bauxite freight from the Caribbean, cut east and north through the Caribbean, and north again along the east coast of North America, before finally moving up the St. Lawrence to the

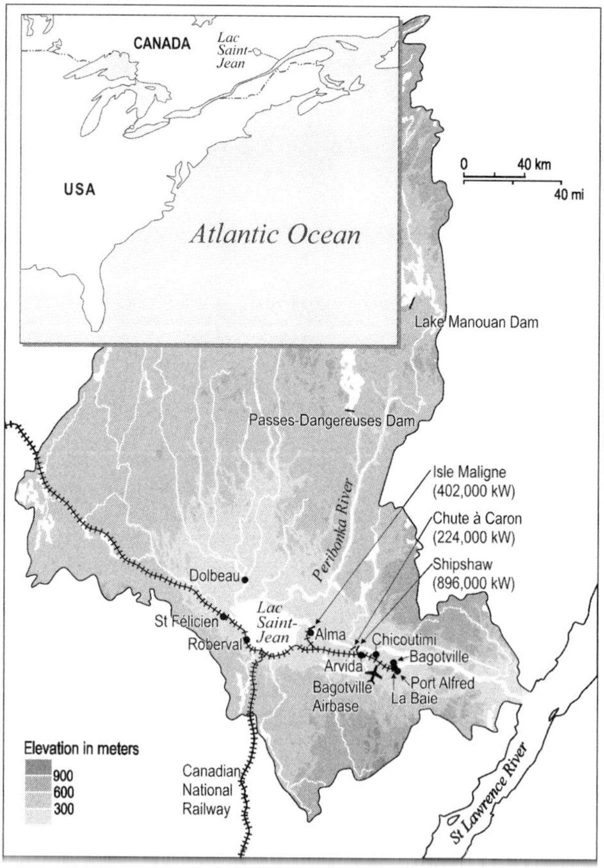

Figure 11.2. The Saguenay basin, Canada

Saguenay. At Port Alfred/La Baie, located at the head of the Saguenay fjord, bauxite was loaded onto rail cars for the final leg of the journey to the upper Saguenay, where raw bauxite was crushed, washed in a chemical solution, and converted into alumina. During winter, this shipping schedule was disrupted owing to ice formation in the Saguenay fjord, and Alcan had to use banked bauxite supplies or rely on rail shipments. For military reasons, this primary shipping arrangement was considerably disrupted over the course of the war.[19]

Alcan's Quebec smelters introduced the second major point of transformation in the commodity chain (see fig. 11.2). Located in Shawinigan and Arvida, a cluster of smelter facilities employing a primarily French-Canadian working class from the region, and overseen by a substantially American and English-Canadian managerial class, converted alumina into aluminum.[20] Although these company towns did not bear the sharp color or colonial lines of British Guiana, their linguistic divides corresponded closely to the social structure.[21] In the vast smelter facilities, alumina was placed in cryolite baths in smelter pots at temperatures reaching 900°C to produce molten aluminum. The temperatures were maintained through huge deliveries of electrical current supplied by lines connecting with powerhouses farther north near Lac Saint-Jean, as well as neighboring power systems to the west. The primary locational advantage of the upper Saguenay region lay in its power potential, close to tidewater. During the war, the original dams and smelters were enhanced and expanded under considerable pressure from the Canadian government. In 1943, Alcan completed massive additions to its hydroelectric generating capacity at Shipshaw, amounting to 650,000 horsepower. This single project produced more hydroelectricity than all the wartime hydro additions of the province of Ontario.[22] Expansions to the Arvida works and smelter rooms created a vast complex. An Australian reporter breathlessly noted in 1944 that the processing plant was the largest of its kind in the world, stretching over one and a half miles in length and three-quarters of a mile wide.[23]

These new facilities took time to build, particularly the dams. The Shipshaw project would not be completed until 1943. Before that time, aluminum production expanded on the basis of diverted current from neighboring power systems. River mobilization on the Saguenay, in short, implicated a much wider territory in which separate systems were integrated, and with them different rivers and flow regimes.

Figure 11.3 gives some sense of the scale of transmissions, reaching west toward Ontario and deepening existing points of transfer in Quebec. Three large central systems in Quebec were tied into closer cooperation in the process: the Montreal Light Heat and Power system centered on Montreal,

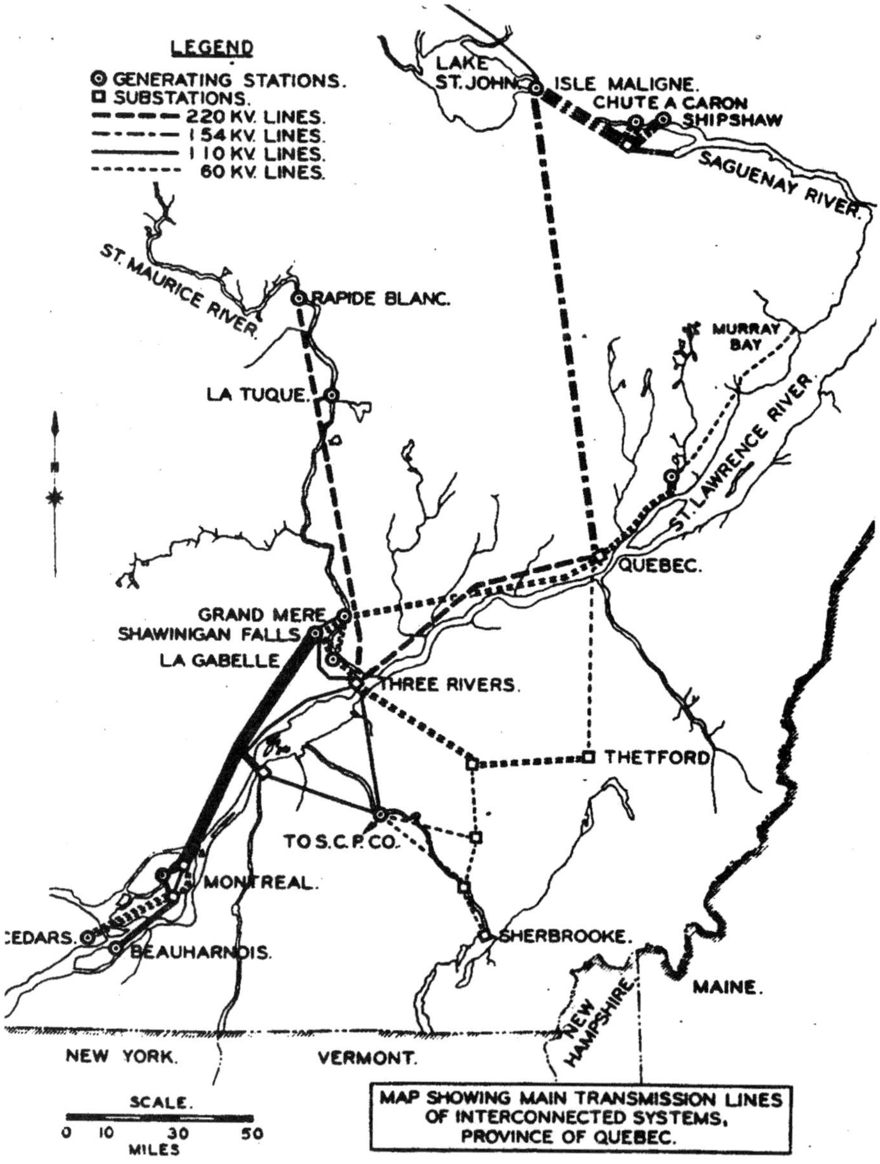

Figure 11.3. Wartime transmission lines in Quebec. Source: Extracted from W. R. Way, "Québec Rivers Unite for Victory: Notes sur l'interconexion des réseaux électriques du Québec," *Shawinigan Journal* I (5) (May 1944), 8–12.

the Shawinigan Water and Power Company system based in the St. Maurice Valley, and the Alcan system farther east on the Saguenay.[24] Such ties allowed the diversion of current to the central Shawinigan system, which could then lower its own generation and store water in its reservoirs to offset future demands in other systems. When Montreal demand fell below capacity, for example, current was delivered east so that the Shawinigan system could bank water in expectation of wintertime demands from Alcan.[25] The federal state mandated a range of power delivery schemes such as this and forced a wide-ranging conservation program to cancel unnecessary demands in pulp and paper plants, and for commercial and domestic uses.[26] In the meantime, the transformation of the Saguenay continued apace as an already-regulated river was built up, and tributary rivers like the Peribonka were dammed for the first time to supply additional storage. The resulting power complex that came online in 1943 provided massive new capacity for aluminum production and outlined the limits of a power transfer system across Quebec that would only be deepened in the postwar context.[27]

Following ingot production, the aluminum commodity chain splintered, and in several directions. Within Canada, aluminum ingots were transshipped to rolling mills in other parts of Quebec, and to a fabricating plant in Toronto. In 1943, a British-owned but Alcan-managed fabricating facility was also built in Kingston, Ontario (see fig. 11.4). Ingots were also loaded onto supply vessels to enter the convoy shipments across the North Atlantic, across the Pacific to Australia, and onto trains moving south to rolling mills in the United

Figure 11.4. Unloading aluminum ingots from Arvida, Quebec, in Kingston, Ontario. Source: Library and Archives Canada, Ronny Jaques, National Film Board, January 1943, Kingston, ON, Mikan 3196473.

States. Whatever the endpoint, the ingots were remelted, turned into sheets or forgings, and mixed with cooper and zinc to be strengthened as alloys. At a host of factory locations in the United Kingdom, Canada, Australia, and the United States, aluminum from the Saguenay smelters was pulled and shaped into bombers, air transporters, and naval parts. The final point in this commodity chain occurred at dispersed locations mainly in the North Atlantic but also beyond, where planes were either destroyed in the course of conflict or, somewhat later, decommissioned, disassembled, and either recycled or disposed of at various military facilities.

The aluminum commodity chain thus connected several discrete and distant locations, drawing on comparative locational advantages, in terms of bauxite supply and cheap hydropower. What linked these places was a superstructure of corporate organization, capital investment, and complementary political jurisdictions within the British Commonwealth, as well as a substructure of long-distance bulk shipping, which connected railroads on the landward ends of the commodity chain through a network of freighters that made it cost effective to ship lightly processed tropical soils over 4,000 kilometers overseas before reduction and smelting. In this way, the economies and ecologies of the Demerara and the Saguenay were practically conjoined.

MILITARIZING THE COMMODITY CHAIN

To this point, we have considered the sites and links in the aluminum commodity chain, and implicitly pointed to the importance of war as a driver. War shaped the commodity chain also in the militarization of the production process. Before the war, the aluminum commodity chain had been produced by corporate initiative in two world regions with a lax regulatory structure and an open-door policy toward investment. During the war, as demand for aluminum rose exponentially, its strategic position became ever more apparent. This made it subject to potential attack and attracted a wide-ranging defensive strategy at several points along the chain.

The first stage in the militarization of the commodity chain occurred in early 1941, as some of the cascading military consequences of the fall of France became apparent. In Canada, air force officials worried that German bombers flying from western France could lead suicide missions as far as the aluminum facilities and in a single strike ground the British air war. In that year, the Canadian war cabinet seriously considered shifting all anti-aircraft guns from Halifax to Arvida to defend the region, but abandoned the plan after reassuring intelligence from the United States alleviated their fears. Nevertheless, a small team of military guards was employed at the aluminum smelter, a Canadian airbase was constructed at Bagotville, close to the bulk transfer point at Port Alfred/La Baie, and expansion plans at Shipshaw remained secret.[28] No bombing raids on the smelters or dams materialized, but their potential was taken with great seriousness.

The southern links in the commodity chain became similarly militarized after the fall of France. As Britain entered the destroyers-for-bases agreement with the United States in 1940, various British Caribbean possessions came under the practical military control of the United States. In the context of European instability and uncertain outcomes in European Caribbean colonies, the United States sought to build a chain of bases to defend resources for the Americas against potential German incursions. On the northwest coast of Trinidad, the United States established a naval base, within which in 1942 Alcan sited its bauxite transfer facility.[29] In British Guiana in the middle Demerara basin, the United States also built Atkinson field, a landing strip and air base, to serve as one point in the transatlantic route to supply military operations in North Africa, but also to protect the exploitation and flow of bauxite down the Demerara and in surrounding waters. As Harvey Neptune argues in his revealing examination of the American presence in wartime Trinidad, such military bases produced diverse social consequences. In the case of Trinidad, a former recreational region was practically annexed for the base site; labor costs on the island rose, leading to labor unrest in domestically oriented sectors; complex race and gender dynamics emerged as imported American ideas about race engaged the island's racially coded social hierarchy; and American military men developed sexual relations off base.[30] Similar processes likely occurred in British Guiana, where wartime bauxite expansion and the building of new military facilities helped stimulate small-scale agriculture at the same time as the sugar plantation economy experienced a series of challenges and contracting production.[31]

While the mining and smelting components of the commodity chain were defended with air bases, the primary threat came at sea (see fig. 11.5). The most dangerous section of the commodity chain occurred in the eastern Caribbean, where German U-boat attacks threatened to disrupt the flow of bauxite, particularly in 1941 and 1942. One Alcan official recorded that in 1942 Saguenay Terminals lost one-tenth of its ships in transit, or twenty-five of 233 ships.[32] US Admiral John Hoover, commander of the Caribbean Sea

Frontier, noted later that the bauxite ships were "just easy meat, slow small vessels cruising along east of Trinidad."[33] In response, British officials mandated smaller tonnage (five hundred tons) to spread the risk of shipping losses over more vessels. Changes to shipping routes also occurred as bauxite-laden ships fled under threat to accessible North American ports. This meant that in 1941 and 1942 a considerable portion of the bauxite shipments was diverted to ports ranging from New Orleans to Portland, Maine to New York City. This freight moved north by rail, or by barge along the Hudson to the New York State canal to Oswego and Lake Ontario. After the erection of the Trinidad transshipment station, the consolidation of convoys and the diversion of U-boat attacks to other sections of the Atlantic, the pressure decreased, but not entirely. When the US War Shipping Administration assumed authority over the Caribbean area from the British Ministry of Shipping in 1943, it mandated that bauxite ships should not move east of New York. Until this precaution was relaxed in 1944, bauxite moved north through New Jersey and New York City by rail.[34]

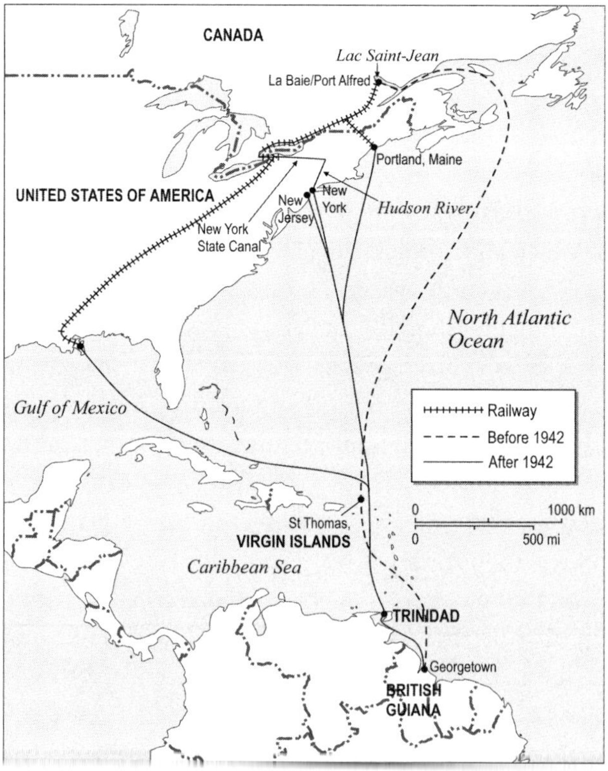

Figure 11.5. A bauxite shipping gauntlet.

The smelter operations depended not only on bauxite deliveries, but also on the availability of cryolite, a scarce mineral used as a bath in the alumina smelter pots and critical to the electrolytic process. In 1939, the only known, accessible location of cryolite in the world was the Ivigtut mine on the west coast of Greenland. After the fall of Denmark in 1940, Alcan officials feared that this supply might be canceled or complicated. The British and Canadian governments immediately considered a military occupation but were strongly advised against this course by the still-neutral United States, fearing a precedent for possible Japanese occupations in Dutch and French Southeast Asia. Canadian and Alcan officials instead pursued a diplomatic approach by offering supplies to Greenland as a way to open discussions about cryolite contracts for the war. Danish officials in Greenland cooperated with this approach and used Alcan as its chief supply agent through the war. After the American occupation of Greenland in April 1941, the cryolite supply was more firmly secured.[35]

The endpoint of the commodity chain—consumption of aluminum for military aircraft—delivered the final point of militarization, in which the metal was formed and deployed entirely for military ends, to facilitate bombing exercises and air fights, the ripping of metal and bodies in the annihilation of total war.

ENVIRONMENTAL DIMENSIONS OF THE COMMODITY CHAIN

Some of the environmental dimensions of the aluminum commodity chain were already evident before the war. Aluminum required massive material and energy inputs from two world regions in the extraction of tropical soils from the Demerara region and the damming of rivers for hydropower on the Saguenay. These critical links in the supply chain were bound by fossil-fueled long-distance transportation system. Aluminum's waste stream encompassed the razed tropical forests of British Guiana, the red mud and air pollutant byproducts of the Bayer process (which converts bauxite to alumina) and smelting, as well as the discarded and destroyed shells of airplanes. What difference did war make to aluminum's environmental history?

The pressures of war drove a general expansion in aluminum production that had immediate effects in the two most important supply and processing regions. In the case of British Guiana, Demba oversaw a fourfold increase in bauxite exports during the course of the war, from 470,147 long tons in 1939 to the wartime peak of 1,941,813 long tons in 1943.[36] Because the bauxite originated in surface mines in the middle Demerara, additional volumes of silicate soils implied a lateral extension of mining activity into tropical landscapes, as well as the construction of rail lines. Such activity worked out of existing facilities, but also involved the creation of new mining sites both east and west of the Demerara. Although the precise acreage brought into production is impossible to know given the paucity of evidence, it represented a profound environmental impact. The legacy landscapes of bauxite mining are best described as lunar: pocked, mineralized surfaces that are devoid of topsoil, flora, or fauna. Such landscapes take years (even with active rehabilitation programs, which did not exist during the war) before they can be colonized by plants and animals. The wartime expansion thus increased the nature and extent of aluminum's environmental effects.

Environmental change in the upper Saguenay was neither as blunt nor as focused as that along the Demerara. Its most immediate effect was hydrological. For the aluminum expansion to succeed, Alcan required more electricity to drive its smelters. Its own development program imposed a new set of main stem structures on the Saguenay that altered the seasonal flow regime, changed the sediment load, and affected the aquatic ecology. To keep such dams running at full potential during low-flow months, Alcan also dammed upper basin tributaries, Lac Manouan and the Peribonka River, which displaced ancestral Innu hunting and gathering territory and flooded a wide section of boreal forest.[37] The hydrological impact spread further still. Because Alcan required electricity diversions from neighboring systems, a policy of interconnection helped deliver the consequences of aluminum's demands over a wider and wider territory. High-voltage power lines became the connective tissues of large technological systems that sought to balance supply and demand over vast spaces and reduce the risks associated with single-river generation. In the process, river flows across the St. Lawrence basin became coordinated to meet regional peak loads. Power exchanges as much as seasons affected the flow and course of rivers.

The scope of these war-driven transformations on fish habitat was profound, though little mentioned in the war. In the upper Saguenay, storage dams interrupted migration pathways and changed the nature of fish habitat from cool, fast-flowing rivers to deep, slow-moving reservoirs. Across the basin, the increased coordination of power systems likely also had effects, as water-release schedules followed power priorities, not fish needs. But the nature of these effects is difficult to judge. The limited scientific research conducted on fish-dam problems on the St. Lawrence during war was sponsored by power companies seeking to avoid litigation.[38] They therefore imposed narrow parameters on research, and focused on applied, not fundamental, research questions. Although dams were understood to affect fish habitat, the assumption seemed to be that the effects were unavoidable collateral damage in the mobilization of rivers for war.

Other environmental effects have accumulated over time. Polynuclear aromatic hydrocarbons (PAHs), which were released in the smelting process through smokestacks, spread through the Saguenay basin as air pollution, and then entered the hydrological system. As the river entered the Saguenay fjord, PAHs, attached to sediments, accumulated on the fjord floor year by year. Studies of the chemical composition of the depositions on the Saguenay fjord floor find a sharp increase in PAH accumulations after 1940, directly in line with the smelter expansion of the war years.[39] PAHs have been linked to cancer occurrence in mammals, particularly St. Lawrence beluga whales. Daniel Martineau and his colleagues have argued that PAH air pollution in the Saguenay region may also account for a higher incidence of cancer in the Saguenay region's human population compared with other parts of Quebec and Canada.[40] Although the installation of scrubbers in smelter stacks since the 1970s has reduced the dispersal of PAHs, in the fjord bed they linger still.

War also imposed its own peculiar pressures on the politics of environmental exploitation. In British Guiana and the broader Caribbean realm, resource exploitation was enfolded, as we have noted, into militarized landscapes, flying at once British and American flags. Because of the fears provoked by the U-boat attacks in 1941 and after, Alcan also sought to diversify its strategic bauxite holdings. As it expanded production along the Demerara, its surveyors simultaneously obtained mineral rights in Jamaica. Using defense regulations to overcome landowner opposition, the Jamaican governor seized bauxite lands for the Crown and then granted access to Alcan. Although the British government granted preference to Alcan as a commonwealth firm, by war's end, two US firms—Reynolds and Kaiser—had also entered the fray with the support of the State Department.[41] These properties would not go into production until after the war, but the war had made them seem necessary, had driven forward the acquisition, and had arguably weakened regulatory scrutiny.

In Canada, the wartime emergency and the strategic role of aluminum also elevated the commodity and Alcan to a favored political position. Because aluminum represented in the language of the industry "baked electricity," the

Figure 11.6. Natzab Airfield, New Guinea (inland from Lae), c. 1945. Source: US National Archives, Reference # 111-SC-242003.

aluminum production program commanded the attention of federal electricity policy. It is not an exaggeration to state that Canada's wartime electricity generation and distribution decisions circulated around Alcan's needs.[42] Alcan received tax depreciation certificates from the Canadian government in aid of construction and priority in terms of access to strategic materials and labor, and brokered no opposition to its developments at the local, provincial, or federal levels. When Alcan moved ahead on its Peribonka dams, for example, the Quebec Ministre de la Chasse et des Pêcheries was left to collect a small fine because of the anticipated destruction of fish habitat.[43] The ministry had no intention of calling the nature of the project into question; it merely sought symbolic compensation.

Perhaps one of the most important environmental legacies of the aluminum production program was its success. The war created, to borrow the language of Thomas Hughes, a solution in search of a problem.[44] In demonstrating the effectiveness of the material, and expanding massively the infrastructure of supply, the war gave profound impetus to aluminum as a metal that promised more than military potential. Aluminum became closely aligned in advertising copy and public discourse with modernity. Alcoa launched a wide-ranging design program and publicity campaign to promote aluminum as a reliable and desirable domestic material useful in a host of consumer goods, while the UK government employed aluminum surplus as housing material, drawing tightly together the associations among social reform, housing, and modern materials.[45] All this activity helped avert the wartime fears of aluminum producers that massive new capacity would face an austere postwar market. To the contrary, domestic consumption rose to absorb the new surplus, and the expansion of military spending in the immediate postwar period helped by 1948 to return demand toward wartime levels. Although a good deal of focus is currently placed on the role of domestic consumption in driving commodity chains, this wartime case suggests the extent to which some facets of domestic consumption emerged incidentally from wartime production priorities.

CONCLUSION

The aluminum revolution identified by Siegfried Moos in 1942 pointed to the impact of world war on a strategic commodity, which drove a dramatic increase in demand, reorganized geopolitics, and reconstituted the social and environmental linkages within and between sites along a transcontinental commodity chain. Figure 11.6 shows one example of the detritus of the aluminum commodity chain in its final phase, a collection of American planes left as waste at Natzab Airfield, Papua New Guinea, and photographed at war's end from the air.[46] It reminds us of the unprecedented capacity of the Second World War to gather and scatter materials with untold human and environmental consequences, linking diverse locations with no necessary former connections. Those connections did not all wither once the military threat subsided. The aluminum revolution entered housing markets and kitchens and became the spine of the Cold War military buildup. Although revolutions tend to end quickly, aluminum's became permanent.

Notes

This essay is a revised and updated version of "Aluminum, Commodity Chains and the Environmental History of the Second World War," *Environmental History* 16, no. 1 (January 2011): 69–93. The author gratefully acknowledges the financial support of the Social Sciences and Humanities Research Council of Canada.

1 Siegfried Moos, "The Aluminum Revolution," *Bulletin of the Oxford Institute of Statistics* 4, no. 11 (August 1942): 209.

2 Ibid., 209–14.

3 I have offered a wide-ranging reflection on the concept and historiography of the commodity chain in environmental history in "Aluminum, Commodity Chains and the Environmental History of the Second World War," *Environmental History* 16, no. 1 (January 2011): 69–93. Readers interested in the commodity chain concept should begin by consulting Terence K. Hopkins and Immanuel Wallerstein, "Commodity Chains in the World Economy prior to 1800," *Review* 10, no. 1 (1986): 157–70. I also recommend Jennifer Bair, "Global Capitalism and Commodity Chains: Looking Back, Going Forward," *Competition and Change* 9, no. 2 (June 2005): 152–80; and Jennifer Bair, "Global Commodity Chains: Genealogy and Review," in *Frontiers of Commodity Chain Research*, ed. Jennifer Bair (Palo Alto, CA: Stanford University Press, 2009), 1–34.

4 Paul A. C. Koistinen, *Arsenal of World War II: The Political Economy of American Warfare, 1940–1945* (Lawrence: University Press of Kansas, 2004), 136.

5 David Massell, *Amassing Power: J. B. Duke and the Saguenay River, 1897–1927* (Montreal: McGill-Queen's University Press, 2000).

6 Duncan C. Campbell, *Global Mission: The Story of Alcan* (Montreal: Alcan, 1985), 251.

7 Ibid.

8 Nigel Balchin, *The Aircraft Builders: An Account of British Aircraft Production, 1935–1945* (London: His Majesty's Stationary Office, 1947), 53.

9 League of Nations, Geneva, *World Economic Survey, 1942/44* (Princeton, NJ: Princeton University Press, 1945), 73.

10 "The Economic Consequences to the Allies and to Germany Respectively of the Seizure by Germany of Denmark and Norway," Statement by Lord Stamp's Survey of Economic and Financial Plans, April 13, 1940, UK National Archives, documents online, Catalogue Reference cab/67/6/3, War Cabinet.

11 H. Duncan Hall, *North American Supply* (London: His Majesty's Stationary Office, 1955), 367.

12 Koistinen, *Arsenal of World War II*, 136; Gerald D. Nash, *World War II and the West: Reshaping the Economy* (Lincoln: University of Nebraska Press, 1990), 91–121; George David Smith, *From Monopoly to Competition: The Transformation of Alcoa, 1888–1986* (Cambridge: Cambridge University Press, 1988), 191–249.

13 David Massell, " 'As Though There Was No Boundary': The Shipshaw Project and Continental Integration," *American Review of Canadian Studies* 34 (Summer 2004): 187–222.

14 J. L. Granatstein, *Canada's War: The Politics of the Mackenzie King Government, 1939–1945* (Toronto: Oxford University Press, 1979), 143. On the importance of Canadian-American links around aluminum and electricity during the war, see Massell, "As Though There Was No Boundary."

15 Carlo Lamur, *The American Takeover: Industrial Emergence and Alcoa's Expansion in Guyana and Suriname, 1914–1921* (Dordrecht: Foris, 1985).

16 Bradford Barham, "Strategic Capacity Investments and the Alcoa-Alcan Monopoly, 1888–1945," in *States, Firms, and Raw Materials: The World Economy and Ecology of Aluminum*, ed. Bradford Barham, Stephen G. Bunker, and Denis O'Hearn (Madison: University of Wisconsin Press, 1994), 87, 101.

17 Campbell, *Global Mission*, 319.

18 For a nostalgic and interesting account of Mackenzie in the prewar and wartime period, see Patricia Wendy Dathan, *Bauxite, Sugar and Mud: Memories of Living in Colonial Guyana, 1928–1944* (Ste-Anne-de-Bellevue, Quebec: Shoreline, 2006); see also Campbell, *Global Mission*, 313–14. Although Campbell's is the official history of Alcan, it nevertheless acknowledges the racial and colonial politics of the settlement.

19 Campbell, *Global Mission*, 295–97.

20 Jose Iguartua, "Les origines des travailleurs de l'Alcan au Saguenay, 1925–1939," *Revue d'histoire de l'Amérique française* 37, no. 2 (1983): 291–308. The second-largest groups of workers were Eastern European immigrants and Finns.

21 José Iguartua, *Arvida au Saguenay: Naissance d'une ville industrielle* (Montreal: McGill-Queen's University Press, 1996).

22 Matthew Evenden, "La mobilisation des rivières et des fleuves pendant la Seconde Guerre mondiale: Québec et l'hydroélectricité, 1939–1945," *Revue d'histoire de l'Amerique francaise* 60, no. 1–2 (2006): 125–62.

23 Don Whitington, "Canada Is Aluminum Nerve Center," *Canberra Times*, August 7, 1944.

24 Two major interties linked the three systems: the first, a 220,000 voltage transmission line, completed in November 1940, ran between Trois-Rivières and Québec City, linking Alcan's Saguenay system with those of the Shawinigan system; the second, a 120,000-volt line, completed a year later, ran under the island of Montreal, connecting the Montreal and Shawinigan systems. See D. M. Farnham and O. W. Titus, "Design, Manufacture, and Installation of 120-Kv Oil-Filled Cables in Canada," *Transactions of the American Institute of Electrical Engineers* 61 (December 1942): 881–88.

25 Evenden, "La mobilisation des rivières et des fleuves," 125–62.

26 "Restriction of Newsprint Paper Production," July 29, 1941, RG 179, Stack Area 570, Row 65, Compartment 10, Shelf 05, Entry 162, Box 1, File: Joint Economic Committee Memoranda, 1941, Memo 12, US National Archives; "Canada Begins Stricter Curb on Newsprint," *Christian Science Monitor* September 3 (1942). The wartime power conservation order is contained in *Order No. P.C. 5* (Power Shortage Areas), Montreal, September, 20, 1942, RG 28, Vol. 251, File 196-11-2, Department of Munitions and Supply, The Power Controller, Library and Archives of Canada (LAC). On wartime power conservation, see Matthew Evenden, "Lights Out: Conserving Electricity for War in the Canadian City, 1939–1945," *Urban History Review* 34, no. 1 (Fall 2005): 88–99.

27 W. R. Way, "Québec Rivers Unite for Victory: Notes sur l'interconexion des réseaux électriques du Québec," *Shawinigan Journal* 5 (May 1944): 8–12; and "Power-System Interconnection in Québec," *Transactions of the American Institute of Electrical Engineers* 61 (December 1942): 841–47.

28 Minutes of the War Committee of Cabinet, March 12, 1941, RG 2, 7C, Volume 3 [reel C-11789], LAC; Memorandum from Chiefs of Staff Committee to Ministers re aluminum plant at Arvida, March 3, 1941, Documents of the Cabinet War Committee; Extracts of Minutes of Defence Council Meeting of March 7, 1941; Volume 4, Minutes of the War Committee of Cabinet, April 21, 1941, May 21, 1941. The British position held that the likelihood of attack was remote. See "B.A. 330 TAUT 397," n.d., Beaverbrook Papers, BBK/D/109 British Parliamentary Archives. On security and sabotage fears, see John MacFarlane, "Agents of Control or Chaos? A Strike at Arvida Helps Clarify Canadian Policy on Using Troops against Workers during the Second World War," *Canadian Historical Review* 86, no. 4 (December 2005): 619–40. The Alcan works manager, Alexander Charles Johnston, noted before the Royal Commission charged with investigating the circumstances of the 1941 strike, that there would only be about ten men on armed patrols at any given time, but that there was also a home guard contingent with anti-aircraft guns on site: Fonds de la Commission royale charge d'enqueter sur les evenements qui se sont produits a Arvida, PQ en julliet 1941, RG 33, Series 56, Volumes 1–3, p. 49, LAC.

29 Fitzroy Andre Baptise, *War, Cooperation, and Conflict: The European Possessions in the Caribbean, 1939–1945* (New York: Greenwood Press, 1988), 87. On the politics behind this decision, see Daryl F. White, "Multinational Patriots: Business-Government Relations in the Canadian Nickel and Aluminum Industries, 1914–1945" (PhD thesis, University of Western Ontario, 2006), 274.

30 Harvey R. Neptune, *Caliban and the Yankees: Trinidad and the United States Occupation* (Chapel Hill: University of North Carolina Press, 2007).

31 Jay R. Mandle, *The Plantation Economy: Population and Economic Change in Guyana, 1838–1960* (Philadelphia: Temple University Press, 1973), 113.

32 Paul Clark's estimate cited in Campbell, *Global Mission*, 299.

33 Quoted in Baptiste, *War, Cooperation, and Conflict*, 143.

34 Duncan Campbell has produced the most complete picture available of Alcan's wartime shipping problem. See *Global Mission*, 295–303.

35 Ibid., 329–35.

36 Imperial Institute, *The Mineral Industry of the British Empire and Foreign Countries, Statistical Summary, 1938–1944* (London: His Majesty's Stationery Office, 1945), 10.

37 David Massell, "Power and the Peribonka, a Prehistory: 1900–1930s," *Quebec Studies* 38 (2004–5): 87–103; and idem, *Quebec Hydropolitics: The Peribonka Concessions of the Second World War* (Montreal: McGill-Queen's University Press, 2010).

38 B. W. Taylor, second report on the fish in the St. Maurice watershed, Fonds Shawinigan Water and Power Company, Box 4106, F1/1296, Archives Hydro-Quebec (AHQ); Arthur Surveyer, "Report on La Croche River Investigation," 1945, F1/1297; "Groundwater Investigations in the Vicinity of Beauharnois Canal," January 14, 1942, Fonds Beauharnois Light, Heat and Power Company, Box 3553, F2/2215/2, AHQ; Professor Marie-Victorin, University of Montreal, "Report on a Physiological and Ecological Reconnaissance along the Shores of the St. Lawrence River between Lake St. Louis and Lake St. Francis (August–Dec 1940)," 3 vols. Box 4195, F2/2215 #1, AHQ.

39 John N. Smith and Eric M. Levy, "Geochronology for Polycyclic Aromatic Hydrocarbon Contamination in Sediments of the Saguenay Fjord," *Environmental Science and Technology* 24, no. 6 (June 1990): 874–79; "Contamination," Saguenay Fjord Virtual Museum, Virtual Museum of Canada, accessed July 3, 2009, http://www.virtualmuseum.ca/Exhibitions/Fjord/english/e_hydro_e.html.

40 Daniel Martineau, Karin Lemberger, André Dallaire, Philippe Labelle, Thomas P. Lipscomb, Pascal Michel, and Igor Mikaelian, "Cancer in Wildlife, a Case Study: Beluga from the St. Lawrence Estuary, Québec, Canada," *Environmental Health Perspectives* 110, no. 3 (March 2002): 285–92.

41 Denis O'Hearn, "Producing Imperialism Anew: The United States, the United Kingdom, and Jamaican Bauxite," in Barnham et al., *States, Firms and Raw Materials*, 154–55.

42 Matthew Evenden, "Mobilizing Rivers: Hydro-Electricity, the State and World War II in Canada," *Annals of the Association of American Geographers* 99, no. 5 (December 2009): 940–47.

43 *Rapport Général du Ministre de la Chasse et des Pêcheries* (Quebec: Rédempti Paradis, Imprimeur de Roi, 1944), 13.

44 Thomas P. Hughes, *Networks of Power: Electrification in Western Society, 1880–1930* (Baltimore: Johns Hopkins University Press, 1983).

45 Brian Finnimore, "The A.I.R.O.H. House: Industrial Diversification and State Building Policy," *Construction History* 1 (1985): 60–71; Dennis Doordan, "Promoting Aluminum: Designers and the American Aluminum Industry," *Design Issues* 9, no. 2 (Spring 1993): 44–50.

46 I am grateful to Judith Bennett for sending me a copy of this photograph when she read and commented on an early draft of the paper for the World Congress of Environmental History meetings. She came across it in the course of researching her book *Natives and Exotics: World War II and Environment in the Southern Pacific* (Honolulu: University of Hawai'i Press, 2009), which has much to say about the environmental legacies of war, and all the planes and material left behind.

Chapter 12
Crisis Utilization in Mexican Forests

CHRISTOPHER R. BOYER

Mexico was not a major theater of conflict in World War II. No battles were fought on its soil, and its military played a largely symbolic role in the conflict. No bombs fell on Mexican cities, nor did its leaders flirt with the Axis powers, as in the case of countries such as Argentina. Nevertheless, the war constituted a Mexican watershed in both political and environmental terms. President Lázaro Cárdenas (1934–40) strongly opposed fascism, but his administration's 1938 enforcement of Mexican subsoil rights and expropriation of the petroleum industry (followed by establishment of a Mexican state-owned petroleum company, Petróleos Mexicanos, or Pemex) sparked a tense diplomatic standoff with the United States and Britain. Cárdenas succeeded in "encouraging" the United States and the United Kingdom to come to the negotiating table by selling exactly one load of oil to the German Navy; this small act was enough to catch the attention of the major powers. The conflict was resolved in 1941 when the Roosevelt administration invoked the Good Neighbor policy to reach a comprehensive agreement with Mexico.[1]

Much changed with his successor, Manuel Ávila Camacho (1940–46), whose developmentalist agenda focused less on peasant communities and workers' cooperatives than on capitalist development and urban-centric modernization. Ávila Camacho maneuvered the nation into an economic and military alliance with the Allies and above all the United States. His administration treated the war not so much as a military emergency but as a means to implement a regime that I call "crisis utilization." I intend this term to be understood in two ways. On the one hand, the state authorized extraordinary measures to extract Mexican natural resources at the disposal of the Allies for utilization in the war effort. On the other hand, the state utilized the reality of the World War II crisis to put in place new policies and institutional forms that favored the increasingly corporatized model of development that Mexican postwar leaders favored. This regime of "crisis utilization" emerged in the early 1940s, when Mexican leaders argued that the extraordinary demands imposed by wartime demanded an equally extraordinary—and ostensibly temporary—abrogation of legal structures intended to conserve forests and ensure that rural communities would have a measure of control over their own property.

Oil was perhaps the nation's single most valuable contribution to the war effort (apart from the thousands of Mexican citizens who fought in the war as part of the US Armed Forces). After overcoming US and British resistance to the 1938 expropriation of the petroleum industry, Mexico soon became a reliable source of fuel for the Allied war machine.[2] Yet wartime oil production did little to change the legal framework or overall practice of petroleum extraction. Mexico sped production as much as possible and made oil available to the Allies, but even so, there were few long-term consequences of this shift. Not so in the case of forest products, which the Allied war machine also craved. In the woodlands, the war-induced crisis utilization led to profound changes within rural communities and in terms of the regulatory structures that governed timber production.

Crisis utilization had both short- and long-term implications for Mexican forests. In the short term, the federal government placed nearly all conservationist measures in abeyance. This situation led forestry officials to abandon the already-ailing system of community forestry structured around producers' cooperatives and to turn definitively to commercial production on a much larger scale. State-led development projects consolidated in the postwar era often functioned to deprive rural communities of the ability to determine how forestland that they had received through the agrarian reform would be used. Simultaneously, the process of commodification—which had been temporarily forestalled by the revolution, the initial phases of land reform, and Cardenismo—galloped forward. The net result of this war-inspired developmentalism was the radical curtailment of environmental protections, local autonomy, and the state's former commitment to balanced economic development in rural areas.

In the longer term, the war can also be credited with putting in place an extractive regime that endured in some cases well into the 1980s. Prior to 1940, Mexico imported most of its newsprint and writing stock from the United

States; the venerable Mexican paper mills Loreto y Peña Pobre, Atenquique, and San Rafael primarily produced upscale stationery. As the United States redirected its industrial capacities to war making and commandeered available stocks of paper, Mexico suddenly faced a shortage. The Mexican paper producers offered to retool their plants and ramp up domestic production, but there was a catch: they demanded access to a steady supply of wood pulp. The paper companies and several leading naturalists suggested an elegant solution to the problem: the creation of a regime of forest concessions not unlike the one that had prevailed during the administration of the disgraced Porfirio Díaz three decades earlier. These postrevolutionary concessions were known as Industrial Forestry Units (Unidades Industriales de Explotación Forestal, better known by their acronym, UIEFs). Initially granted only to paper companies, they were soon sought by logging companies and at least one mining corporation. Ironically, none of these were established until the war had effectively come to an end. Born out of wartime exigencies, they far outlived the context of their creation.

This chapter opens with an examination of the extraordinary measures authorized by the Alemán administration (1946–52) to institutionalize and expand measures originally intended to produce forest products for the Allied war effort, establishing the productive logic of "crisis utilization." This chapter also examines the longer-term consequences of crisis utilization. It argues that the creation of a concessionary regime intended to promote the industrial use of forests originated with the "extraordinary" wartime measures, but that this productivist logic quickly became entrenched as both corporate and policy interests resisted any return to the prewar status quo. I conclude with some observations about the implications of crisis utilization for natural resource exploitation more generally.

CRISIS UTILIZATION

Instead of an explicit policy shift, the overall effects of World War II were primarily responsible for putting an end to the 1930s regime of local production and the ideal of peasant-led conservation championed by the Cárdenas administration. Cárdenas had promoted a policy of "social landscaping" that manages forests and rural communities in tandem in a bid to rationalize the use of resources and promote the economy of the land-reform sector and rural development generally.[3] He had sought to mobilize the popular classes under the aegis of state-controlled organizations such as peasant confederations and workers unions, with the intention of achieving social justice and creating a balanced modernization agenda. The heart of his agenda for rural areas depended on a rapidly expanding agrarian reform project. Although most land-reform communities received agricultural land, Cárdenas recognized that many *pueblos,* and particularly indigenous ones, had lost forestlands to developers during the nineteenth century. Overriding the concerns of conservation-minded scientists, Cárdenas included forests within his land-reform initiative but decreed the establishment of producers' cooperatives in forestlands as a means to regulate peasant production and guarantee that land-reform beneficiaries had access to jobs and economic development that took place in their woods.[4] This brand of paternalism was a hallmark not only of Cardenismo but also of the postrevolutionary agrarian reform in general: the land reform functioned in a juridical sense by first nationalizing property and then turning it over to rural communities for perpetual usufruct. Nevertheless, this meant that the federal government retained ultimate control of these properties. Cooperatives represented a means to boost peasant production while allowing regulators to manage peasant logging and to mitigate unregulated and ecologically destructive logging in the land-reform sector.

In fact, Cárdenas's paternalistic and highly bureaucratic forestry regime had begun to unravel even before Cárdenas left office. He demoted the Departamento Forestal to a subsecretariat soon before leaving office and publicly broke with the excessively hierarchal and rigid approach to forest management that characterized most conservationist thought at the time. Nevertheless, it was the crisis utilization associated with World War II that ultimately doomed the Cardenista vision. The wartime imperative to produce as much construction material as possible swamped the Forestry Department's once-scrupulous oversight of logging operations and justified the promotion of large-scale commercial production over and above community-based forestry.

These changes were not long in coming. President Manuel Ávila Camacho deepened ties with the United States immediately after taking office in 1940. He ordered an easing of export restrictions on such primary goods as oil, minerals, and forest products, and he consummated Mexico's alliance by declaring war on the Axis powers two years later.[5] Particularly in the north, US and Mexican companies raced into the woods, where they built roads and makeshift logging camps at a furious pace. The federal government also sped the construction of the western branch of the Northeast Railroad (a line known as the Chihuahua al Pacífico) and dedicated three labor gangs to the task.[6]

Nearly all of the logging for these initiatives was done using "temporary" permits that were granted without the paperwork normally required before any cutting could be done, such as a professionally written forest inventory of the area to be logged and multiyear management plans approved by federal foresters. Instead, the Ávila Camacho administration opened the way for forestry officials to issue temporary permits without even the approval of authorities in Mexico City. In many cases, these permits made no mention of how much wood loggers could extract, and many did not even specify where loggers were authorized to cut. It appears that federal officials billeted in the countryside made some of these authorizations orally, meaning that no paper trail existed for their superiors in Mexico City (or historians) to review. We do know that nearly all of this production was exported to the United States, which as one forester put it, "absorbed whatever the country produced" (*absorbía toda la producción tal como saliera*).[7] Logging companies streamed into forests, especially in the northern parts of the country. Some of the nation's largest timber companies—including the venerable Bosques de Chihuahua—trace their origins to this moment of bonanza, when commercial logging regained a preeminence it had not held since the revolution.

It is all but impossible to determine the extent of damage that Mexican forests sustained during the war years. In theory, most of these operations took place on federal territory such as parks, forest reserves, or other public lands. Given the spotty documentation during the early 1940s and foresters' subsequent complaints about overlogging, it seems clear that most operations simply cut every tree they could reach within a convenient radius of a sawmill or railroad track, sometimes without the knowledge of its proper owner, and then moved on. Semiarid pine-oak forests can theoretically recover from this sort of clear-cutting, but their slow rate of growth and the tendency of rural people to convert cleared forest land into cattle pastures mean that such logging operations result in a permanent loss of forest cover. On the other hand, there was a limit to how much new forest loggers could hope to reach. The war made it virtually impossible to acquire logging equipment, rails to extend existing lines, and sawmill parts. As a result, timber interests tended to intensify projects they had already established, or to revitalize projects they had abandoned during the Cárdenas era, rather than opening up completely new logging frontiers.

THE LEGACIES OF CRISIS UTILIZATION

In 1942—the same year that Mexico declared war on the Axis powers—legislators passed a new forestry code that endorsed many of the pro-development propositions made during a conference of industry experts and forestry officials held in Mexico City the previous year. Most delegates agreed that peasant production would never meet the nation's need for forest products and that modern corporations would hew more closely to forest management plans. The new law committed the government to what can be called a "muscular" program of conservation, proclaiming that all Mexicans should cooperate with the protection, restoration, and propagation of forests. It defined unoccupied unclaimed (*baldío*) and federally owned woodlands as national forest reserves, and even made it possible to expropriate the property of landowners who logged their woods too intensively or failed to replant trees thereafter. Perhaps because some of the nation's top industrialists weighed in on the legislation, however, these ostensibly conservationist provisions cloaked a loophole that opened public lands to private use on an unprecedented scale. The new law authorized the forestry department (now demoted to a subsecretariat of the Department of Agriculture) to grant logging companies permission to work inside forest reserves. The legislation also made it easier for private enterprises to sign joint venture agreements (*contratos de asociación en participación*) with communities that had received grants of forest land through the postrevolutionary land-reform process. These agreements theoretically obliged timber companies to provide communities with machinery, technical assistance, and in some cases sawmills in exchange for a share of profits from local logging operations.[8] In this, they resembled the long-term rental contracts between logging companies and rural communities that the Cardenistas had vigorously assailed a few years earlier, and they were technically illegal because they allowed outsiders to work on property granted through the agrarian reform.[9] The 1942 forestry code was superseded by a fundamentally similar law seven years later that reiterated "conservationist" goals but tellingly pledged that all logging projects needed to follow "a technical plan . . . to obtain the maximum return from the forest."[10] This emphasis on maximum yield contrasted with previous legislation's stated goal of sustainable use over the long term. While the new laws did express concerns about hyperexploitation and deforestation, the 1949 code in particular sought to speed development by reasserting federal sovereignty over woodlands and encouraging private corporations to push ahead with the industrialization of forest products.

To achieve these goals, legislators first needed to address the legacy of Cardenismo, which had delivered around a third of the nation's forests to rural communities by that point. Most fundamentally, the new forestry codes displaced forest management from the local to the regional level by progressively dismantling the embryonic regime of community forestry. The 1942 law contained a vague injunction that logging on collectively held lands—such as land-reform parcels known as *ejidos* and indigenous commons—should "benefit the relevant groups of population" and allowed for the continued existence of producers' cooperatives.[11] Most foresters interpreted this to mean that existing cooperatives could continue to operate but no new ones should be authorized. As a result, the cooperatives slowly disappeared over the course of the decade as officials rescinded their charters for alleged irregularities such as ignoring the terms of their logging permits, falling under the control of outsiders, or failing to meet the needs of the community.[12]

The 1949 code stripped its predecessor's rhetorical commitment to local production and formalized the postwar state's progressive abandonment of community production. Its authors extended the authority of federal officials and private corporations over woods in public land, private properties, and most notably *ejidos*. Whereas the 1942 code allowed the president to grant logging companies permission to operate inside national forest reserves, for example, and the 1949 legislation created a National Forestry Council composed of government and business leaders charged with oversight of commercial forestry on a national scale, again, the land-reform sector was noteworthy for its absence. The 1949 code also ratified and expanded two other powerful tools first mentioned in the 1942 law. First, it gave the president the authority to declare it "in the public interest" (*de utilidad pública*) to grant long-term concessions (the aforementioned UIEFs) to private corporations, effectively allowing the president to override the rights of landholders and land-reform communities in favor of the concessionaires. Second, it allowed authorities not only to declare logging bans but also to grant exceptions to UIEFs and other producers wherever necessary.[13] These new regulations and institutional forms significantly diluted local control of forests. They displaced the primary locus of forest management from land-reform communities and handed far more authority to a network of official and private-sector decision makers—many of them based in national or state capitals—tasked with the industrialization of forest landscapes on a regional scale.

THE RETURN OF THE CONCESSION

World War II shortages had choked off Mexico's access to newsprint. The forestry code enacted in 1946 gave the president authority to grant UIEF concessions to any corporation dedicated to "mining, paper manufacture, construction, the production of war materiel, etc.," but did not spell out how they should function in practice.[14] The 1949 forestry code cleared the air. Companies authorized to form a UIEF received the right to purchase wood products from all private landowners, indigenous communities, and *ejidos* within a territory demarcated by presidential decree. In exchange, the concessionaires would file a comprehensive management plan, replant trees wherever they cut, hire their own foresters, maintain tree nurseries, and pledge not to consume more wood than needed. From a management perspective, it seemed that UIEFs could simultaneously modernize the woodlands while achieving maximum sustained use of natural resources. By the mid-1950s, some experts concluded that the concessions had industrialized the forestry sector while maximizing productivity, minimizing waste, and more efficiently distributing forest products.[15]

In addition, UIEFs were promoted as a tool to create jobs and build a more robust market for locally produced forest products. Most UIEF contracts directed concessionaires to hire people living within their territories as road builders, sawmill operators, and lumberjacks. In practice, relatively few jobs materialized. Most companies preferred to use existing employees rather than hire villagers who typically lacked the skills and labor discipline.[16] Nor did woodland communities experience UIEFs as a boon to their local economies. While the law directed concessionaires to pay a "just price" (as well a stumpage fee known as *derecho de monte*) for the wood harvested on *ejidos* and other properties in their boundaries, it remained silent about how to arrive at such a price. This was a particularly thorny question because the concessions distorted the market by removing other (legal) buyers from the scene and depressing the value of forest products, except in the rare case that a community management plan allowed it to produce more timber (or pine resin or pulpwood) than the concessionaire was able to consume. In many instances, UIEFs aggravated the situation by continuing to pay the same price for forest products from year to year, even when inflation began to erode rural incomes.

The first three concessions went to paper companies in the central highlands, beginning with the 1945 creation of a UIEF for the Atenquique Paper Company of Jalisco. Two years later, the Loreto y Peña Pobre paper mills in the southern reaches of the Federal District received control of approximately 80,000 hectares. The following year, the nation's largest paper company, San Rafael y Anexas, was awarded jurisdiction over 117,275 hectares made up of federal lands, indigenous commons, and *ejidos* in eight municipalities in Mexico State and parts of Morelos and Puebla. These early concessions had terms in the range of fifty-five to sixty years, though subsequent presidential decrees reduced them to thirty-five years.[17] Their recipients agreed to plant ten trees for each one they cut, to establish schools and hospitals for workers, and to organize corps of fire fighters and armed forest wardens in their respective jurisdictions. Some companies did hire the required personnel, but few if any fulfilled the other obligations.

Some UIEF concessions no doubt contributed to socioeconomic development. Loreto y Peña Pobre opened a paper mill just before receiving a concession in 1942 and made a major investment in machinery built in Mexico. Likewise, San Rafael constructed a small company town and installed new water lines for its modernized plant in Mexico State.[18] Most notably, every company that received UIEF concessions employed workforces numbering between several hundred and two thousand people who worked as loggers, road builders, truck drivers, and sawmill operators, not to mention professional and clerical staffs.

The UIEF concessionaires also took responsibility for managing forests on a scale that posed unprecedented challenges. One key problem was that rural people typically wanted a predictable annual income, which meant either leasing to a company that would send crews to fell trees over a period of years or logging the land themselves on a more or less permanent basis. Yet resource managers objected to such arrangements on both logistical and technical grounds. Logistically, companies lacked the capacity to build roads and saw wood simultaneously on every parcel in their domain; it was much easier to build roads, temporary sawmills, and logging camps in one area and work it until moving on to the next. Rotations of this sort also conformed to what foresters regarded as the best scientific practice. Experts working in Atenquique devised a system in 1951 known as the "Mexican management method for uneven-age forests" (*método mexicano de ordenación de montes*), which soon became the norm for logging in most temperate forests. It entailed dividing the landscape into individual stands of trees or properties and selectively cutting 35 percent of the total volume of wood in each of these plots, rotating from one to the next until the earliest-cut woods had recovered and the rotation could start once again.[19] In theory, such a technique ensured sustainable harvests; over the long term, however, it favored broadleaf species over more commercially attractive and heavily harvested conifers.[20] More importantly, it imposed a high social cost on landowners and land-reform beneficiaries who might wait two, three, or more years before they perceived any benefits from "scientifically" managed UIEFs, and then would have to endure another five years or more before the company returned to their land.

These structural problems were exacerbated by UIEFs' inability or unwillingness to fulfill other elements of their social contracts. Atenquique failed to establish a resin-tapping program that promised to employ hundreds of *campesinos,* for example, and San Rafael dithered for nearly fifteen years before opening any schools or medical facilities, or even organizing a fire brigade. The Forest Service's systematic lack of oversight made it easy for corporations with UIEF concessions to ignore the obligations in their charters, making them lightning rods for political opposition and rural discontent. As one of the nation's leading newspaper declared in 1962, it seemed that UIEFs' real mission was merely "to protect private interests."[21]

Even so, presidents approved twenty-one UIEFs—including three in Michoacán and one in Chihuahua—in the developmentalist heyday of the 1940s and 1950s. The concessions gave paper mills and logging companies privileged access to no fewer than 2.2 million hectares of the nation's most desirable timber reserves.[22] Unlike the early grants made to paper companies that had been in business for decades, most subsequent UIEFs went to newly formed timber companies that investors cobbled together for the express purpose of requesting a logging concession. The federal government usually approved these requests so long as concessionaires agreed to make substantial investments in sawmills, transportation, and social services. True to form, however, nearly all of these UIEFs chronically failed to fulfill their contracts. Some observers concluded that concessionaires simply did not want to share the "fabulous wealth" they made in the forests, but few if any logging companies profited enough to meet all the financial and social obligations to which they had agreed. Whatever the case, nearly half of the UIEFs managed their territories so poorly—or provoked such vehement opposition from peasant communities—that forestry officials ended up canceling their concessions altogether.[23]

Local opposition to the UIEFs soon became a standard feature of rural politics. Communities that had received land-reform grants of forestlands

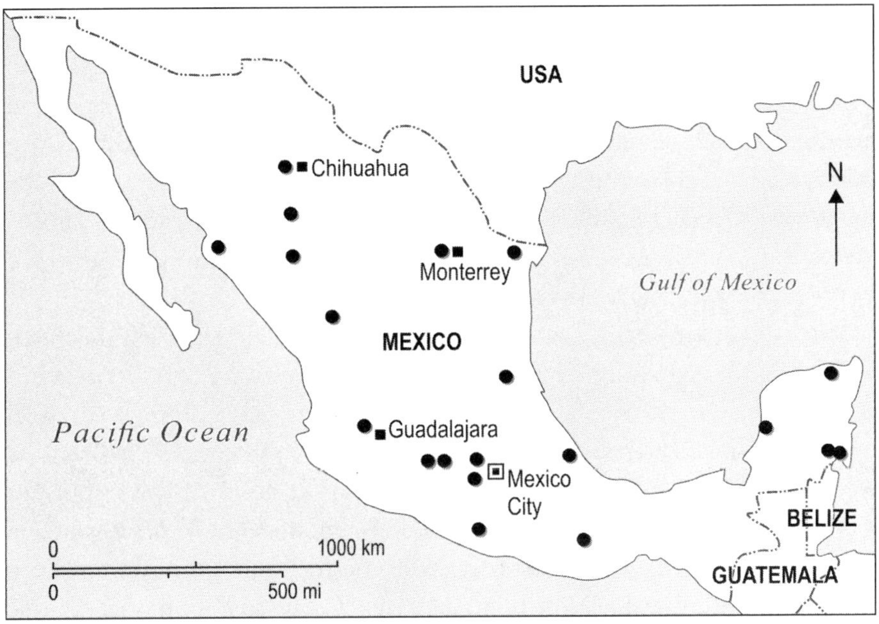

Figure 12.1. Location of UIEFs in post-war Mexico. Source: Karel Chambille, *Atenquique: Los bosques del sur de Jalisco* (Mexico City: Instituto de Investigaciones Económicas, UNAM, 1983), plate following page 16.

sometimes did not learn that their property fell within the jurisdiction of a UIEF until lumberjacks appeared in the hills. Villagers desperately sought to keep their lands segregated from logging concessions. As competitive markets vanished, many communities stood to receive much less from their products whether they worked the woods themselves or simply received royalties from the concessionaires. To make matters worse, concessionaires rarely followed the forestry policies that required selective harvesting of timber (rather than clear-cuts), which would have mitigated environmental damage. They often ignored the production limits altogether and overcut woodlots, with devastating consequences for their owners. Village leaders responded by complaining to the forestry department or sympathetic leaders in the officially sponsored peasants' union (the National Confederation of Campesinos, or CNC) whenever UIEF managers failed to pay royalties into the bank accounts held in trust by the agrarian development bank (Banco Ejidal) or encroached on land they had no right to work. Some communities took resistance a step further and refused to let loggers onto their lands or threatened to kill company foresters who scouted stands of timber for harvest.[24]

CONCLUSION

The crisis utilization associated with World War II had immediate effects in Mexico, such as speeding the death of Cárdenas-era community forestry and favoring the reemergence of commercial logging after six years of intensive management and land reform had placed unaccustomed restrictions on the industry. Especially in some of the northern regions closest to the United States, some forests were no doubt permanently converted to agricultural use or simply to marginal cattle lands.

It is the longer-term consequences of crisis utilization that left the deepest mark on the environment: the logic of crisis-justified institutional innovations in the form of new forestry codes and the establishment of UIEF concessions that had much greater impact in the long term far outlived the war itself. The wartime crisis had already ended before the most important institutional innovations were even up and running; the logic of wartime scarcity had ceased to be an issue by the time that UIEF concessions were made or the 1949 legal code ratified their juridical existence. In other words, the war effectively served as a justification for institutional shifts that political leaders hoped to make anyway—it functioned in this sense as a pretext to increase commercial production at the expense of local, *ejido*-centric logging projects that were becoming more commonplace during the Cárdenas era.

The long-term consequences were profound. The demise of scientific paternalism and the explosion of unregulated markets for forest products also undermined any real possibility that community-based forestry could meet even domestic needs at any point in the foreseeable future. The demands of global warfare favored the rough logic of quick and efficient production, and neither the ideals of rational use nor local production posed much of a barrier. By the time foresters began to survey the situation after the wartime crush, the tide had already turned in favor of large-scale production and scientific planning on a regional scale that left little room to consider the interests of would-be local producers.

Analysts such as Naomi Klein have shown that neoliberal reformers have taken advantage of political crises to advance their economic agenda.[25] The analysis presented here suggests that a similar logic functions in terms of environmental policies. The crisis utilization of 1940–45 imposed unprecedented demands on Mexican forest production, yet the effort to meet these demands had a relatively transitory impact on the environment. The same cannot be said of the institutional structures put into place as a consequence of crisis utilization, however. Legislators took advantage of emergency conditions not only to open the doors to unfettered commercial use of the forests, but also to codify these new conditions in a way that benefited these same interests over the long term. In essence, they institutionalized crisis utilization and converted it into the predominant form of forest use in Mexico for decades to come.

Notes

1 Lorenzo Meyer, *México y los Estados Unidos en el conflicto petrolero, 1917–1942*, 2nd ed. (Mexico City: El Colegio de México, 1972); Friedrich E. Schuler, *Mexico between Hitler and Roosevelt: Mexican Foreign Relations in the Age of Lazaro Cardenas, 1934–1940* (Albuquerque: University of New Mexico Press, 1998).

2 See Catherine E. Jayne, *Oil, War, and Anglo-American Relations: American and British Reactions to Mexico's Expropriation of Foreign Oil Properties, 1937–1941* (Westport, CT: Greenwood, 2000).

3 Christopher R. Boyer and Emily Wakild, "Social Landscaping in the Forests of Mexico: An Environmental Interpretation of Cardenismo, 1934–1940," *Hispanic American Historical Review* 92, no. 1 (February 2012): 73–106; Christopher R. Boyer, "Revolución y paternalismo ecológico: Miguel Ángel de Quevedo y la política forestal, 1926–1940," *Historia Mexicana* 57, no. 1 (July–September 2007): 91–138.

4 Christopher R. Boyer, *Political Landscapes: Forests, Conservation, and Community in Mexico* (Durham, NC: Duke University Press, 2015), 101–21.

5 Mauricio Cruz García, "Gobierno y movimientos sociales mexicanos ante la segunda guerra mundial," *Foro internacional* 51, no. 3 (July–September 2011): 458–504.

6 Fernando Alcocer Patiño, *Un siglo en el bosque (1887, 1987)* (Chihuahua: Unión de Productores e Industrias Forestales, 1987), 78.

7 "Proyecto de Ordeniación para la UIEF Bosques de Chihuahua: Serie de Explotación El Largo" (December 1954), Caja 255, Archivo General de la Nación, Ramo Secretaría de Agricultura y Recursos Hidráulicos (hereafter AGN-SARH).

8 Información Aduanera de México, *Nueva ley forestal y reglamento* (Mexico City, 1943), 5–6. On industrialists' involvement in writing the new law, see Héctor Agustín Salvia Spratte, *Los laberintos de Loreto y Peña Pobre* (Mexico City: Ediciones El Caballito, 1989), 72.

9 Manuel Hinojosa Ortiz, *Los bosques de México: Relato de un despilfarro y una injusticia* (Mexico City: Instituto Mexicano de Investigaciones Económicas, 1958), 91–97.

10 National Congress of Mexico, *Ley forestal y su reglamento*, annotated by Enrique González Flores (Mexico City: Ediciones Botas, 1951), 7.

11 Información Aduanera de México, *Nueva ley forestal*, 3.

12 *Novedades*, August 9, 1950; *El Universal*, June 16, 1951. A list of presidential orders quashing the cooperatives can be found in José Luis Calva Téllez, ed., *Economía política de la explotación forestal en México: Bibliografía comentada, 1930–1984* (Mexico: Universidad Autónoma de Chapingo, 1989), 418–35.

13 Mexico, *Ley forestal*, 2–3, 6–8, 11–12, 59–60.

14 Información Aduanera de México, *Nueva ley forestal*, 3; Roberto Barrena, "Unidades de Ordenación Forestal," *México Forestal* 35, no. 4 (June–August 1961): 12.

15 Dictamen Pericial by Ings. Juan Manuel González, Ricardo Tecanhuey, Miguel Esquivias, and Emilio Flores Calderón, April 15, 1957, Caja 246, Exp. 54603, Leg. 3, AGN-SARH.

16 Mexico, *Ley forestal*, 12, 19–20.

17 Felipe Birgos Martínez, "Proyecto General de Ordenación Forestal: Unididad Industrial de Explotación Forestal de las Fábricas de Papel San Rafael y Anexas, S.A." (unpublished

typescript, 1965), 3; Karel Chambille, *Atenquique: Los bosques del sur de Jalisco* (Mexico City: Instituto de Investigaciones Económicas, UNAM, 1983), 35–39.

18 Spratte, *Los laberintos*, 69–70; Ings. Rodolfo Rodríguez Caballero et al., *Algunas prácticas de ordenación de montes* (Morelia: Comisión Forestal del Estado de Michoacán, 1960), 13; Emmanuel Raufflet, "Institutional Change and Forest Management: The Case of Tlalmananco, Mexico" (PhD diss., McGill University Faculty of Management, 2002), 6–7. See also Chambille, *Atenquique*.

19 On Atenquique's role in establishing the norms for the Mexican management method, see Rodríguez Caballero et al., *Algunas prácticas*; and Ing. Tirso Gutiérrez Jarquín, "Proyecto General de Ordenación Forestal, Unidad Ejidal de Ordenación Forestal "Melchor Ocampo" (typescript; Ocampo, 1969), 243–45.

20 Broadleaf species tend to reproduce more rapidly after logging in Mexico; loggers usually cut more pines than oaks to begin with. In either case, this led to high grading. This was not usually a serious problem, however, since so few loggers actually followed the guidelines. See David Barton Bray and Leticia Merino Pérez, *La experiencia de las comunidades forestales en México: Veintcinco años de silvicultura y construcción de empresas forestales comunitarias* (Mexico City: Instituto Nacional de Ecología–SEMARNAT, 2004), 157, 222–23; and Laura C. Snook and Patricia C. Negreros, "Effects of Mexico's Selective Cutting System on Pine Regeneration and Growth in a Mixed Pine-Oak (*Pinus-Quercus*) Forest," in *Current Topics in Forest Research: Emphasis on Contributions by Women Scientists. Proceedings of a National Symposium* (Washington, DC: US Forest Service, 1986).

21 *El Universal*, January 31, 1962. See also Union Nacional de Resineros to Dámaso Cárdenas, July 25, 1951, Caja 6, Exp. 3, AGN-SARH.

22 Roque Oscar Aguilar Espinoza, "Organización forestal en México" (paper presented at Seminario de Titulación, Universidad Autónoma de Chapingo, 1990), 36.

23 *Excelsior*, June 11, 1956, and February 9, 1957.

24 Spratte, *Los laberintos*, 114–16; *Excelsior*, November 3, 1962.

25 Naomi Klein, *The Shock Doctrine: The Rise of Disaster Capitalism* (New York: Picador, 2007).

Chapter 13
Food Disruption and Agricultural Policy in Tanganyika

GREGORY MADDOX

Scholars of African history, particularly in east Africa, have long debated the long shadows of the World War II era. John Iliffe and John Reader, among others, have argued that the demands of the war led to increased emphasis on both agricultural and industrial production, particularly in the British colonies and in the Union of South Africa.[1] Iliffe in particular argues that the war years capped a crisis in colonial rule in British Tanganyika that led to what he, following other scholars, labeled the "second colonial occupation"[2] as the British created a development infrastructure in its east African colonies. I have elsewhere argued that the war era marked for one region closer integration of local populations into the colonial economy as a labor reserve.[3]

Other scholars have critiqued this perspective, arguing that changes during the war years consolidated earlier policy initiatives and economic changes.[4] David Anderson and David Throup challenged the notion of the war years as a turning point. At the same time, though, they do provide evidence for it as at least an inflection in an ongoing process of the integration of rural communities in Kenya into a colonial economy, both as primary producers and as labor on white-owned estates. In Kenya, the colonial state intensified efforts to encourage production of foodstuffs, especially maize, during the war by easing previous marketing restrictions intended to favor production on white estates. The drive for increased production led British officials and settlers to argue after the war that African ignorance had caused massive land degradation in "native reserves." Continuing a discourse that had developed in the 1930s, the colonial state promoted soil conservation programs that sparked local resistance.[5] This resistance helped feed both the Mau Mau insurgency and the general rise in opposition to colonial rule.

Different regions in sub-Saharan Africa faced several interconnected demands during the war. Fighting in sub-Saharan Africa was limited to the rapid liberation of Ethiopia in 1941 and minor skirmishes between French colonial forces loyal to Vichy and Free French and Allied forces in west Africa. British, Belgian, and French colonies loyal to the Free French (all of them after 1943) served as reservoirs of manpower and suppliers of raw materials. Several hundred thousand Africans served in the colonial armies of the Free French, the British, and the Belgians, seeing action in Europe and in Asia.[6] Mining operations expanded dramatically in many colonies, with Africa becoming a leading source of several critical minerals.[7] South Africa saw rapid industrialization and urbanization during the war as industry passed mining as the most important sector of the economy.[8] In South Africa and Southern Rhodesia (Zimbabwe), wartime expansion of mining also led to increased industrialization. In contrast, the explosive expansion of mining operations in the Belgian Congo (Democratic Republic of the Congo) and in Northern Rhodesia (Zambia) remained enclaves with little in the way of economic linkages to other sectors in those colonies.[9]

If the war saw in a sense the raiding of African mineral resources and the conscription of African manpower for the Allied war effort, it also saw colonial regimes seek to increase agricultural output, whether on settler farms and estates or by African primary producers, at low cost and with a lack of resources such as transport and inputs. In many cases, output expanded as prices rose from Depression-era lows, even in the face of colonial attempts at price controls. But the expansion of agricultural production, especially by African primary producers, faced limits in the form of a shortage of manpower in many colonies owing to conscription for the military and forced labor on private estates as well as public works.[10] At the same time, the South African state sought to impose onerous conservation controls in the black reserves that began to spark protests against cattle culling and conservation work.[11] The demand for output coupled with the shortage of inputs, including labor, resulted in subsistence crises in a number of areas as climatic conditions varied.[12] Famine struck northwestern Africa from 1940 to 1943[13] and the Sahel in 1943–44.[14] Across the eastern half of the subcontinent, food shortages and famines occurred in the early 1940s and again in 1944–45.[15]

In Tanganyika the situation developed along similar lines. Westcott suggests the war "stimulated the growth of managerial government" in Tanganyika that brought greater government control over production and distribution and until the later 1940s limited consumption of imported goods.[16] He notes that efforts to increase production of food for export to war zones along with increased conscription helped contribute to widespread shortages in 1942–43.[17] Giblin has argued that before World War II, British administrators took a passive view toward economic intervention in the colonial economy, limiting themselves to propaganda. After the war, administrators turned their attention to investment for development.[18] D. M. P. McCarthy has maintained that the scope of British initiatives changed from manipulating markets to manipulating production decisions following World War II.[19] John Iliffe says that the end of the war saw the beginning of a "second colonial occupation" as the British Empire, reeling from the Depression and the costs of the war, tried to exploit its colonial dependencies more efficiently.[20] Iliffe argues that the war marked a turning point and as a result the culmination of a crisis of colonialism that began with the Depression. He goes on to argue that an "agricultural crisis" existed by 1945 with "four main causes: the concentration of economic activity in export-producing areas; changing European perceptions of African agriculture; continuing ecological change; and the special pressures of depression and war."[21] He suggests that the colonial state emerged from the war years reinvigorated with both a new drive for economic development and a new mechanism for funding it through the provisions of the Colonial Development and Welfare Act of 1940.[22]

OUR CONTRIBUTION IS TO FEED OURSELVES: THE FOOD CRISIS IN TANGANYIKA

These generalizations do not do justice to the variety of outcomes faced by people in different regions. This essay examines one region, Ugogo[23] in central Tanzania, made up of three districts—Dodoma, Mpwapwa, and Manyoni—and administered as part of the Central Province under the British. The region sits on the high plateau around one thousand meters above sea level. It receives on average around five hundred millimeters of rain annually, with more in the east and west and less in the center. The people of the region have survived by combining rain-fed agriculture and extensive stock keeping. The region suffers from frequent drought and, in the late nineteenth and early twentieth centuries, repeated famines.[24]

For the people of Ugogo (the Wagogo), the beginning of World War II brought increased coercion from the government. Not only were they told they had to plant more crops, but starting in 1940, they were also told to sell more cattle and send their young men either to the army or to Tanganyika's rubber and sisal plantations. During the first few months of the war, the people of Ugogo felt it mostly through an intensified drive to produce more crops and sell more cattle. The director of agriculture in Tanganyika, A. J. Wakefield, summed up wartime policy when he announced through a press release in December 1939 that the goals for the 1939–40 growing season would be to "PRODUCE TO THE LIMIT: EXPORT ALL WE CAN."[25] In a confidential report to the chief secretary, he expanded on his proposals, arguing that each region should concentrate on producing export crops. The sisal plantations would have to use more local labor while more traditional migrants, such as the Nyamwezi of western Tanganyika, stayed home to grow groundnuts and cotton. Any shortfalls in food production, including in Central Province, would then be made up from production in the southern provinces of the territory.[26] His mention of food shortages proved only too prophetic.

In Central Province, local administrators tried to offer incentives along with propaganda as a means of encouraging increased production. For the 1938–39 and 1939–40 growing seasons, the administration distributed groundnut seeds to farmers discovered to be short. The government paid for the seed, and the Indian Association, which represented most merchants in the province, distributed them to farmers. As it turned out, poor rains in Ugogo early in the year meant that groundnut exports were only 1,500 tons from all of Central Province, and areas in Mpwapwa and southern Dodoma already had mild food shortages by the end of 1939.[27] To supply food, the administration had the Native Authorities pay for work on soil conservation measures with grain from other parts of the province.[28] The 1939–40 growing season proved adequate, however, and the people of Ugogo took advantage of higher prices and good rains to produce large crops of both groundnuts and grain for sale.[29]

The government also began to order sales of cattle for the war effort. The armed forces wanted beef for both the Ethiopian and Middle Eastern fronts. The military simultaneously entered into a contract with Liebig's, Hornby and Lowe, a Kenya-based meatpacking firm, for tinned beef. During 1940, the government for the most part contented itself with buying on the open market. Liebig's also entered the market. As a result, cattle prices rebounded sharply, and sales reached a new record.[30]

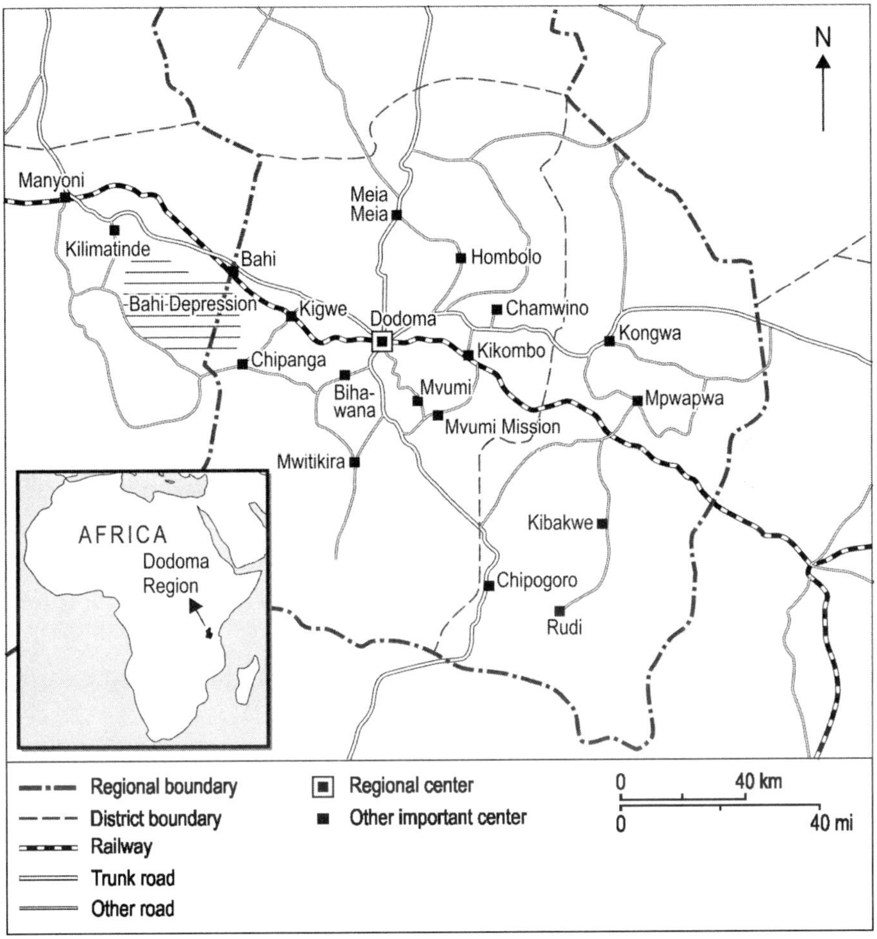

Figure 13.1. The Ugogo region of Central Tanzania comprising the administrative districts of Mpwapwa, Dodoma, and Manyoni.

During 1940, the government also began to conscript men for the King's African Rifles, the East African Labour Brigade, and civilian labor on estates.[31] Conscripts first underwent a physical exam. Those deemed fit enough were immediately inducted into one of the military units. Less fit men were held in camps locally and then sent north to the estates. The British, with tribal stereotypes in full force, did not deem the Wagogo suited to be soldiers, and few entered the military. The Wagogo were too small and thought to be likely to look after their own survival first and worry about obeying orders later. Most conscripts from Ugogo were sent first to the sisal plantations to replace Nyamwezi and other "tribes" thought to be better soldiers, and, after the fall of Southeast Asia to the Japanese, to hastily revived rubber plantations near Muheza in Tanga Province. During 1940, numbers of conscripts remained low, and men were often pressed into local civil defense work for only a short period. But conscription both for the military and for work on the estates to replace other draftees increased steadily throughout the next four years.[32]

Many people in Ugogo remember World War II as the time the government began to apply pressure directly to producers to get what it wanted. During 1939 and 1940, this pressure remained at the level of exhortation.[33] As food supply conditions worsened, by 1942, increasing numbers of men volunteered for work on the estates.

European colonial government employees also had to make direct contributions to the war effort. A large number either volunteered for military service or were drafted. As a result, many administrative functions had to be carried out by African employees of the government for the first time. Because of the war, the number of Europeans in government service was cut back as men entered the military, and no one replaced them. Hence power devolved to African employees. Many used their new powers to increase their standing in local society; however, their new power brought with it new dangers, for district officers could find more reasons to dismiss them.

In Ugogo, the increased level of coercion used by the colonial government succeeded in closely tying the producers to the colonial economy. People without cattle, a majority of households in Ugogo, had to either produce cash crops on their own or become migrant laborers. Cattle owners were now forced to sell cattle instead of hiring labor. Their production of agricultural surpluses declined, and the flow of cattle to households without livestock slowed. Those households then became more dependent on the colonial economy and the colonial government for their subsistence when they failed to grow enough food for themselves, and less dependent on cattle owners.

RELIEF ON A SCALE NEVER BEFORE SEEN

By 1941, Tanganyika Territory had been prepared to help win *Vita vya Uhuru*, or "the War for Freedom," as British propaganda called it. Measures to encourage not only crop production but also increased wage labor and cattle sales had been introduced. From 1941 on, production was coordinated at the highest levels by the intercolonial Economic Control Board for East Africa that directed the distribution of all marketed production throughout the region.

Famine, starting in Ugogo in 1941 and then eventually spreading throughout all of Tanganyika by 1943, disrupted these arrangements. Caused in part by drought but mostly by the drain on local resources imposed by the British for the war effort, this famine helped push the people of Ugogo into a downward cycle of poverty and famine.

In March 1941, the district officer in Dodoma, A. H. LeGeyt, began to report that the harvest had failed in the western part of the district.[34] In August, a survey by LeGeyt and R. R. Staples, the provincial agricultural officer, convinced the provincial commissioner, R. A. J. Maquire, that the lack of local food supplies meant the potential for a killing famine loomed.[35] The local administration succeeded in obtaining some food and money from the central administration and organized forced cattle sales to give at least some people the means to pay for the aid.[36] Despite these efforts, the first of many deaths from hunger over the course of the war years occurred in Kintinku in the Rift Valley in August.[37]

By the end of September, the special cattle sales were in full swing. European officers made spot checks on the markets, reporting that most cattle owners were willing to sell at least some of their herds. Local stock traders buying cattle for resale in other parts of the territory and buyers for Liebig's competed to drive prices up to thirty shillings a head, the highest price since 1930.[38] In some areas, however, especially those where the food shortage was slight to nonexistent, the limits of voluntary cattle sales had already been reached. Administrative officers resorted to force, as they had anticipated, in an effort to coerce money out of cattle owners to pay for the relief food brought into the province.[39]

Many informants recalled the war as a time when the government wanted them to sell their cattle. Mzee Magagi of Handali remembered that the British "called on the people" to sell their cattle.[40] Other informants said the government took cattle to feed its soldiers. Mzee Matao of Hombolo recalled that when the government wanted more cattle, it would send a messenger to the Mtemi. The Mtemi would then send his *ruga ruga*, roughly his police, to seize cattle. When the Mtemi had enough, he would send them on to Dodoma. He pointed out, however, that once the famine started, "the people were forced by hunger alone to sell their cattle."[41]

By the end of October, the administration had a rough estimate of the numbers needing relief. Le Geyt estimated that between 24,900 and 32,600 people would need food.[42] Slowly, the administration tried to get food for sale to the areas that needed it. They faced numerous problems, as supplies were often delayed and new areas in the province began to report shortages.[43]

Finally, a black market sprang up within the region as a result of controlled prices. Low prices made farmers in Mpwapwa and northern Dodoma reluctant to sell their surplus grain at official markets. Merchants in Dodoma quickly realized they could make profits by using their lorries to pick up grain in the countryside and then transfer it to areas in the Eastern Province, where estates paid prices higher than the official ones in Ugogo.[44] Although Le Geyt accepted that this black market grain helped food supplies, it left the administration with more stocks of grain than it could store in Dodoma during the middle of the rainy season.[45] Transport away from the railway remained a problem. Despite the efforts of Le Geyt and Maquire, food often ran out at the storehouses in the villages. Le Geyt had to write to V. Allen, the district works foreman in charge of a food depot at Kigwe in November, promising a fresh supply as soon as transport became available.[46]

The local name for this famine reflects the importance of African government employees in the administration of relief. Many people identified the famine with Jioni, for John, referring to John Mbogoni, a Native Authority clerk.[47] Mzee Mtopochi of Zuzu described how John began to handle the relief: "During the Jioni famine, John, who was in charge of giving out famine food, at first only sold the food to the people. Then, when he saw how bad the famine was, he began to give it out free."[48] John Mbogoni not only supervised the delivery of famine relief to rural areas from Dodoma, but also paid for food bought with Native Authority funds and paid bonuses awarded for the tails of wild pigs, a measure designed to control destruction of crops. More than anyone else in the local administration, he represented the government in famine relief.[49]

Mbogoni's growing power ran afoul of several important people. In late 1943, Mtemi Mazengo of Mvumi and District Officer Z. E. Kingdon accused Mbogoni of failing to pay for food ordered from Kondoa by the Native Authority and of falsifying the number of pig tails brought to his office. Although Mbogoni claimed that the charge was "through malice for my clan" on the part of Mazengo, he was fined and sentenced to a short jail term. Mazengo seized sixty of Mbogoni's cattle, took fourteen sacks of grain from his house, and ordered him to repay 1,241 shillings and 65 cents. The amount of property seized during the midst of a killing famine is indicative of the power of Native Authority employees.[50] If Mazengo thought he was ridding himself of a rival, he was wrong. After his short prison term, Mbogoni returned to his home area in Kinyambwa and eventually became a Mtemi in his own right.

As the famine progressed, and more people became dependent on government stores for food, the number of Africans in control of food distribution increased. Mzee Daniel Lyacho's father was a clerk at Matumbulu south of Dodoma during the 1940s. In 1943 and 1944, when many people had exhausted their cash and cattle reserves, the government organized public works on which people had to labor to get food. Lyacho described his father's work, saying "The Government organized the camps where people worked on roads and digging trenches, but the clerks ran them. My father was a clerk at the camp in Matumbulu. He gave out food one day at a time, for each day's work. Those with cattle or who didn't want to work got food where they could."[51]

The administration used African employees not only to distribute food, but also to enforce measures designed to ensure the famine of 1941 would not be repeated. In particular, during the planting season in November and December 1941, the administration tried to enforce early planting. This strategy grew out of the findings of the provincial agricultural officer, Staples. He wrote that late planting had caused much of the shortages, with farmers who had planted early getting sufficient harvests.[52] M. Kirge, the district agricultural officer for Dodoma, echoed his superior's view when discussing a survey in Kinyambwa.[53] Given this reasoning, Staples and the agricultural staff instituted a strong campaign to plant early. He charged that late planting had occurred because of "the customary indolence" of the people and an ample supply of grain for beer from the good crops reaped the previous year.[54] Staples seemed in this case to have given in to the urge to blame the victims; only a few months previously, he had defended the agricultural practices of the Wagogo during the Increased Production campaign.

Staples's campaign turned out to be futile, if not absolutely detrimental. Maquire reported irregular rains and "unprecedented" losses to "Abyssinian weaver birds," meaning that the 1942 harvest was mostly lost.[55] Staples, an entomologist by training, did not seem to understand the rationale behind staggered planting. Farmers had long spread their plantings over November, December, and January to ensure that they got at least some crops. Given the extreme variability in the rains, in some years they would get caught not planting enough to take full advantage of good early rains such as in 1941, but they would have seed left for replanting after a break in the rains such as in 1942. As it turned out, the big early plantings almost all failed throughout Ugogo, and some areas did not have enough seed for all the replanting required.[56]

By March, Maquire and his officers realized the famine would continue after only a small break after the harvest in May or June.[57] Maquire marshaled his arguments for a skeptical audience in Dar es Salaam. He wanted to prevent both potential unrest and wide-scale, unmanaged migration from Ugogo. The secretariat, however, was more concerned with keeping expenses down. He argued that "the people have suffered a steady attrition of their wealth and, possibly, of their physical strength." He even suggested that the colony should put local needs above even requests from the military for exports of food.[58]

The famine in 1942–43 would cover almost the whole of Ugogo, striking also in Mpwapwa, which had produced a small surplus the year before. Despite the gravity of the situation, the suggestion that territorial needs should be put above imperial ones did not sit well in Dar es Salaam. As Maguire noted, many areas in Tanganyika, throughout eastern and central Africa, and almost the entire Indian Ocean basin faced food shortages in 1942–43. These production shortfalls were complicated by both the reduction in shipping available to import food and by the fall of some food-exporting regions in East Asia to the Japanese. Given the enormity of the Bengal famine in India, the African colonies received low priority for the supplies that were available.[59] In Tanganyika, the difficulty in obtaining supplies meant that the central government had to ration what grain it had between the estates; the areas in the northern, eastern, and western regions that had shortages; and Ugogo.[60] This problem faced colonial governments throughout the region. In Zanzibar, areas alternated between feast and famine as the government shifted available food around.[61] In Kenya, surpluses produced in some areas were constantly rerouted as a food riot broke out here, the military needed food there, or another colony made an urgent request for help.[62]

CONTINUATION OF FAMINE

But Maquire's warnings about a second year of famine did not fit into the plans of the Economic Control Board. R. W. R. Miller, director of agricultural production, opined that Maquire seemed "rather prone to famine," and planning for food deliveries lagged. In June, Maquire, both to defend his name and to put relief plans into action, polled missionaries and officers about the famine.[63] Father George Joye of the Passionist Fathers Mission in Bahi wrote that the only difference between this famine and the Mtunya famine of World War I was the availability of food from relief.[64] Canon R. Banks of Kilimatinde was even more emphatic. After over two decades of work at Kilimatinde, he knew that the colonial government, and especially the central government, often blithely ignored the fate of its subjects when their interests conflicted with imperial ones. He wrote an impassioned letter with an eye toward the relief that would be required in the coming year. He noted that the administration

> arranged for stocks of grain to be brought into the district. These were bought up like hot cakes and there was nearly a riot here one day when the stock in the store was sold out and no more expected for some days. Now the Mugogo never buys a single kilo of grain more than he actually needs, and on the theory that relief was unnecessary it is hard to explain why these people should have bought such huge quantities, in small quantities at a time, if they had corn hidden away in some inaccessible place.

He estimated that hundreds would have died without the food imported, perhaps up to 5 percent of the population, or four thousand deaths. Failure to act would have "damned the British government forever in the eyes of all decent people" and forced "the most patriotic missionary" to speak out.[65] Unfortunately for the people of Ugogo, Banks's words fell on deaf ears. Maquire failed to convince the Economic Control Board or the secretariat of the urgency of the situation. Instead, they worried about the cost of continued relief and the state of the food supply in the military.

In October 1942, the effects of indecision at the higher levels began to be felt. Food from Kenya originally ticketed for Central Province was rerouted to Mombasa because of riots caused by the lack of maize in that city.[66] R. W. Varian, who had replaced Le Geyt as district commissioner in Dodoma, countered by embargoing five hundred tons of rice produced in the Rift Valley and collected in Bukene in western Manyoni. Despite explicit orders to release the rice to the military, he wrote, "much as I dislike the thought of feeding starving Wagogo on rice," that he would confiscate the rice unless the Maize Control Board released more grain to him.[67] With famine also facing the Western, Lake, and Eastern Provinces, and the sisal and rubber estates cutting their daily ration to less than twenty-four ounces of grain meal a day, a starvation diet, little help was available locally. What grain that did make it into the province was quickly consumed.[68] The lack of relief supplies during the critical planting period meant that people had to be away from their fields at the most important time of the year. Even when food did arrive, they had to go to the railway to get it and carry it back home because of a lack of transport.[69]

Despite increased cattle sales,[70] money was much scarcer during the 1942–43 growing season than the year before. Crop sales had fallen in 1942, and cattle prices had stabilized. Many owners of small herds had exhausted their stock and no longer had a means of raising cash. The administration authorized the issue of food on loan against repayment at the next harvest. For the season, the price of grain was raised to three shillings per debe of grain or two-thirds debe of rice or beans because of higher prices charged to the government.[71] Many people recall with bitterness having to repay loans from previous years in the midst of shortages. Mzee Nonya of Handali recalled that "if you could not repay the loan at the end of the famine, the British would go to your relatives and take their cattle."[72] Most informants recall the *mchele* (rice), or *mhuzi* (soup), famine as the time when men began to go work on *manamba* (contracts, derived from the word "number," for the number of tasks a laborer had to complete before being paid) on sisal and rubber estates. For the whole province, Maquire estimated that thirteen thousand people went to the estates either voluntarily or as conscripts in 1942.[73] Often, young men did not want to go to the estates. They preferred the better pay and conditions of military service. The sisal estates claimed that the Wagogo arrived in no condition to work because of their hunger and took as few as possible.[74] The newly reopened rubber plantations found the Wagogo acceptable, however, and preferred them over any other "tribe" in Tanganyika Territory. By the time the rubber estates opened, the famine was in its zenith in Ugogo. The men recruited now had a strong motive for keeping their jobs.[75]

The men of Ugogo retain vivid memories of being forced to take up *manamba*. Mzee Magabo of Matumbulu recalled, "I tried to join the army, but I was told I had bad eyes. Instead I went to Muheza to work gathering rubber."[76] When the men left, according to Bibi Akisa of Handali, "sometimes the men sent back money. Otherwise, the women had to work in the fields of others to get food."[77] But several men claimed they never sent money back home while working. Mzee Yakongo remembered, "Our families stayed at home, and we men went to work on the estates alone. We brought all our money back with us when we finished. . . . It was up to the people at home to get food. they could work in the *shamba*, but we got *posho* [rations] on the estates."[78]

The absence of men created hardships for women even if their husbands owned stock. In this case, a wife could not sell any animals without the permission of her husband or his closest male relative, and she could not get food on loan from the stores because she owned stock.[79] The experience of going to work *manamba* became so common that the name of the age set for men who underwent circumcision during World War II was *manamba*.[80]

Only a few Wagogo made it into the King's African Rifles (KAR) or the East African Labour Brigade. These men often came from wealthy and well-connected families. Hassan Biringi, son of Mtemi Biringi Lusambo of Dodoma, served in the KAR throughout the war. He recalled that the men from Central Province became agitated when they heard news of the famine while training in Ceylon for action in Burma. Eventually, Mtemi Nolo of Makutupora took to BBC radio broadcasts to assure the troops that their wives and children had food.[81]

Despite Hartnoll's belief to the contrary, there can be no doubt that conscription helped perpetuate the famine during World War II. Recruiting disrupted areas, and the loss of labor at a time when demand for field labor was at its peak undoubtedly reduced the harvest in 1944. Because of the increased need for cash with which to buy food, men did not stop migrating at the end of the war. Instead, regular labor migration among men became part of the bundle of strategies employed to ensure the reproduction of the household. By 1944, people no longer ran when recruiters came. As Mzee Zazalo of Bahi said, they learned they could get money: "[Labor migration] started during the war. Then young men were forced to go to Muheza to work on rubber estates and to Tanga to work on sisal. At first the people did not want to go. Then, after the first group came back, they told everyone about it. Young men started to go by themselves."[82]

Taken as a whole, Ugogo had become unable to reproduce itself. Whereas before, trade and redistribution of food had countered local shortages, now poorer people had to look outside Ugogo not only to earn money for food during years of dearth but also just to meet their day-to-day requirements for survival.

Despite the efforts of the Provincial Administration, food began to run out in December 1942. People began to leave their fields to search for wild food or to congregate in the towns, hoping for first call on any relief shipments. Everyone in Ugogo from the provincial commissioner to farmers in Chipogoro foresaw imminent death. By January of 1943, Maquire had reached a state approaching panic. The short rains had failed again, there was little seed grain left for replanting, and Dodoma had run completely out of food. Maquire feared that people moving into larger settlements looking for food could raise the possibility of serious unrest and had the police make preparations to suppress it.[83] As bad as the situation had been, it deteriorated even further over the next few weeks. When the district commissioner for Singida wrote requesting a small amount of food for parts of his district, Maquire had to reply that he and the people of Singida were on their own.[84]

Provincial Agricultural Officer N. V. Rounce summed up the position at the end of January when he wrote that food supplies ran out in many places during the middle of the planting season, seriously reducing the area cultivated as people left the fields to search for food.[85] Thereafter, the death toll mounted. Between 1943 and 1945, when the harvest proved adequate, thousands died. Relief efforts in 1943 proved wholly inadequate despite the confiscation of large numbers of cattle. In 1944, more food came to the province, much from within the territory, but the colonial government's insistence on dispensing aid only in famine camps where people had to provide unpaid labor meant many stayed away and suffered.[86]

CONCLUSION

The crisis of the war years caused distinct changes in local practices to manage hunger. Local practices continued to balance immediate subsistence needs with the imperative of preserving assets such as cattle for recovery;[87] cattle owners increasingly sold stock rather than investing it in reciprocal relationships that brought labor to their fields. Cattle sales improved above levels before the war and stayed there, rising higher during years of shortage and falling as owners rebuilt their herds in good years. Farmers in Ugogo also increasingly began to produce maize as both a food crop and a cash crop. Maize produces higher yields per acre during a shorter growing season than the local varieties of sorghum and millet. The variety used locally, mostly Proofromsten Pearl, yielded around seven hundred pounds per acre compared with about six hundred pounds per acre for bulrush millet and five hundred pounds per acre for sorghum.[88] As everyone in Ugogo realized, however, maize requires more rain than sorghum and millet. Many informants noted that people began to switch to maize after the war because they received higher prices for their surplus than they would have for sorghum or millet.[89] Mzee Masai recalled, "I remember that people began to grow a lot of maize in around 1952. There were times when we tried to grow maize and we did not get a harvest. We started to grow maize because that was the type of food that we got as famine relief. Also we were paid a better price for maize than for millet. Those living here in the area would buy only maize. This was always true of the Swahili."[90]

Labor migration, though, became the major response of local populations to the shortages during the war years. Before the war, locals had worked for cattle owners in return for access to cattle as loans or payment and for food during hard times. But during the war conscription, especially for sisal and rubber estates in eastern Tanganyika, migration took many men out of the region. Even when the rains returned in 1944–45, rates of labor migration stayed high, with up to 30 percent of the population migrating.[91]

Broadly speaking, the famine in central Tanganyika during the war years can be seen as a crisis in food entitlement.[92] In terms of production, at the level of the specific colony or of the east African region, even in the worst of the period of shortage, food existed and was exported. At the level of central Tanganyika, the famine resulted from both a collapse in entitlement and a food availability decline.[93] The decline in entitlements to food in the region occurred, in ways similar to those outlined by Sen for Bengal, in the appropriation of resources by the colonial state during the crisis of the war. The food availability decline occurred both because of climatic conditions and because of the loss of labor in the region.[94]

In the aftermath of the war and of the killing famine in the region, the colonial government faced pressure both internally and externally to do something to end the drain on territorial resources represented by famine relief and to promote viable economic production in the region. Externally, two factors pushed the colonial government toward action. First, the Colonial Office expected each colonial government to develop a plan under the terms of the Colonial Development and Welfare Act of 1940 for increased investment in development. Iliffe outlined the planning process that occurred in Dar es Salaam in 1944–45, which resulted in a plan heavy on roads and "land rehabilitation."[95] In the provinces, local administrations developed plans for investment in conservation and improved agriculture. In Central Province, the several plans of the 1940s all got lost in a series of food shortages and famines between 1946 and 1950, and by the implementation of the other major postwar initiative in the region, the infamous Groundnut Scheme.

The Gogo Development Scheme marked the last attempt by the colonial government to take the initiative in solving the problems of poverty and food security in Ugogo. In essence, in the aftermath of the crisis generated during the war years, the colonial government tried two distinct methods of promoting what it called development in the region. Both failed, one spectacularly. One sought to exploit land and to a lesser extent local labor in an externally financed and oriented large-scale project. The other sought to reinforce local autarky and a social order that no longer existed.

Central Tanganyika's continuing marginalization within the late colonial economy after World War II demonstrates the differential impact of colonial economic development that occurred across the continent. The war years saw an intensification of this process in many ways. The famines in Ugogo brought death and impoverishment not just because of a lack of rain, but also because the colonial state starved the region of resources that it felt had to go to areas producing goods deemed critical for the war effort. While those areas had value forcibly transferred to the colonial metropoles in the form of repressed commodity prices and wages, they also became the areas that saw economic growth in the decades after the war. For regions like Ugogo, impoverishment during the war years set the stage for continued poverty in the future.

Notes

1 John Reader, *Africa: The Biography of a Continent* (London: Hamish Hamilton, 1997), 634–37; and John Iliffe, *A Modern History of Tanganyika* (Cambridge: Cambridge University Press, 1979), 342–54.

2 Iliffe, *Modern History*, 436.

3 Gregory H. Maddox, "Famine, Impoverishment and the Creation of a Labor Reserve in Central Tanzania," *Disasters* 15, no. 1 (1991).

4 David M. Anderson and David Throup, "Africans and Agricultural Production in Colonial Kenya: The Myth of the War as Watershed," *Journal of African History* 26, no. 4 (1985).

5 David M. Anderson, "Depression, Dust Bowl, Demography, and Drought: The Colonial State and Soil Conservation in East Africa During the 1930s," *African Affairs* 83, no. 1 (October 1984); John Lonsdale, "The Depression and the Second World War in the Transformation of Kenya," in *Africa and the Second World War*, ed. David Killingray and Richard Rathbone (London: Macmillan, 1986), 121–24.

6 David Killingray and David E. Omissi, *Guardians of Empire: The Armed Forces of the Colonial Powers, c. 1700–1964* (New York: Manchester University Press, 1999); Nancy Ellen Lawler, *Soldiers of Misfortune: Ivoirien Tirailleurs of World War II* (Athens: Ohio University Press, 1992); Myron J. Echenberg, *Colonial Conscripts: The Tirailleurs Sénégalais in French West Africa, 1857–1960* (Portsmouth, NH: Heinemann, 1991).

7 Raymond Dumett, "Africa's Strategic Minerals during the Second World War," *Journal of African History* 26, no. 4 (1985), 381–83.

8 William Beinart, *Twentieth-Century South Africa* (Oxford: Oxford University Press, 2001), 129–30.
9 Dumett, "Africa's Strategic Minerals during the Second World War," 389–97.
10 Elias C. Mandala, *Work and Control in a Peasant Economy: A History of the Lower Tchiri Valley in Malawi, 1859–1960* (Madison: University of Wisconsin Press, 1990), 191, 214; Michael Crowder, "The 1939–1945 War and West Africa," in *History of West Africa*, ed. J. F. A. Ajayi and M. Crowder (New York: Columbia University Press, 1973), 606–11; David Killingray, "Labour Mobilisation in British Colonial Africa for the War Effort, 1939–46," in Killingray and Rathbone, *Africa and the Second World War*, 69–71.
11 Beinart, *Twentieth-Century South Africa*, 134–36; idem, "Soil Erosion, Conservationism and Ideas about Development: A Southern African Exploration, 1900–1966," *Journal of Southern African Studies* 11, no. 1 (1984): 71–74.
12 Killingray, "Labour Mobilisation in British Colonial Africa," 69.
13 Gervase Clarence-Smith, "The Impact of the Spanish Civil War and the Second World War on Portuguese and Spanish Africa," *Journal of African History* 26, no. 4 (1985): 319.
14 David Killingray and Richard Rathbone, "Introduction," in Killingray and Rathbone, *Africa and the Second World War*, 17; Michael Watts, *Silent Violence: Food, Famine and Peasantry in Northern Nigeria* (Berkeley: University of California Press, 1983), 326–36.
15 Clarence-Smith, "Impact of the Spanish Civil War," 321–22; Ian Spencer, "Settler Dominance, Agricultural Production and the Second World War in Kenya," *Journal of African History* 21, no. 4 (1980): 506, 512–14.
16 Nicholas Westcott, "The Impact of the Second World War on Tanganyika, 1939–49," in Killingray and Rathbone, *Africa and the Second World War*, 157.
17 Ibid., 147–49.
18 James Giblin, "Famine, Authority, and the Impact of Foreign Capital in Handeni District, Tanzania, 1840–1940" (PhD thesis, University of Wisconsin, 1986), 527–30.
19 D. M. P. McCarthy, *Colonial Bureaucracy and Creating Underdevelopment: Tanganyika, 1919–1940* (Ames: Iowa State University Press, 1982), 111–14.
20 Iliffe, *Modern History*, 436–37.
21 Ibid., 347.
22 Ibid., 436–40.
23 Ugogo means "the land of the Gogo people." Bantu languages form nouns by adding class prefixes to roots, so "U-gogo" is "the land of the Wa-gogo people."
24 Gregory H. Maddox, "*Mtunya*: Famine in Central Tanzania, 1917–1920," *Journal of African History* 31, no. 2 (1990); idem, "'Leave, Wagogo! You Have No Food!': Famine and Survival in Ugogo, Central Tanzania 1916–1961" (PhD thesis, Northwestern University, 1988); Peter Rigby, *Cattle and Kinship among the Gogo: A Semi-Pastoral Society of Central Tanzania* (Ithaca, NY: Cornell University Press, 1967); Michael Patton, *Dodoma Region, 1929–1959: A History of Famine*, Bralup Research Reports No. 44 (Dar es Salaam: BRALUP, 1971).
25 A. J. Wakefield, press release, December 13, 1939, Tanzania National Archives (TNA) 26298 Vol. II.
26 A. J. Wakefield to Chief Secretary, December 14, 1939, TNA 26298 Vol. II.
27 G. J. Partridge to Chief Secretary, June 6, 1939, TNA 26298/6 Vol. II.
28 A. V. Hartnoll, *Central Province*, Annual Reports of the Provincial Commissioners 1939 (Dar es Salaam: Government Printers, 1940), 7-8.
29 R. R. Staples, "Central Province Agricultural Report, 1941," TNA 46/A2/1/1941.
30 R. A. J. Maguire, *Central Province*, Annual Reports of the Provincial Commissioners 1940 (Dar es Salaam: Government Printers, 1941), 7.
31 Meshack Owino, "The Discourse of Overpopulation in Western Kenya and the Creation of the Pioneer Corps," in *The Demographics of Empire: The Colonial Order and the Creation of Knowledge*, ed. Karl Ittman, Dennis Cordell, and Gregory H. Madddox (Athens: Ohio University Press, 2010), 157–73.

32 Maguire, *Central Province*, 6–7.

33 I/5/5A Father Steven. Father Steven was the first African in the region to be ordained in the Catholic Church and was already serving as a priest at the time of the war. Full transcripts of all interviews cited can be accessed by index number and the name of one of the informants listed in Maddox, "Leave, Wagogo!," appendix.

34 A. H. LeGeyt to Provincial Commissioner, March 31, 1941, TNA 46/A20/1 Vol. III. Since 1938, Dodoma District included Manyoni and Mpwapwa.

35 TNA 46/A20/1 Vol. III contains the relevant correspondence.

36 R. A. J. Maquire to J. E. S. Lamb, Administrative Secretary, September 1, 1941, TNA 46/A20/1 Vol. III.

37 Jumbe Filipo to District Commissioner, Dodoma, August 28, 1941, TNA 46/A20/1 Vol. III.

38 Minutes by R. A. J. Maquire, November 4, 1941TNA 46/18/2 Vol. II.

39 Maquire, *Central Province*, 2.

40 I/38/87A-95A Magagi etc.

41 I/34/54A-60A Lechipiu etc.

42 R. A. J. Maquire to Chief Secretary, October 27, 1941, TNA 30077 Vol. I.

43 See TNA 46/A20/I Vols. III and IV for daily reports on the food situation throughout the Province.

44 A. H. Le Geyt to Provincial Commissioner, December 14, 1941, TNA 46/A20/1 Vol. IV; and R. A. J. Maquire to Chief Secretary, December 15, 1941, TNA 20908 Vol. II.

45 A. H. Le Geyt to Provincial Commissioner, December 29, 1941, TNA 46/A20/1 Vol. IV.

46 A. H. Le Geyt to A. Allen, November 10, 1941, TNA 46/A20/1 Vol. IV.

47 I/6/6A,7A Sikana and Lyacho, I/32/52A Biringii, I/29/42A Kaka, and I/31/46A-51A Musa etc.

48 I/31/46A-51A Musa etc.

49 R. W. Varian to the High Court, October 17, 1944, TNA 13879. For more on Mazengo, see Gregory H. Maddox with Ernest M. Kongola, *Practicing History in Central Tanzania: Writing, Memory, and Performance* (Portsmouth, NH: Heinemann, 2006), 115–36.

50 J. Mbogoni to High Court, October 2, 1944, and R. W. Varian to High Court, October 17, 1944, TNA 13879/83.

51 I/6/6A,7A Lyacho and Sikana.

52 R. R. Staples, Monthly Agricultural Report for August, September 16, 1941, TNA 46/20/31.

53 McK. Kirge to District Commissioner Dodoma, September 22, 1941, TNA 46/A20/1 Vol. III.

54 R. R. Staples, Monthly Agricultural Report for October, November 21, 1941, TNA 46/20/31.

55 Maquire, *Central Province*, 1.

56 R. A. J. Maquire to Chief Secretary, March 11, 1942, TNA 30077 Vol. I.

57 Ibid.

58 R. A. J. Maquire to Chief Secretary, April 15, 1942, TNA 30077 Vol. I.

59 Iliffe, *Modern History*, 351–53. Amartya Sen, *Poverty and Famines: An Essay on Entitlement and Deprivation* (Oxford: Clarendon Press, 1981); Mark B. Tauger, "Entitlement, Shortage and the 1943 Bengal Famine: Another Look," *Journal of Peasant Studies* 31, no. 1 (October 2003). See also Richard Tucker, "Environmental Scars in Northeastern India and Burma," chapter 7 in this volume.

60 Gregory Maddox, "*Njaa*: Food Shortages and Famines in Colonial Tanzania, 1919–1920," *International Journal of African Historical Studies* 1, no. 19 (1986): 13–17.

61 Timothy Welliver, personal communication, 1986.

62 See the correspondence in TNA 30077 Vol. III between the Economic Control Board in Dar es Salaam and the Maize Control Board and Kenya Farmers Association in Nairobi.

63 R. A. J. Maquire to J. E. S. Lamb, "Confidential," June 22, 1942, TNA 46/A20/1 Vol. V.

64 Father George Joye to Provincial Commissioner, June 26, 1942, TNA 46/A20/1 Vol. V.

65 R. Banks to Provincial Commissioner, June 15, 1942, TNA 46/A20/1 Vol. V. Maquire allowed that Banks might have been "a little emphatic."
66 R. S. Griffith, Maize Control Board Nairobi to Provincial Commissioner, October 19, 1942, TNA 46/A20/1 Vol. IV.
67 R. W. Varian to Economic Control Board, October 31, 1942, TNA 46/A20/1 Vol. VI. I suspect he was being sarcastic about his regrets.
68 R. A. J. Maguire to Economic Control Board, October 14, 1942, TNA 46/A20/1 Vol. VI.
69 I/31/46A-51A Musa etc.
70 Maquire, *Central Province*, 8.
71 D. S. Troup to Provincial Commissioner, "Brief Report on Famine Administration," March 9, 1943, TNA 46/A20/1 Vol. VIII.
72 I/38/87A-9sA Magagi etc.
73 Maquire, *Central Province*, 14.
74 K. C. Charron, Medical Officer Department of Labour to Provincial Commissioner, May 31, 1943, and R. A. J. Maguire to K. C. Charron, June 9, 1943, TNA 30077 Vol. I.
75 Hartnoll, *Central Province*, 14.
76 I/40/102A-106A Chipanga etc.
77 I/38/87A-95A Magagi etc.
78 I/40/102A-I06A Chipanga etc.
79 I/38/87A-95A Magagi etc.
80 H. A. Fosbrooke "A Note on the Dating of Age-Sets and Famines in *Historia, Mila na Desturi–*Wagogo–Tanganyika," October 18, 1954, TNA 435/A2/24.
81 I/32/52A Biringi; and R. A. J. Maguire to Chief Secretary, January 22, 1943, and Chief Secretary to Provincial Commissioner, January 23, 1943, TNA 30077 Vol. I.
82 I/39/96A-10lA Luangi etc.
83 R. A. J. Maquire to all District Officers, "Strictly Confidential," January 12, 1943, TNA 30077 Vol. I.
84 J. A. Armstrong, District Commissioner Singida to Provincial Commissioner, January 27, and R. A. J. Maquire to District Commissioner, January 28, 1943, TNA 46/A20/1 Vol. VI.
85 N. V. Rounce, Monthly Agricultural Report for January, February 9, 1943, TNA 46/20/31.
86 Maddox, "Leave, Wagogo!," 271–80.
87 Alexander de Waal, *Famine That Kills: Darfur, Sudan, 1984–1985* (Oxford: Clarendon Press, 1989), 7.
88 H. S. Senior, "Gogo Development Plan," July 1943, TNA 42245/1.
89 I/3/3A Kongola I/6/6A, 7A Sikana and Lyacho; I/33/53A Sah Ali; and I/43/113A Chilega.
90 I/13/9B, l0B, 11A Masai etc.
91 Tanganyika Territory, *Statistical Abstract 1951* (Dar es Salaam, 1952), 4; and H. S. Senior, "Gogo Development Plan," June 1953, TNA 42245/1.
92 For the general definition, see Sen, *Poverty and Famines*.
93 Stephen Devereux, *Theories of Famine* (New York: Harvester/Wheatsheaf, 1993), 76–81.
94 Tauger, "Entitlement, Shortage and the 1943 Bengal Famine," 53–57.
95 Iliffe, *Modern History*, 438–40.

Chapter 14
Japanese Imperialism and Marine Resources

WILLIAM M. TSUTSUI AND TIMO VUORISALO

In this essay we review the long-term legacies of what has been termed Japan's "pelagic empire"—the sprawling realm of commercial fisheries, state institutions, and policies—and strategies for marine resource exploitation and market dominance established in the decades before World War II. Today, as for most of the past century, the Japanese fishing industry and its practices are controversial internationally. Foreign fishing interests and governments have long been threatened by Japan's highly efficient and wide-ranging fleets. Meanwhile, although Japan was ranked twenty-sixth in the Environmental Performance Index of 2015[1] and hosted the historic United Nations (UN) conference in 1997, where the Kyoto Protocol was signed, it continues to be a special target of environmental groups such as Greenpeace. Japanese persistence in whaling, in particular, continues to receive unfavorable attention in the European and American media.[2] Our assertion is that Japanese imperialism in the Asia-Pacific region from the late nineteenth century until 1945 developed patterns of exploitation of marine resources that are still present, and contribute to the overharvesting of marine fish and the continuing conflicts between Japanese fishing interests and overseas competitors and conservationists. Thus the long shadows of the pelagic empire still influence Japan's environmental reputation and policy today, more than half a century after defeat in World War II ended the nation's formal imperial ambitions.

THE MAKING OF JAPAN'S "PELAGIC EMPIRE," 1868–1945

The Japanese people have long practiced intensive coastal fishing and whaling.[3] Japan only developed extensive offshore fisheries in the late nineteenth century, however, as the modernizing nation became increasingly integrated into the global economy. From the 1870s, for instance, Japanese fishermen gained rights in Russian territory and rapidly established operations in Sakhalin and along the coast of the Sea of Okhotsk.[4] In the wake of the Russo-Japanese War of 1904–5, additional concessions were won off of Kamchatka, and Japanese enterprises aggressively developed them. Advanced technologies introduced from the West also played a significant role in expanding the reach and the productivity of Japanese fishermen. The first modern otter trawler was imported from Britain in 1908 and was enthusiastically copied by Japanese shipbuilders; within only a few years' time, over a hundred new vessels were pulling catches from the seas around Japan.[5] In 1906, Japan launched its first motor-powered tuna boat, and soon Japanese fishermen were eagerly using them to pursue skipjack through the Bonin (Ogasawara) Islands.[6] Significantly, and not coincidentally, these first steps toward establishing offshore fisheries paralleled the growth of Japan's terrestrial empire. With new territories won in the Sino-Japanese (1894–95) and Russo-Japanese Wars, and with the annexation of Korea in 1910, Japanese fishing operations moved progressively beyond home waters and through the neighboring seas of East Asia.

By any measure, Japanese fisheries grew rapidly in the first decades of the twentieth century. The number of motor-powered fishing boats, for example, increased from fewer than three dozen in 1908 to almost six thousand in 1920. Over the same period of time, the offshore catches of Japanese fishermen expanded fivefold. By the early 1930s, as fisheries production in Japan and the empire swelled to over five million tons a year, Japan could legitimately claim to be the world's leading fishing nation.[7] From the perspective of the Ministry of Agriculture and Commerce, Japan's progress seemed almost inevitable: "With the steady increase of population . . . the demand for fishing products is showing a striking advance, a condition still further accelerated by

Figure 14.1. Japanese war fleet at home in the early 1920s. Source: J. E. Rosberg and V. Tolvanen, eds., *Maapallo. Maantieteellinen tietokirja. Seitsemäs osa* (Helsinki: Kustannusosakeyhtiö Otava, 1927)

the increasing demand from abroad. Under these circumstances the fishermen can no longer remain satisfied with coasting work alone, but are obliged to a greater extent than ever before to venture into the open sea and even to the distant coasts of Corea [sic] and the South Sea Islands."[8] And indeed, while the yield from Japan's coastal waters did increase steadily prior to World War II, the major growth in Japanese production came in offshore and deep-sea fisheries.[9] Japanese trawlers first worked the rich grounds of the East China and Yellow Seas in 1921, began sweeping the South China Sea in 1927, and expanded operations to the Gulf of Tonkin in 1928. After Japan claimed the mandated islands of Micronesia from Germany as the spoils of World War I, Japanese boats ranged across the southwestern Pacific in search of tuna. Meanwhile, in northern waters, the Japanese developed and dominated the crab fishery: Japanese crabbers started working the Kurils in 1909, and fleets of Japanese factory ships moved up and down the Siberian coast starting in 1922. The results from the crab grounds, as from Japan's other new fisheries, were stunning: in 1931, for example, Japan produced 407,542 cases of canned crab, more than eighty times as many as had been processed only a decade earlier.[10]

The drive to expand production in Japan's fisheries continued to intensify in the 1930s. The impetus for these heightened efforts was not growing concern for food security, although Japanese policy makers did recognize the need to provide ever more marine protein to the home population and the growing ranks of imperial subjects.[11] Nor was it motivated primarily by a desire to maintain and grow employment in Japanese fisheries at a time of global economic distress.[12] Instead, Japan's fishermen were mobilized after the Manchurian Incident as part of Japan's imperial project. What the nation's trawlers, factory ships, purse seines, and pelagic longlines were expected to deliver was not food for Japanese tables or jobs for Japan's unemployed, but rather foreign exchange—the dollars, pounds, and reichsmarks necessary to keep Japan's growing empire and military machine fed with imported oil, iron ore, and rubber. Thus Japanese fisheries were focused as never before on exports, on developing new sources of marine products, and on finding markets for them in Europe, North America, and around the world.

Canned goods, over 80 percent of which were exported, were an important part of this ambitious strategy: production of canned salmon and crab, long staples in British and American kitchens, was promoted; Japanese fleets combed the Pacific for albacore and yellowfin to meet the seemingly insatiable American appetite for canned tuna; sardines in tomato sauce were packed specially for the hungry consumers of India and the Philippines. Frozen tuna was sold by the ton to canners in California, prawns caught off the coast of Mexico by Japanese ships were targeted at the US market, and inexpensive dried and salted fish poured into China and Southeast Asia. At the same time, trawlers and processing ships swept the Japan Sea, the Sea of Okhotsk, and the Bering Sea, producing large quantities of fishmeal and liver oil, most of it sold to German consumers. And in the colonial states of Southeast Asia—Malaya, Singapore, the Dutch East Indies—Japanese interests gained a predominant role in the supply and distribution of fresh fish.[13] Import substitution was also pursued: to limit the purchase of leather from Western suppliers, eager Japanese entrepreneurs promoted the domestic production of sharkskin, whale leather, and tanned salmon skin.[14] As one commentator proclaimed in 1940, at the peak of Japan's prewar fisheries, "The fishing industry serves Japan as an important source of foreign exchange. Her exports of marine products amount annually in value to between ¥150,000,000 and ¥160,000,000 and thus rank third after raw silk, and cotton yarn and piece-goods exports. Also, the fish cost practically nothing, whereas raw cotton must be imported for the textile industry and vast mulberry plantations must be maintained by silk-growers."[15]

The growth of Japan's fisheries—especially overseas, export-oriented operations—was vigorously supported by the Japanese state. Both fishermen and policy makers long recognized their shared interest in expanding catches and extending Japan's fishing grounds: "fisheries build up the nation" (*suisan rikkoku*) was a slogan promoted during the Sino-Japanese War, and, as one 1900 fisheries textbook boldly declared, "Fishery development is the basis of a rich nation, as well as the basis of a strong army."[16] Japanese diplomats, for example, were tenacious in bargaining for fishing rights in treaty negotiations, especially with the Russians, to ensure Japanese boats access to the rich grounds off the shores of Russia's maritime provinces of Sakhalin and Kamchatka. The ties between Japan's armed forces and its fishermen were always intimate: canned seafood production was first nurtured by the government as a source of rations for soldiers and sailors; decommissioned warships often entered the fishing fleet and private fishing vessels were frequently drafted into use by the military in wartime; and the Imperial Navy was regularly called upon to safeguard Japanese fishing operations in distant and disputed waters.[17] Japanese fisheries interests could also count on substantial financial support from the state. From as early as 1897, when the Pelagic

Fisheries Encouragement Law was enacted, government subsidies supported the construction of technologically advanced boats for expanded offshore and deep-sea fisheries. In the early twentieth century, state aid underwrote the development of factory ships, the provision of advanced refrigeration and ice-making facilities, and the growth of Japanese fishing concerns in Southeast Asia.[18]

Significantly, Japanese authorities recognized that fishing was an industry and as such needed to be nurtured, guided, managed, and mobilized just as much as other strategic economic sectors.[19] One can argue that fisheries were an important early proving ground for Japan's nascent "industrial policy" that Chalmers Johnson traced back to the late 1920s and 1930s.[20] Government bureaucrats, especially those in the Fisheries Bureau of the Ministry of Agriculture and Forestry, took an active role in coordinating the growth of Japan's fishing industry, working particularly closely with the handful of large corporations that dominated Japan's modern, export-oriented fisheries. As one official put it in 1939, "The Government controls and supervises the high seas fisheries in a prudent manner [so that] demand and supply of aquatic products [may be] perfectly adjusted."[21] To maintain order in the rapidly growing high-seas fisheries, an elaborate licensing system was established and a variety of trade organizations, from cooperatives to producers' federations, were set in place. When mounting competition threatened to bring disorder to the fishing grounds, the state was not hesitant to step in, forcing consolidation of firms or limiting licenses in waters that were growing depleted.[22] The authorities were also assertive in directing the industry toward strategic goals, especially with regard to opening new fishing grounds to Japanese producers. Thus the state encouraged and funded numerous reconnaissance expeditions, where Japanese research vessels or small groups of trawlers were sent into distant, unexploited waters. In the 1920s, such exploratory missions laid the foundation for large-scale Japanese trawling in the South China Sea, off the Russian coast of the Sea of Japan, and in the Bering Sea; in the 1930s, expeditions worked the Mexican Pacific coast, the Bay of Bengal, and even the Arabian Sea. Had World War II not intervened, large-scale Japanese fishing operations would likely have followed in many of these promising offshore locations.[23]

Japan's rapid rise as the world's foremost fishing nation was also supported by a range of educational and research institutions established by the central government and local authorities. The Imperial Fisheries Institute, which provided overall coordination for fisheries research and training initiatives, was founded by private interests in 1889 and brought under the Ministry of Agriculture and Commerce in 1897. A number of prefectural fisheries experiment stations and local fisheries high schools were established starting in the 1890s. Marine science degree programs were offered at the imperial universities in Hokkaidō (from 1872) and Tokyo (from 1910). Numerous public institutions operated their own high-seas training, and research vessels and the government oversaw a far-flung network of experiment stations, marine laboratories, and oceanographic observation posts around the coasts of Japan and throughout the empire.[24] This infrastructure for scientific research and professional training in marine fisheries was internationally renowned; in fact, the School of Fisheries at the University of Washington was modeled on the Imperial Fisheries Institute.[25] It was also specifically focused, at all levels, on leveraging scientific knowledge to enhance the productivity and profitability of Japan's fisheries. Just as prewar Japanese fishermen were mobilized for the needs of the state and the demands of empire, oceanography and fisheries education was also mobilized in the support of ever-greater catches and ever-growing export returns.

Surveying the progress of Japanese fisheries in the early twentieth century, Nomura Masuzō, president of the Imperial Fisheries Association, concluded, "As a result of aggressive measures taken by the Government to promote the industry, coupled with application of scientific provisions, both the scale and technical skill of the industry have made marked progress in official and public undertakings."[26] While rapid growth in fisheries may have been good for Japan's food supply, employment rate, and (above all) its balance of payments, the range and efficiency of Japanese fishermen put significant pressures on marine resources across the Pacific Ocean and its marginal seas. Overexploitation of coastal fishing grounds was almost certainly an important factor in Japan's assertive expansion into offshore fisheries in the first two decades of the twentieth century. According to one Japanese fisheries expert, "At present we are often told of the devastation of the coastal fisheries or of the scarcity of fish along the coast. We are unable to prove the patent facts of these reports, due to the imperfection of statistics, but the decrease of fish and the deterioration of their breeding rate are felt everywhere."[27] As Japan then dispatched its state-of-the-art trawlers and high-capacity factory ships into offshore waters, similar reports of overexploitation began to surface. From the 1920s, the yields of key demersal species (such as bream and croaker)

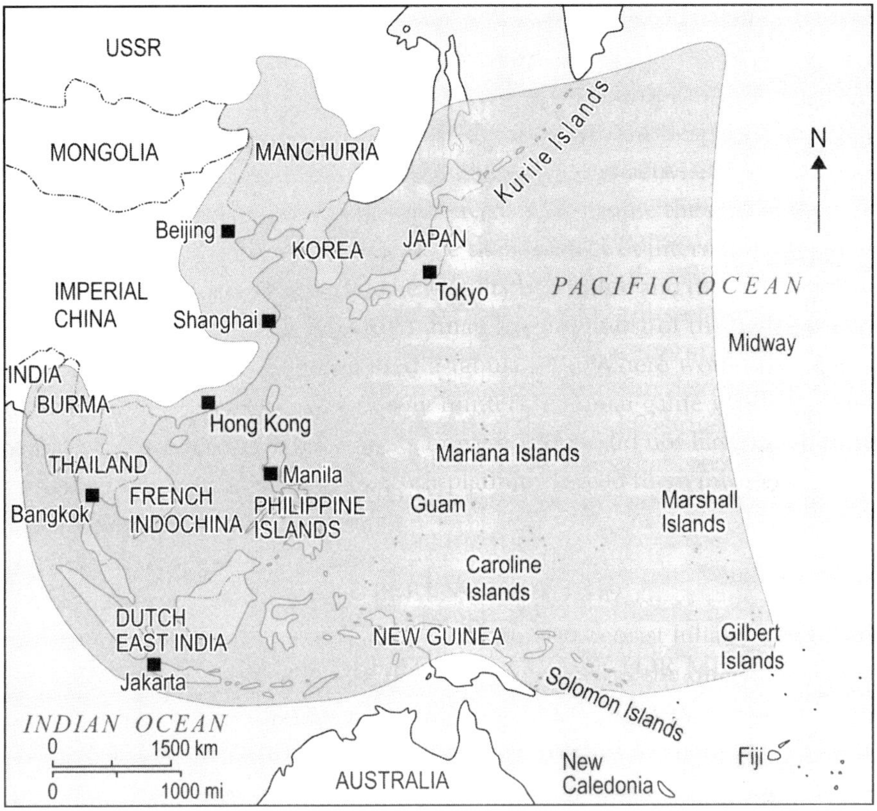

Figure 14.2. Japan's marine realm at its greatest extent.

had begun to decline in the East China and Yellow Seas, causing Japanese fishermen to look farther afield for productive trawling grounds; a precipitous drop in catches of sardines and herring in the early 1930s led Japanese boats to sink their nets in new waters and in search of new species; and Japan's crab, salmon, and tuna fishermen found themselves regularly seeking increasingly distant resource frontiers to maintain peak production.[28]

The seemingly relentless expansion of Japanese fisheries and the apparent lack of interest in conservation among Japan's fishermen, officials, and marine scientists earned Japan an international reputation as a lawless predator and ruthless exploiter of the seas. Russians were long highly sensitive to Japan's marine intrusions, as were the Chinese after Japanese trawlers displaced local fishermen along the East China Sea coast.[29] In the United States, the threat from Japan first gained widespread attention in 1906, when a group of Japanese poachers raided the fur seal rookery on St. Paul Island in the Bering

Sea.[30] In 1937, when Japan sent its first exploratory expedition to harvest salmon in Bristol Bay in southwest Alaska, American observers condemned the intrusion as "a damaging invasion" and an "attack."[31] A correspondent for the *New York Times* wrote, "Just as the Nipponese military juggernaut rolls across China without regard for the amenities of civilization, so do these fishing vessels from Tokyo completely overlook conservation rules and principles in their quest for the finny wealth of the ocean."[32]

The efficiency and zeal of Japan's modern fisheries also characterized its whaling activities. Japanese coastal whaling—practiced for centuries using relatively unsophisticated technology and focused on harvesting gray, right, fin, humpback, and Baird's beaked whales—began modernizing in the late nineteenth century. Diminishing catches forced the Japanese to use larger vessels, adopt more advanced European technology, and expand whaling farther out to sea.[33] Japan's territorial expansion before World War II also benefited whaling: Japanese whalers had access to waters from the Bonin Islands and Taiwan in the south to the Kuril Islands in the north, meaning whaling could become a year-round activity, as the vessels sailed the southern seas in the winter and the northern ones in the summer.[34] Japanese whalers did not venture into the rich waters of the Antarctic until 1934, three decades after the Norwegians, but they made rapid progress there as well. This assertive expansion was very much part of Japan's drive for pelagic empire and was focused, just like crabbing in the Bering Sea or tuna fishing in the South Pacific, on landing rich catches and delivering strong export returns. By 1938, six Japanese whaling fleets were active off Antarctica, their oil destined mainly for European margarine and soap factories. Japan's share of the world's whale catch grew from 1 percent in 1930 to 12 percent in 1938. By then, Japan was already the third-largest whaling country, after England and Norway.[35] Whale meat production, which unlike the oil was consumed in the home market, rose to 45,000 tons in 1939.[36] Significantly, Japan did not join the first international whaling convention in 1931, with its cautious steps toward regulation, although Tokyo did later did impose some modest restrictions on the Japanese harvest.[37]

On the eve of Pearl Harbor, Japan commanded a pelagic empire as well as a terrestrial one. Japan had twice the fisheries production of any other nation on earth, and its 1.4 million fishermen, processors, and aquaculturalists pulled heavy catches from the Bering Sea to the Antarctic Ocean, from the Bay of Bengal to the continental shelf off the coast of Argentina. Fisheries were more than just a "vehicle for empire" or a "platform," as Carmel Finley has described them: the oceans were a central component in Japan's imperial vision, and Japan exploited them with all the thoroughness that it exploited the soybean fields of Manchuria, the mines of the Korean peninsula, or the forests of Taiwan.[38] Ironically, though, just as Japan's terrestrial empire would reach its height in the first months of the Pacific War, its empire of fisheries would begin to retrench and fade. Deep-sea and offshore operations, including pelagic whaling, ground to a halt as most of Japan's large fishing vessels were drafted into the war effort. Many of Japan's experienced seamen were also called into naval service. Even coastal fisheries contracted significantly as endemic shortages of cotton yarn, Manila hemp, and, most critically, petroleum depressed wartime production. In any case, as Japan embarked on an autonomous course in 1941, the primary impetus behind the pelagic empire—generating trade surpluses to fuel Japanese expansion—lost all momentum. The wartime oceans would have to be won with carriers and cruisers rather than trawlers and factory ships.

In the end, although the Japanese pelagic empire collapsed during the Pacific War as dramatically as the terrestrial imperium, it left important legacies for the fisheries that would regenerate after Japan's defeat. Among these were

1. A history of exploiting the "freedom of the seas" (mare liberum) for Japan's economic benefit and venturing far from home waters to tap the riches of the world's oceans.
2. A worldwide reputation for aggressive, predatory fishing practices and a lack of respect for international norms.
3. A legacy of distant-water fisheries that contributed to domestic food needs but were often oriented primarily toward export markets.
4. A pervasive government presence in all aspects of the fishing industry.
5. An infrastructure of research and educational institutions designed to support the commercial advance of Japanese fisheries.
6. An economic model based on highly mechanized, large-scale deep-sea fishing operations using low-wage crews to assure competitiveness in international markets.
7. A pattern of continuously expanding fishing grounds and enhancing fishing technology as existing waters are overfished and depleted.

BUILDING AN INFORMAL FISHERIES EMPIRE, 1945–77

At the time of Japan's defeat in 1945, the nation's once-proud fisheries were at lowest ebb. Forty percent of Japan's fishing fleet (including virtually all the large, modern vessels) was destroyed during the war; nets, lines, and other fishing gear were ragged from the years of deprivation; and many of Japan's most productive offshore and deep-sea fishing grounds were lost with the empire. Japan's American occupiers, however, quickly realized the critical importance of fisheries, and not just to the pressing problems of a hungry Japanese populace and the hundreds of thousands of fishermen who needed work. Acknowledging the prominence of the fishing industry in the prewar economy and Japan's international trade, American officials in Tokyo prioritized the rapid rebuilding of Japanese fisheries as essential to restoring the nation's economic self-sufficiency and jump-starting postwar exports.[39] Within weeks of the defeat, the occupation authorized the return of Japanese fishermen to coastal waters and, in 1946, the resumption of Japanese whaling operations in the Antarctic Ocean.[40] At the same time, scarce raw materials (steel for ships, diesel fuel for engines, cotton for nets) were targeted to the rebuilding effort, financial resources flowed from the Japanese government to fishing interests, and new fisheries technology from abroad was introduced. The resulting resurgence was quick and strong: by 1947, Japan had more large fishing vessels than at the prewar peak, and the volume of fisheries production exceeded two-thirds of what had been achieved in the 1930s.[41]

Across the board, the immediate postwar recovery of Japanese fisheries was only too successful. Excess capacity was soon apparent, with legions of fishermen demobilized or repatriated from the empire; naval shipyards eager to churn out modern, efficient fishing boats; and coastal fish stocks rapidly showing signs of stress from overfishing. The occupation authorities responded much as the prewar Japanese state had, by facilitating the exploitation of ever-expanding fishing grounds by Japan's growing fleet.[42] Although many of America's wartime allies—the Soviet Union, China, Australia—vehemently opposed the return of Japanese fishermen to their coastlines, the occupation progressively loosened restrictions on the grounds open to Japanese ships. As early as 1946, Japanese fishermen had access to waters where 80 percent of the prewar catch had been harvested, and by 1950, Japan's fishing vessels could range across much of the western Pacific.[43] Bowing to political pressure from fisheries' interests in the Pacific Northwest, who feared a reprise of Japan's 1937 "invasion" of Bristol Bay, American authorities did extract some voluntary concessions from Japan. In the International Convention for the High Seas Fisheries of the North Pacific Ocean, negotiated in the final months before Japan regained its independence in 1952, Japan agreed to an "abstention" from harvesting salmon, halibut, and herring from a broad swath of ocean worked by American and Canadian fishermen. At the same time, however, the United States supported Japan's efforts to secure "freedom of the seas" for its fishing fleet elsewhere in the Pacific and around the world.[44]

Even before the end of the occupation, Japanese fishermen had surpassed the largest annual catches prior to World War II, and Japan could rightly claim once again to be the world's top fishing nation. Between 1950 and 1960, Japanese fisheries production doubled; by 1973, it had doubled again.[45] Japan would lead the world in whaling by 1960, its catch reaching its historic peak of about 226,000 tons in 1962.[46] And although fishing and whaling are seldom credited with being a major component in Japan's "miraculous" postwar growth, they very much contributed—just like the steel, automobile, and electronics industries—to Japan's unprecedented and sustained economic expansion. Some of the soaring catch statistics can be credited to advances in technology: stronger nets made from synthetic materials, fish finders, powerful fish lamps, radar, improved refrigeration facilities, and new mother ships that dwarfed prewar floating canneries in size and sophistication.[47] But the rapid resurgence of Japanese fisheries after the war was a result, in large part, of the legacies of the pelagic empire of the 1930s. Using a familiar repertoire of practices and strategies developed to support Japan's imperial ambitions

Figure 14.3. A captured Sowerby's beaked whale. Source: H. Wolf in Alfred Brehm, *Djurens liv. Däggdjuren, andra bandet* (Stockholm: Aktiebolaget Familjeboken, 1921).

and its march to war, the Japanese fishing industry would be mobilized once again in the wake of defeat and occupation as an important element in the new national drive for wealth and economic power.

Scarcely a decade after World War II, Japanese fishermen had returned to most of the grounds they had worked regularly in the 1930s and were eagerly seeking out new waters even farther afield.[48] Tuna fishing began in the Indian Ocean in 1952, and by 1957, Japanese vessels were chasing yellowfin and albacore in the Atlantic. Bases for Japanese ships were built throughout the South Pacific, and Japanese fishermen worked intensively down the coasts of New Zealand and into the Tasman Sea. Joint ventures were established in Guatemala and Chile as well as in Mozambique, Madagascar, and Ghana. Japanese trawlers plied the Persian Gulf, and Japanese fishmeal factory ships spent summers in the Bering Sea and winters off the coast of Angola. By the mid-1960s, over two hundred Japanese fishing vessels were based in foreign ports, including Tahiti, Port-au-Prince in Haiti, and Recife in Brazil. Japanese tuna boats were active throughout the Mediterranean as well as the Caribbean, and from 1962, Japanese fishermen harvested cod off the west coast of Greenland and along the Grand Banks of Newfoundland. As one British observer wrote of Japan's fisheries, with a clear sense of awe, "Since 1955 [their] geographic expansion has been almost breath-taking—one of the truly great and significant happenings on the world's food front."[49]

Fisheries products were even more important to Japan's export economy in the immediate postwar decades than they had been in the 1930s. From 1949, when Japanese canned tuna was allowed to reenter the American market, Japan soon rose to become the world's leading exporter of fisheries products. This rapid growth in both volume and value was driven by many factors, including the liberalization of the canned food trade in Britain, rising global demand for frozen fish, and a healthy international appetite for seafood products after World War II. Japanese canned salmon had a strong market in the United Kingdom, canned tuna in brine and crab were top sellers in the United States, and tinned sardines and saury were exported mainly to developing nations. Japanese dried fish found ready buyers throughout Asia, fish liver oil was sold to America and Europe, and US consumers increasingly sought out frozen halibut, swordfish, and oysters from Japanese fishermen. Frozen tuna from Japanese ships supplied not just the canneries of California, but also packers in American Samoa and from the newly opened fisheries of the Atlantic, in Puerto Rico, France, Italy, Portugal, and Spain.[50] By

Figure 14.4. A captured baleen whale on the beach. The importance of whale meat increased in Japan in the postwar years. Source: J. E. Rosberg and V. Tolvanen, eds., *Maapallo. Maantieteellinen tietokirja. Toinen osa* (Helsinki: Kustannusosakeyhtiö Otava, 1923).

the mid-1950s, for example, canned, frozen, and dried fish made up nearly 6 percent of Japan's total exports, far exceeding the prewar peak of 3.9 percent in 1936. Among all Japanese export goods, fish ranked a strong fourth after ships, cotton textiles, and artificial fibers. Fisheries products were the United States' second-largest import from Japan by value, behind only silk. And by 1960, Japanese fisheries exports totaled almost $200 million a year.[51]

As had been the case since the dawn of modern fisheries in Japan, the state played a central role in the postwar expansion of the fishing industry. The experiences of war and occupation had little effect on the established strategies of Japanese industrial policy, on the prerogatives of central government bureaucrats, or on the close ties linking Japanese fisheries interests to the state. Government licensing of fishing operations after the war, as before it, served to regulate competition and minimize conflicts with other nations; state financial institutions supported targeted investments in new technology and larger, more efficient vessels; the new Fisheries Agency, established in 1948, subsidized insurance for fishing interests and underwrote the development of infrastructure like harbors and shore facilities.[52] Educational and research activities funded by the central and local governments continued to provide critical services to Japanese fishermen. Government research ships in the late 1960s were one example of the economic viability of the new technology of purse seining for tuna. A decade earlier, exploratory work by vessels from the Fisheries Agency and the Kanagawa Prefectural Fisheries Experiment Station had encouraged Japanese fishing interests to enter the Atlantic. Much of the geographical expansion of postwar Japanese fisheries to distant waters around the globe was first charted by government research and training ships or by private vessels subsidized by the Japanese state.[53]

Such prospecting for untapped stocks of marine resources became increasingly challenging—and ever more important to the growth of Japanese fisheries—in the years after World War II. "The search for new grounds is unceasing," one observer noted, as waters intensively exploited by Japan's fishermen for decades showed unmistakable signs of depletion and distress.[54] Although fish stocks around Japan seemed to regenerate during the war, Japan's rebuilt and technologically upgraded fleet soon faced stagnant or declining yields for some species in coastal waters and the trawling grounds of the East China Sea.[55] And while the industry managed to squeeze increasing harvests from the coastal and offshore fisheries (the catches between 1960 and 1975 increased by 43 and 77 percent, respectively), the biggest growth (126 percent over those fifteen years) came from distant waters.[56] By the mid-1960s, however, catch productivity was even falling in the king crab grounds of Bristol Bay and in the tuna fisheries of the southwest Pacific and South Atlantic. As had been the case during the 1930s, plummeting postwar yields of some commercially important fish (such as herring, sardines, and yellowfin sole in the Bering Sea) led Japanese fishermen to focus on different species (including squid, mackerel, and especially pollock) that seemed, for the time being at least, to be relatively plentiful.[57] But even as Japanese research ships and trawlers continued to hunt for underutilized fish stocks in the remotest corners of the world's seas, it was becoming apparent by the 1970s that Japan's time-tested strategy of exhausting one fishery and simply moving on to another may have reached its inevitable limits.

For all the achievements of the Japanese fishing industry in the three decades after World War II—from unprecedented catches to gleaming new fleets to a dominant presence in global markets—Japan's fisheries also faced a number of new challenges in the postwar world. First, growing domestic demand for aquatic products put pressure on Japanese fishermen and threatened the industry's long-standing prioritization of exports. Japan's population increased steadily after the war (from seventy-two million in 1945 to over one hundred million in 1970), as did the nation's prosperity. Per capita daily consumption of fish ballooned from only 52.9 grams in 1952 to 145 grams in 1965 and 191 grams in 1975.[58] Japanese consumers wanted not just more fish to eat, but higher-quality fish as well: urbanites lost their taste for species like saury, rural residents grew unsatisfied with exclusively dried and cured fish, and items once considered luxuries (like canned crab) became everyday products. In the 1950s, Japanese fishing interests found more of their canned pack going to domestic buyers than ever before; by the 1960s, the demand for fresh fish soared, and with the invention of flash-freezing technology, the Japanese tuna fleet increasingly shifted from supplying canneries to producing *sashimi*-grade fish for the home market. In the 1960s, for the first time ever, significant imports of marine goods began to enter Japan, and by the mid-1970s, imports exceeded exports by volume.[59]

Second, the Japanese fishing industry, and especially its offshore and deep-sea operations, faced mounting challenges to its business model. The time-tested strategy of high-technology, large-scale, low-wage fisheries served Japan well, at least through the 1960s. In the decades after World War II, however, Japanese fishermen encountered more acute competition in the world's fishing grounds, both from nations with new, highly mechanized fleets (like the Soviet Union) and from developing countries with labor costs far lower than Japan's. Even with extensive subsidization by the Japanese state, Japan's fishermen suffered from rising labor expenses, a shortage of workers (as fewer young people entered the industry), and skyrocketing fuel costs (especially as Japanese vessels had to venture ever farther afield to find their catches).[60]

Third, the postwar decades brought a progressive narrowing of mare liberum, the freedom of the seas that had made possible the rise of Japan's pelagic empire in the early twentieth century and allowed for the global revival of Japanese fisheries after World War II. Prior to the war, of course, Japanese fishermen faced only minimal restrictions; after 1945, however, fisheries management agreements proliferated. In addition to the 1952 Fisheries Convention, by 1970, fourteen other international agreements regulated fishermen in the North Pacific alone.[61] This new regime was often less rigorous and comprehensive than aggrieved fishing interests claimed: restrictions on halibut and herring were soon lifted from the 1952 convention, and scholars have observed that 90 percent of the catch in the North Pacific still remained outside international management.[62] Far more threatening to Japanese fishing interests over the long run was the growing oceanic enclosure movement, which accelerated after World War II. Although the United States and major fishing powers like Japan sought to defend free access to the world's seas, many nations (especially in Latin America and Asia) that felt exploited by foreign distant-water fleets increasingly sought to extend their jurisdiction over neighboring oceans.

Fourth, despite Japan's active postwar participation in international fisheries management agreements, Japanese fishermen continued to be regarded as ruthless and aggressive predators of marine resources. Japan's widespread, efficient, and highly visible fishing activities were blamed by many observers for the perceived depletion in critical fish stocks around the globe. The nascent environmental movement also came to regard Japan as a dangerous outlaw devastating the world's oceans through overfishing, the indiscriminate use of driftnets on the high seas, and continued whaling operations.[63]

After 1945, Japanese fisheries no longer served an explicitly imperial project, as they had in the years leading up to Pearl Harbor. But because of the scale, scope, and clout of the postwar fishing industry in global waters and global markets, the three decades after World War II can be seen as a new period of the fisheries empire—this time an informal one—for Japan. The new informal imperium looked much like the pelagic empire of the past—in its practices and strategies, its impact on the marine environment, and its image in the larger world—and played an important part in Japan's peaceful postwar rise to economic superpower status. By the 1970s, however, the signs of an impending and profound transition in Japanese fisheries had become clear. Even as Japan continued to lead the world in fishing, Japanese fishermen were pressing the limits of their imperial reach, increasingly constrained by tangles of international agreements and creeping territorial waters, hampered by the decline of competitive advantage, and frustrated by the ecological reality of finite fishing grounds and finite fish stocks. The legacies of the pelagic empire had buoyed Japanese fisheries through a time of unparalleled expansion, but a sea change was about to transform Japan's fishing dominance.

AN EMPIRE TRANSFORMED, 1977 TO PRESENT

With the opening of the third UN Conference on the Law of the Sea (UNCLOS) in 1973, the institutionalization of the global enclosure movement began in earnest. Coastal states had long been chipping away at the freedom of the seas, seeking to gain control over the marine and mineral resources of expanded territorial waters: by 1967, only twenty-five countries held to the traditional three-mile limit, while sixty six claimed twelve-mile territorial seas and eight had established two-hundred-mile limits. In 1976, the United States abandoned its long commitment to mare liberum and declared a two-hundred-mile fishery conservation zone; the Soviet Union and Canada soon followed suit. Japan felt it had little choice but to join the trend of the times, not least so that it would be in a stronger bargaining position with other fishing nations: on July 1, 1977, Japan extended its territorial seas to twelve miles and its fishing area to two hundred miles. These new "lines in the water" were formalized as Exclusive Economic Zones (EEZs) in the Law of the Sea Convention drafted in 1982.[64]

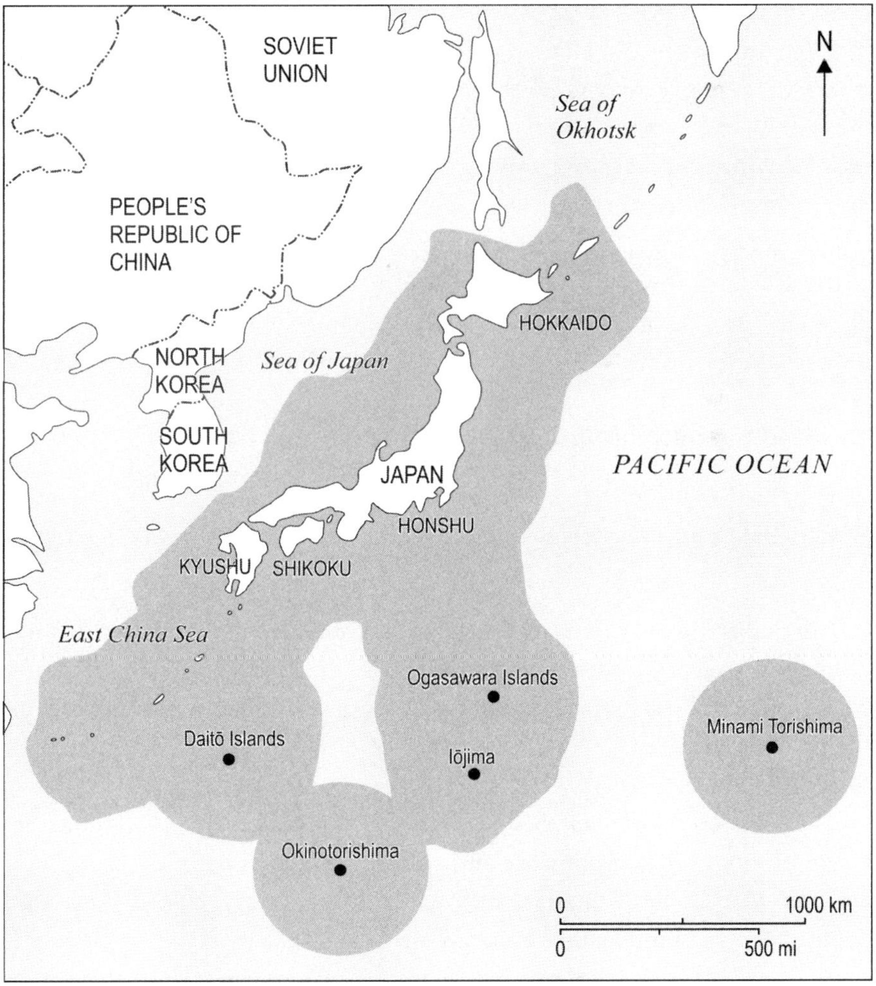

Figure 14.5. Japan's Exclusive Economic Zone (EEZ).

The enclosure of the world's oceans had profound effects on Japanese fisheries. Prior to the 1970s, Japan's tuna fleet regularly operated in the coastal waters of more than fifty countries. In 1975, almost half of Japan's total catch came from within two hundred miles of the shores of other nations, and most of that was from what would soon become the EEZs of the United States and the Soviet Union.[65] The Japanese fishing industry and the government in Tokyo responded assertively to the extended jurisdictions. By the mid-1980s, Japan and private fisheries interests had concluded agreements with eighteen states allowing Japanese fishermen access (for a fee) to tuna in their coastal seas.[66] Japan had a long history of joint ventures with fishing enterprises abroad, stretching back to the prewar period, but the establishment of such relationships accelerated in the 1970s. These arrangements were increasingly made in developing nations (with Japan supplying the bulk of the capital and technical expertise, the partner providing labor and the all-important access to restricted waters) and, compared with earlier joint ventures, were focused more on supplying Japan's domestic seafood market than on producing export goods.[67] The proliferation of EEZs also pushed Japanese fishermen to look elsewhere for their harvests: in the wake of UNCLOS, Japan's coastal fisheries enjoyed a brief boom (as local operators reaped the benefits of Japan's extended territorial claims), and Japanese distant-water fleets focused more on the high seas, which remained beyond the jurisdiction of individual states.[68]

With all the changes buffeting Japanese fisheries, the role of the Japanese state in guiding the industry's development gradually changed. After decades of orchestrating the expansion of fisheries as an economic engine for Japan, the state perceptibly shifted its emphasis from adjustment and rationalization to managing a soft landing for an industry that was beginning to decline. Tokyo was generous in providing subsidies to the struggling (but still politically powerful) fisheries, in facilitating strategic reductions in the size of the fleet, and in assisting financially and technologically with the transition to more fuel-efficient and less labor-intensive vessels and practices.[69] The Japanese government also used its large foreign aid budget to advance the interests of Japanese fishermen and ensure adequate seafood supplies for Japanese home consumption.[70] At the same time, Japan's large fishing companies, long the bureaucrats' closest partners in industrial policy, smoothly accommodated to the new realities, thanks in no small part to state support. Over the 1970s, sprawling enterprises like Nippon Suisan moved from direct harvesting operations to an emphasis on the import trade, distribution, the manufacture of high-value-added seafood products, and a diversified range of other activities (including industrial chicken farming).[71]

Global environmental anxieties also took their toll on Japanese fisheries. One target of activists and governments concerned with the conservation of marine resources was Japan's extensive use of large driftnets, often called "curtains of death" or "maritime strip mining" for their ruthless efficiency in ocean harvests. Japan's salmon fishermen began using large driftnets on the high seas in the 1960s in response to the restrictions of the 1952 Fisheries Convention; they were subsequently deployed in the North Pacific to catch flying squid starting in 1979 and in the South Pacific albacore fishery from 1982, largely as a result of the loss of fishing grounds from the establishment of two-hundred-mile EEZs.[72] Strong opposition to driftnets from the United States and the countries of Oceania put substantial pressure on Japan. Finally, Tokyo bowed to international opinion and announced a voluntary ban on large driftnet use, effective at the end of 1992.[73]

Whaling emerged as another major focus of environmental criticism and political contention. The world's first whaling treaty, the Convention for the Regulation of Whaling, was concluded in 1931 but remained largely unimportant because five major whaling states, including Japan, refused to accede to it. It was not until 1951 that Japan became a party to this treaty's successor, the International Convention for the Regulation of Whaling, which established the still-active International Whaling Commission (IWC).[74] Japan has participated in the activities of the IWC, although usually as one of the countries registering objections to its decisions.[75] As early as 1954, Japan resisted a prohibition on blue whale harvests in the North Pacific, which was among the IWC's first substantial conservation initiatives. In 1981, Japan opposed an IWC decision to ban the "cold grenade" harpoon as a means of killing minke whales. Japan also objected to the historic 1982 IWC moratorium on commercial whaling, and to the present day continues whaling for "the purposes of scientific research" despite ongoing criticism from environmental groups. The convention allows governments to grant special permits authorizing their nationals to kill whales for scientific purposes, and this right has been actively used by the remaining active whaling nations, including Japan. Between 1985 and 2011, Japanese vessels caught a total of 19,182 whales (mostly minke whales).[76] Although the Japanese harvest has declined significantly from its postwar peak, the complete abandonment of whaling demanded by the world's nature conservationists is unlikely in the foreseeable future.

The statistics tell the story of the decline of Japan's postwar fisheries empire. The catch from Japanese high-seas operations fell by more than half between 1980 and 1993. In the critical tuna fishery, Japanese production peaked in both volume and value in 1984. Strong harvests of Japanese pilchard, a highly cyclical fishery off the coast of Hokkaidō, buoyed the industry from the late 1970s to the late 1980s. But 1988 proved to be the last year that Japan could claim to be the world's greatest fishing nation; by 1989, it was surpassed in volume by both the People's Republic of China and the Soviet Union.[77] The deterioration in Japan's international trade in marine products was even more stark. Japanese fishery exports as a percentage of the world total peaked (at almost 10 percent) in the mid-1970s and declined steadily thereafter. US fisheries exports, which benefited greatly from the extension of American territorial waters, exceeded Japan's total in 1978. After 1988, Japanese exports declined steadily in volume and, after a brief plateau, in value as well. Meanwhile, Japanese imports of seafood soared: by 1989, the Japanese market consumed 28 percent of world imports, up from only 8 percent in 1970. Over the same period of time, Japan fell from a situation of essential self-sufficiency in fisheries products to being dependent on foreign sources for 22 percent of domestic seafood consumption.[78] At the turn of a new millennium, the days of a Japanese fishing industry that dominated global waters, catches, and trade seemed increasingly like ancient history.

Yet ironically, just as the Japanese fisheries empire seemed in eclipse, Japan suddenly found itself an oceanic superpower, the master of a sprawling Exclusive Economic Zone:

> With the coming into effect of the Law of the Sea Convention, Japan has come to administer an offshore area spanning 4.47 million square kilometers—12 times more than its total landmass of 380,000 km^2 and the sixth largest in the world. Thanks to its possession of scattered Pacific Ocean islands, Japan's total area (land and sea combined) now covers 4.85 million km^2, which is among the world's top 10. Among the deep-sea areas under Japan's administration are the Japan Trench, Izu-Ogasawara Trench, Nankai Trough, and Ryūkyū Trench, and in terms of cubic volume, Japan's EEZ is estimated to be the fourth largest in the world.[79]

So, it seems, even with the world's oceans progressively closed off to Japanese fishermen, the legacies of past fishing glories rapidly fading, and the biological limits of the sea apparently reached, Japan's once-cherished ambition of a marine empire may yet be realized.

Notes

1 See the Environmental Performance Index, accessed October 30, 2016, http://www.epi.yale.edu/epi/country-rankings.

2 Conrad Totman, *A History of Japan*, 2nd ed. (Oxford: Blackwell, 2005), 515.

3 This section is based closely on William Tsutsui, "The Pelagic Empire: Reconsidering Japanese Expansion," in *Japan at Nature's Edge*, ed. Brett Walker, Julia Thomas, and Ian Miller (Honolulu: University of Hawai'i Press, 2013); and Timo Vuorisalo and Johanna Lahdenperä, "Japanin valloituspolitiikka, luonnonvarat ja ympäristö," in *Sodan ekologia: Nykyaikaisen sodankäynnin ympäristöhistoriaa*, ed. Simo Laakkonen and Timo Vuorisalo (Helsinki: Suomalaisen Kirjallisuuden Seura, 2007), 403–32.

4 John Stephan, *Sakhalin: A History* (Oxford: Clarendon Press, 1971), 76.

5 Supreme Commander for the Allied Powers, *Japanese Offshore Trawling*, Natural Resources Section Report 138 (Tokyo, 1950), 7.

6 Supreme Commander for the Allied Powers, *The Japanese Tuna Fisheries*, Natural Resources Section Report 104 (Tokyo, 1948), 6. On technological change in the Japanese fisheries, see Ninohei Tokuo, *Nihon gyogyō kindaishi* (Tokyo: Heibonsha, 1999).

7 Supreme Commander for the Allied Powers, *Japanese Fisheries Production 1908–46 (A Statistical Report)*, Natural Resources Section Report 95 (Tokyo, 1947), 8, 31.

8 Alfred Stead, ed., *Japan by the Japanese: A Survey by Its Highest Authorities* (New York: Dodd Mead, 1904), 428–29.

9 Association for Liberty of Trading, *The Economic Development of the Japanese Fishing Industry* (Tokyo, 1933), 5.

10 Ibid., 9.

11 Even two of the most knowledgeable experts on Japan's modern fisheries, Carmel Finley and Roger Smith, mistakenly focus on food security as the major impetus behind the rapid growth of the Japanese industry in the mid-twentieth century. See Carmel Finley, *All the Fish in the Sea: Maximum Sustainable Yield and the Failure of Fisheries Management* (Chicago: University of Chicago Press, 2011); Roger Smith, "Japan's High Seas Fisheries in the North Pacific Ocean: Food Security and Foreign Policy," in *Japan at the Millennium: Joining Past and Future*, ed. David Edgington (Vancouver: University of British Columbia Press, 2003); Roger Smith, "Food Security and International Fisheries Policy in Japan's Postwar Planning," *Social Sciences Japan Journal* 11, no. 2 (October 2008).

12 Shozui Sen'ichi, "Our Fishing Industry," *Contemporary Japan* 9, no. 6 (June 1940): 701; John Stephan, *The Kuril Islands* (Oxford: Oxford University Press, 1974), 122; Association for Liberty of Trading, *Economic Development of the Japanese Fishing Industry*, 13.

13 Japan Times and Mail, *Japan's Fisheries Industry 1939* (Tokyo, 1939), 74–75; Haraguchi Takejiro, "Japan's Contributions in the South Seas," *Contemporary Japan* 8, no. 2 (April 1939): 261–62; Mark Peattie, "*Nanshin*: The Southern Advance, 1931–1942, as a Prelude to the Japanese Occupation of Southeast Asia," in *The Japanese Wartime Empire, 1931–1945*, ed. Peter Duus, Ramon Myers, and Mark Peattie (Princeton, NJ: Princeton University Press, 1996), 201; Shimizu Hiroshi and Hirakawa Hitoshi, *Japan and Singapore in the World Economy: Japan's Economic Advance into Singapore, 1870–1965* (London: Routledge, 1999), esp. chap. 4.

14 Japan Times and Mail, *Japan's Fisheries Industry 1939*, 156–59.

15 Shozui Sen'ichi, "Our Fishing Industry," *Contemporary Japan* 9, no. 6 (June 1940): 701.

16 Quoted in Micah Muscolino, "Marine Environmental Encounters between Japan and China, 1900–1960," in Walker et al., *Japan at Nature's Edge*.

17 Japan Times and Mail, *Japan's Fisheries Industry 1939*, 144; Stephan, *Kuril Islands*, 130.

18 Imperial Fisheries Bureau, *Japan, Special Catalogue, Fisheries* (Tokyo, 1915), 101–4; Asia Kyokai, *Japanese Fisheries: Their Development and Present Status* (Tokyo, 1957), 2; Hiroshi and Hitoshi, *Japan and Singapore in the World Economy*, 96–97.

19 Muscolino, "Marine Environmental Encounters between Japan and China," 3.

20 Chalmers Johnson, *MITI and the Japanese Miracle: The Growth of Industrial Policy, 1925–1975* (Stanford, CA: Stanford University Press, 1982).

21 Fujita Iwao, quoted in Japan Times and Mail, *Japan's Fisheries Industry 1939*, 10.

22 Supreme Commander for the Allied Powers, *Japanese Offshore Trawling*, Natural Resources Section Report 138 (Tokyo, 1950), 26–29; Georg Borgstrom, *Japan's World Success in Fishing* (London: Fishing News Books, 1964), 23, 247; Finley, *All the Fish in the Sea*, 22–31.

23 Borgstrom, *Japan's World Success in Fishing*, 252–55.

24 Uda Michitaka, *The Fisheries of Japan* (Nanaimo, BC: Fisheries Research Board of Canada, 1959), 8; Imperial Fisheries Bureau, *Japan, Special Catalogue, Fisheries*, 105–39; Japan Times and Mail, *Japan's Fisheries Industry*, 16–17.

25 Finley, *All the Fish in the Sea*, 77.

26 Japan Times and Mail, *Japan's Fisheries Industry 1939*, 6.

27 Association for Liberty of Trading, *Economic Development of the Japanese Fishing Industry*, 21.

28 Supreme Commander for the Allied Powers, *Japanese Fisheries Production 1908–46*, 32–35; Borgstrom, *Japan's World Success in Fishing*, 44–45; Muscolino, "Marine Environmental Encounters between Japan and China," 12–13.

29 See Micah Muscolino, "The Yellow Croaker War: Fishery Disputes between China and Japan, 1925–1935," *Environmental History* 13, no. 2 (April 2008): 306–24.

30 Finley, *All the Fish in the Sea*, 30.

31 Joseph Bingham (1938), quoted in Julian Minghi, "The Conflict of Salmon Fishing Policies in the North Pacific," *Pacific Viewpoint* 2, no. 1 (March 1961): 77; Anthony Dimond, quoted in Larry Leonard, *International Regulation of Fisheries* (Washington, DC: Carnegie Endowment for International Peace, 1944), 134.

32 Richard Neuberger, quoted in Finley, *All the Fish in the Sea*, 34.

33 "Small-Type Coastal Whaling in Japan," accessed April 19, 2005, http://luna.pos.to/whale/gen_coast.html.

34 Ibid.

35 William Tsutsui, "Landscapes in the Dark Valley: Toward an Environmental History of Wartime Japan," *Environmental History* 8, no. 2 (April 2003): 298.

36 "A History of the Traditional Diet: Japanese and the Whale," accessed September 27, 2006, http:/luna.pos.to/whale/jwa_trad.html.

37 Simon Lyster, *International Wildlife Law: An Analysis of International Treaties Concerned with the Conservation of Wildlife* (Cambridge: Cambridge University Press, 1993), 18.

38 Finley, *All the Fish in the Sea*, 3, 20.

39 See, for example, Henry Esterly, "Overseas Fisheries and International Politics in the Occupation of Japan, 1945–1952," in *The Occupation of Japan: Economic Policy and Reform*, ed. Lawrence Redford (Norfolk: MacArthur Memorial, 1980). Food security concerns seem to have been more important than export prioritization or broader economic objectives in the postwar whaling industry. For the Japanese, whale meat played a significant role as a protein source after World War II. In the 1930s, the proportion of whale meat in total Japanese meat consumption had been about 15 percent. In 1947, however, as much as 47 percent of Japanese animal protein came from whale meat, and in 1964 this figure was still 23 percent. From 1947 to the mid-1950s, whale was the only source of meat for the country's schools. Although meals of pork and chicken gradually grew more common, whale remained the

single most important source of meat for school lunches until the mid-1970s. Even in 1973, about 15 percent of Japan's overall whale meat production went to school cafeterias. "Whale Meat in Japan," accessed May 31, 2006, http://www.upstarts.net.au/site/non_commercial/whales/_conservation/whalecon_main2.html; "Dining Out on a Guilt Trip," BBC, accessed April 19, 2005, http://news.bbc.co.uk/2/hi/sci/tech/2077818.stm; "Small-Type Coastal Whaling," accessed April 19, 2005, http://luna.pos.to/whale/gen_coast.html; "A History of the Traditional Diet."

40 Tsutsui, "Landscapes in the Dark Valley," 302.

41 Wilvan Van Campen, trans., "The Present Condition of the Tuna Fisheries," in *The Japanese Tuna Fishing Industry*, Special Scientific Report: Fisheries No. 79 (Washington, DC: US Department of the Interior, July 1952), 1; Supreme Commander for the Allied Powers, *Fisheries Programs in Japan, 1945–51*, Natural Resources Section Report 152 (Tokyo, 1951), 7, 11. See also *Suisan Nenkan* (Tokyo: Suisansha, 1954).

42 Finley, *All the Fish in the Sea*, 73–75, 106–7.

43 Harry Scheiber, "Postwar Fishery Regimes of the Pacific: Ocean Law, International Rivalry, and Japanese Economic Expansion after 1945," in *Ocean Resources: Industries and Rivalries Since 1800*, ed. Harry Scheiber (Berkeley: University of California Intercampus Economic History Program, 1990), 71. On the "MacArthur Line" and its extensions, see Matsuda Yoshiaki, "Postwar Development and Expansion of Japan's Tuna Fishery," in *Tuna Issues and Perspectives in the Pacific Islands Region*, ed. David Doulman (Honolulu: East-West Center, 1987), 72–73.

44 See Henry Scheiber, *Inter-Allied Conflicts and Ocean Law, 1945–53: The Occupation Command's Revival of Japanese Whaling and Marine Fisheries* (Taipei: Academia Sinica, 2001); and Finley, *All the Fish in the Sea*, esp. chap. 6.

45 UN Food and Agriculture Organization, Statistics Division (FAOSTAT) website, accessed July 15, 2012, http://faostat.fao.org/.

46 "A History of the Traditional Diet."

47 See, for example, Uda, *Fisheries of Japan*, 3–5.

48 On the postwar expansion of Japanese fisheries, see Borgstrom, *Japan's World Success*, esp. chaps. 10–20; Uda, *Fisheries of Japan*; Asia Kyokai, *Japanese Fisheries*.

49 Borgstrom, *Japan's World Success*, 299.

50 Ibid., 97–102, chaps. 10–20; Asia Kyokai, *Japanese Fisheries*, 112*ff*.

51 Finley, *All the Fish in the Sea*, 125; Asia Kyokai, *Japanese Fisheries*, 90, 112*ff*.

52 Anthony Bergin and Marcus Haward, *Japan's Tuna Fishing Industry: A Setting Sun or New Dawn?* (Commack, NY: Nova Science, 1996), 51–60; Fujinami Norio, "Development of Japan's Tuna Fisheries," in Doulman, *Tuna Issues and Perspectives in the Pacific Islands Region*, 65–67.

53 Yoshiaki, "Postwar Development and Expansion of Japan's Tuna Fishery," 81; Uda, *Fisheries of Japan*, 45; Borgstrom, *Japan's World Success*, 109–10. See also *Suisanchō 50-nenshi kankō iinkai, Suisanchō 50-nenshi* (Tokyo, 1998).

54 Borgstrom, *Japan's World Success*, 123.

55 Tsutsui, "Landscapes in the Dark Valley," 303.

56 Toda Yasushi, *The Northern Pacific Fishery: A Case Study of Soviet-Japanese Economic Relations* (Pittsburgh: University of Pittsburgh Center for Russian and East European Studies, 1988), 9.

57 Borgstrom, *Japan's World Success*, 40–41, 274–75; Asia Kyokai, *Japanese Fisheries*, 4–5.

58 Asia Kyokai, *Japanese Fisheries*, 5; FAOSTAT website.

59 Borgstrom, *Japan's World Success*, 82, 100–102; Bergin and Haward, *Japan's Tuna Fishing Industry*, 13; Wilfram Ken Swartz, "Global Maps of the Growth of Japanese Marine Fisheries and Fish Consumption" (MS thesis in Resource Management and Environmental Studies, University of British Columbia, 2004), 10–11.

60 Kate Barclay, "Ocean, Empire, and Nation: Japanese Fisheries Politics," in *Water, Sovereignty and Borders in Asia and Oceania*, ed. Devleena Ghosh, Heather Goodall, and Stephanie

Donald (Abingdon, UK: Routledge, 2009), 42–43; Bergin and Haward, *Japan's Tuna Fishing Industry*, 9, 14, 80–83; Matsuda, "Postwar Development and Expansion of Japan's Tuna Fishery," 86–87.

61 Hiroshi Kasahara and William Burke, *North Pacific Fisheries Management* (Washington, DC: Resources for the Future, 1973).

62 Scheiber, "Postwar Fishery Regimes of the Pacific," 74; Kasahara and Burke, *North Pacific Fisheries Management*, 43–44.

63 Kasahara and Burke, *North Pacific Fisheries Management*, 47; Borgstrom, *Japan's World Success*, 272–73.

64 See Smith, "Japan's High Seas Fisheries," 77–81; Sandra Tarte, *Japan's Aid Diplomacy and the Pacific Islands* (Canberra, ACT: National Center for Development Studies, 1998), 81.

65 Smith, "Japan's High Seas Fisheries," 81.

66 Fujinami, "Development of Japan's Tuna Fisheries," 61.

67 Bergin and Haward, *Japan's Tuna Fishing Industry*, 123–30; Yamaka Junko, "Fisheries in Asia and the Pacific—Japan's Involvement and Its Problems," *AMPO* 16, no. 1–2 (1984): 94.

68 Smith, "Japan's High Seas Fisheries," 81; Tarte, *Japan's Aid Diplomacy*, 128*ff*.

69 Barclay, "Ocean, Empire, and Nation," 42–43; Bergin and Haward, *Japan's Tuna Fishing Industry*, 14; Fujinami, "Development of Japan's Tuna Fisheries," 62–64.

70 Bergin and Haward, *Japan's Tuna Fishing Industry*, 72–75; Tarte, *Japan's Aid Diplomacy*, 83–84, 93–96.

71 John Sproul, "Effects of North Pacific 200-Mile Exclusive Economic Zone Marine Management Policy on Japanese Seafood Production, Trade, and Food Security," *Bulletin of the Faculty of Fisheries, Hokkaido University* 43, no. 3 (1992): 150; Matsuda, "Postwar Development and Expansion of Japan's Tuna Fishery," 87.

72 Bergin and Haward, *Japan's Tuna Fishing Industry*, 95; Miyaoka Isao, *Legitimacy in International Society: Japan's Reaction to Global Wildlife Preservation* (Basingstoke, UK: Palgrave Macmillan, 2004), 50–51.

73 Miyaoka, *Legitimacy in International Society*, chap. 4.

74 Lyster, *International Wildlife Law*, 18.

75 Ibid., 27–35.

76 See *Wikipedia*, s.v. "Whaling in Japan," accessed July 15, 2012, http://en.wikipedia.org/wiki/Whaling_in_Japan#Catches.

77 Poul Holm, "The Global Fish Market: Internationalization and Globalization, 1880–1997," in *Global Markets: The Internationalization of the Sea Transport Industries since 1850*, ed. David Starkey and Gelina Harlaftis (St. John's, Newfoundland: International Maritime Economic History Association, 1998), 250; Bergin and Haward, *Japan's Tuna Fishing Industry*, 17; Sproul, "Effects of North Pacific 200-Mile Exclusive Economic Zone," 126.

78 Sproul, "Effects of North Pacific 200-Mile Exclusive Economic Zone," 124, 132, 134; Bergin and Haward, *Japan's Tuna Fishing Industry*, 6; Holm, "Global Fish Market," 254.

79 Terashima Hiroshi, "On Becoming an Ocean State," *Japan Echo* 34, no. 1 (February 2007): 37–39; Matsuzawa Takatoshi, "What Is the Volume of Japan's 200-nm Exclusive Economic Zone?," *Ocean Policy Research Foundation Newsletter* 123 (September 20, 2005): http://www.sof.or.jp/en/news/101-150/123_2.php.

Chapter 15
Opening the Circumpolar Arctic World

ILMO MASSA AND ALLA BOLOTOVA

This essay tracks general trends in the military use of natural systems and natural resources during and after World War II in northern Canada, the northern Soviet Union, and northern Finland, discussing these areas in the context of the circumpolar north as a whole. Our review spans the period from World War II through the decades of the Cold War. Its scope aligns with the synthetic tradition of environmental history and is based on general geographical, historical, and anthropological texts about the earth's northern regions.[1] While several national, provincial, and local environmental historical case studies exist in this field,[2] we are not aware of any previous environmental historical research that examines the northern zone of the globe as a whole. Only fragmentary information has been gathered on the ecological costs of intensified extractive resource strategies in this zone and their impact on native peoples.[3] The concept of an extractive economy refers to the excessive and unsustainable use of natural systems and natural resources in peripheral areas.[4] War-generated resource scarcity has historically acted as an engine driving political and economic development toward intensive extractive economies.

Global accounts of the ecological history of World War II in the circumpolar north have yet to emerge,[5] but some of the most important wartime trends are identifiable. First, World War II and its various consequences resulted in the increasingly extensive exploitation of the world's northern peripheries. It generated numerous economic opportunities, opened up new communication routes, and stimulated the discovery of new resources. Second, the war alerted the governments of the core belligerent states to the strategic, economic, political, and geopolitical importance of the northern regions. Third, the northern environment suffered direct impacts from combat in the form of scorched-earth tactics and increased levels of pollution. Fourth, during World War II, dwindling resources increased the belligerents' reliance on northern sources of resources. This trend continued after the war; as the southern reaches of the northern zone remained open for extractive industrialization, the ecological costs of this development were completely ignored, during both the war and the postwar decades of reconstruction and rapid economic growth.

By the beginning of the twentieth century, the idea of the Arctic as a zone at the margins of human history was being superseded by a new notion among Western nations of the Arctic as a frontier waiting for industrial colonization. Yet, generally speaking, the intensive extractive economy did not extend deep into the circumpolar north prior to World War II. By the end of the 1930s, the provincial norths of Canada had experienced the initial stages of integration into the rest of the country, but they remained very much on the periphery.[6] In the Soviet Union, extremely poor transportation and communications infrastructures throughout the nation's northern regions meant that northern expansion before World War II was focused primarily on construction of transportation and communications networks.[7] Prior to World War II, considering the northern peripheries of the globe as a whole, the only major industry was situated in Europe.[8] During and after the war, the focus of the world economy shifted significantly northward, giving rise to extractive economies in the belligerent states. What follows is a summary of environmental histories of war-generated extractive development of northern Canada, the northern Soviet Union, and northern Finland.

NORTHERN CANADA

World War II officially began on September 1, 1939, with the German invasion of Poland. Britain and France declared war on the Third Reich on September 3, 1939, and on September 10 of that same year, Canada entered World War II. It wasn't until two years later, after the Japanese attack on Pearl Harbor on December 7, 1941, that the United States entered the war, and this event proved to be a major turning point in the history of the northern territories and provinces of Canada.[9] Canada's long slide into a position of economic dependency on the United States accelerated during the war,[10] when tens of thousands of American military and civilian workers carried out various duties in Canada "with virtually no supervision from

Canadians, forming a kind of friendly army of occupation."[11] An extractive economy took hold. During the war, for instance, Canada's federal government became directly involved in the uranium-mining industry through the creation of the Crown corporation Eldorado Mining and Refining and its auxiliaries. The government also continually intervened in the generation of electric power, mining, transportation, and even fisheries. Wartime technological innovations—such as the development of freight planes, aerial photography, and nuclear weapons—prepared the way for the race for resources that eventually enveloped the Canadian north.[12] War-related projects had an impact across the entirety of northern Canada, from Labrador to the territories bordering on Alaska.[13]

In 1942, Japan attacked the US Aleutian Islands, occupying Kiska and Attu; nearly four thousand American soldiers fell in the recapture of Attu. As a result, the armaments industry and defense investment became the single most important factor influencing the development of Alaska. Anticipating the Japanese assault, the government of the United States had already in 1940 built large military bases at Fairbanks and Anchorage and on the Aleutian Islands, each manned by several thousand personnel. At the time, Alaska was supplied primarily from the south by sea.[14]

In 1942, within weeks of the attack on Pearl Harbor, the United States received permission from Ottawa to construct the Alaskan Highway from Dawson Creek, British Columbia, across northern British Columbia and the Yukon Territory to Fairbanks. The stated purpose of the new road was to provide a new communications and transportation link between the lower forty-eight states and Alaska, one less susceptible to Japanese attack. The 2,500-kilometer highway took only a year to build; at the peak of the project, 40,000 men and substantial amounts of heavy equipment were dedicated to construction. The highway and its adjacent telephone lines and emergency airstrips were opened to military traffic in November 1943.[15] The greatest changes took place in Canada, especially in the Yukon and Northwest Territories, which had previously been thought of as a "wasteland."[16] These two territories make up more than 40 percent of Canada's land area, but contained only about 1 percent of its population at the time.

The Alaska Highway was a prime example of the so-called northwestern defense projects, which included the Canol pipeline, the Northwest Staging Route, and Haines Highway. These defense projects consisted of a series of airstrips, airports, weather stations, radar stations, and radio ranging stations built in British Columbia, the Yukon, and Alaska. The Canol pipeline linked the oil fields at Norman Wells to the refinery at Whitehorse; another three pipelines were built to take the refined oil to Skagway and Fairbanks in Alaska and to Watson Lake in the Yukon.[17] The Northwest Staging Route, built by Canada and the United States between 1940 and 1944, was an air route crossing northwestern Canada from Edmonton to Fairbanks. It was critical for the support of defense projects in northwestern North America and for delivering aircraft to Alaska and the Soviet Union, where it was extended as ALSIB, the Alaska-Siberian air road. Major landing fields in Canada included Grande Prairie, Fort St. John, Fort Nelson, Watson Lake, and Whitehorse.[18] The Northwest Staging Route also provided a diplomatic route between Moscow and Washington, DC. Diplomats, senior political figures, and other government officials shuttled back and forth along it in military transports during the war.[19]

The most important defense project in northeastern Canada was the Crimson Route (also called the Northeast Staging Route). This was a series of joint US and Canadian transport routes planned to protect convoys to the Soviet Union, which was not allied with Great Britain and Canada until June 1941. The project ended in 1943 and was never fully developed. But the newly built airfields of Goose Bay, Fort Chimo, Frobisher Bay, the Pas, Churchill, and Southampton Island opened up the Canadian Arctic Archipelago and Greenland to air traffic in 1941–42, and a telecommunications network was built along with the airfields.[20]

The United States subsequently occupied both Greenland—a Danish colony—and Iceland on April 8, 1941, subjecting Thule Air Base in Greenland and Keflavik Air Base in Iceland to American hegemony. This was a strategic countermove, because Germany had occupied both Denmark and Norway in the early stages of the war. Denmark regained Greenland and Iceland regained its sovereignty in 1945, but the United States maintained its military presence at both Thule and Keflavik.[21]

Rising demand for agricultural products (wheat, flour, and pork) produced a three- to fourfold increase in Canadian exports of those commodities between 1939 and 1945.[22] Demand for canned and cured freshwater and saltwater fish also increased sharply because of wartime exigencies, encouraging further expansion and intensification of fishing operations with the support of the federal government. These trends intensified agricultural colonization in the provincial norths of Canada.[23]

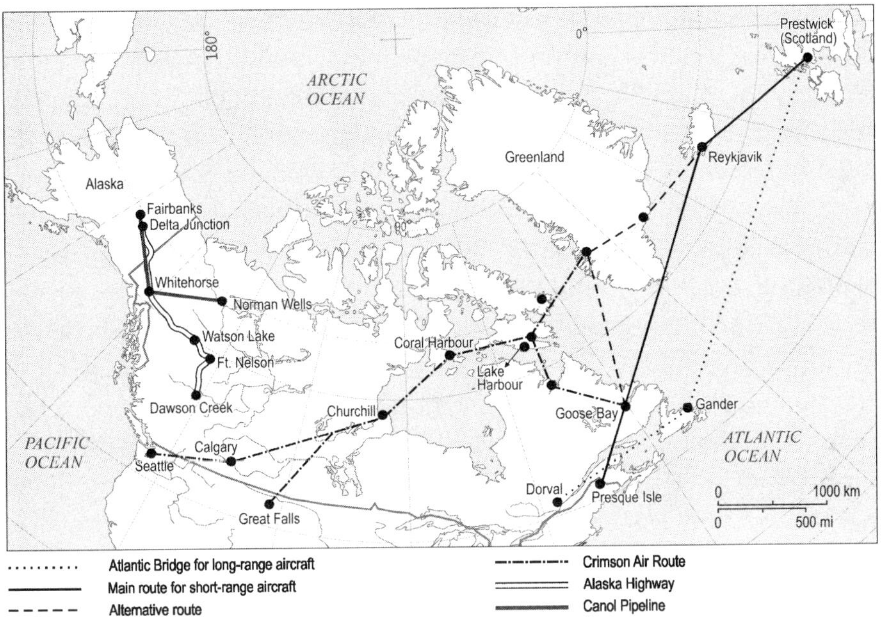

Figure 15.1. Main traffic lines in Canada and Alaska.

This new resource-based industrialization caused numerous local environmental problems in Canada, although no major catastrophes occurred. Graeme Wynn describes the social and environmental effects of highway construction as "enormous," massive projects that left in their wake new diseases, melting permafrost, forest fires, waste, fuel-oil spills, and decreasing stocks of fish and wildlife. Native First Nations peoples were barred from their traditional hunting grounds.[24] It is difficult to overestimate the significance of the highway in opening up regions previously accessible only to canoes, barges, and bush planes. And yet British Columbia benefited from the construction of the Alaska Highway, which made it possible to build the first road to the Peace River region, with the aim of being able to harness hydroelectric power.[25]

Canada's role as a raw material supplier to the United States grew significantly after the war, and the country's abundant natural resources were also needed for European reconstruction. This had significant effects on the Canadian north.[26] In the Paley Report,[27] commissioned by the US government, Canada was considered the most reliable supplier of thirteen of twenty-two indispensable raw materials, minerals, and fuels. In accordance with the policies outlined in that report, Canada became increasingly integrated into the US sphere of influence as a specialized producer of minerals, gas, and oil. Canada's postwar governments also participated in paving the way for "American resource capitalism," in the form of, for example, Prime Minister John Diefenbaker's Roads to Resources program (1959–70).[28] After World War II, almost all Canadian provinces had their own northern development programs, with different emphases and environmental profiles. The goal was to channel the natural resources of this region, called the "Middle North," to the benefit of industry.[29]

In all, the greatest war-generated changes in Canada took place in the Yukon and Northwest Territories. The defense projects built as a result of the war brought a series of weather stations, radar stations, and radio ranging stations across the Canadian north. Other signs of progress included highways, railways, radio and television stations, schools, recreational and cultural facilities, and airline services, creating thousands of new jobs in the resource and service industries, all of which were favorable for modern businesses. But this development was grounded in high-intensity expansion of the wartime extractive economy in these previously peripheral lands. During the war, the changes native peoples faced in the provincial and territorial north were small and localized, but the First Nations subsequently had to endure a great deal of turmoil and upheaval, much of it in the name of progress, charity, and social engineering.

THE NORTHERN SOVIET UNION

On August 24, 1939, the Soviet Union signed the Molotov-Ribbentrop Pact with Nazi Germany. In a secret clause of the pact, Finland, Estonia, Latvia, Lithuania, eastern Poland, and Bessarabia (part of Romania) were included in the Soviet sphere of influence, and the Soviet Union used the early years of the war to strengthen its position in this sphere. On September 17, 1939, sixteen days after Nazi Germany attacked Poland from the west, the Red Army invaded eastern Poland. On November 30, 1939, the Soviet Union attacked Finland, starting the war known in Finland as the Winter War and in the Soviet Union as the Soviet-Finland War of 1939–40. The following year, the Soviet Union annexed the Baltic states of Estonia, Latvia, and Lithuania, as well as parts of Romania.[30] On June 22, 1941, Germany invaded the Soviet Union, effectively ending the Molotov-Ribbentrop Pact; as a result, the Soviet Union joined the Allies. After October 1941, the United States and the Soviet Union struck a barter agreement known as Lend-Lease, according to which the Soviet Union bartered natural resources—such as gold, tin, timber, furs, and caviar—for American deliveries of aircraft, tanks, cars, oil products, railway machinery and locomotives, food and clothing, tanks, and other supplies.[31] In practice, the value of the supplies and services turned over by the United States amounted to approximately $11 billion (in 1947 dollars), which was nearly ten times more than it got from the Soviet Union. On the other hand, the American economic inputs of Lend-Lease were not very high, if we compare them with the enormous human and material losses of the Soviet Union during its "patriotic war" against Germany.[32] The most important air-traffic route created during World War II was the 4,850-kilometer-long Alaska-Siberia route (ALSIB), which was used from 1942 to 1945 to transport fighters from the United States to the Soviet Union. Several thousand prisoners participated in building the airstrips. After the war, information on this air route was classified as highly secret in the Soviet Union.[33]

The evacuation of Soviet industry to the country's eastern and northern peripheries began at the end of June 1941 and continued until the end of December. To keep the country's war industry safe from attack and capture, more than 1,500 large industrial plants were moved eastward during the war, and 3,500 new ones were built. Industrial plants that hitherto produced nonmilitary items were also converted to military use.[34] In general, all production was reoriented toward military interests. Nearly half of the evacuated factories were relocated to the Urals, and the rest to the Volga region, western Siberia, Kazakhstan, and central Asia, while a handful moved even farther, to eastern Siberia. This resulted in a permanent east- and northward shift in the Soviet defense industry's center of gravity. Because of the war, the industrial output of the country's economic heartland plummeted to a fraction of what it had been.[35]

As huge defense and metallurgical factories were relocated to the remote interior, sparsely settled rural communities were required to develop rail and road links, power lines and generating capacity, and homes and services for workers, all in a brief span of weeks or months. Gulag prison labor was used to build transportation networks of roads, railways, and airfields in northern regions. A good example of these new cities of the north is Vorkuta, which was founded in 1932 to develop the Pechora coal basin. Vorkuta's coal became critical during the war because the Don River basin had been lost to the Nazi armies. Another example is the city of Norilsk on the Taymyr Peninsula. This city grew in the Russian Arctic out of the Norillag, which from 1935 to 1956 employed tens of thousands of prisoners mining under the permafrost for the world's largest nickel-copper-palladium deposits.[36]

In the northern periphery, two railway lines were particularly important: the Leningrad-Murmansk railway and the Pechorskaya railway. When the Donbass coalmines in Ukraine were lost to the enemy and new sources of coal were needed for energy production, the Pechorskaya railway was used to transport coal from Pechora to central Russia. The first coal train from Pechora arrived in Leningrad in December 1941.[37] The Kolyma road from Magadan to Ust-Nera, which was more than a thousand kilometers in length, was constructed because of the gold deposits found in and around the River Kolyma. Many people died of cold and starvation during its construction. The road came into use during the war, but it was not fully completed until 1953.[38] The Northern Sea Route was essential to the war effort because it allowed the import of military and food supplies from the West. This shipping route was complemented by river navigation to the interior of Siberia.[39]

This development of military-oriented production required intensification of natural resource extraction, with the added task of finding specific minerals necessary for military production in significant volumes. Gold was the most important ore in the northern Far East (Kolyma, Aldan), termed "metal number one" by the Soviet media. It paid for imported machinery, equipment,

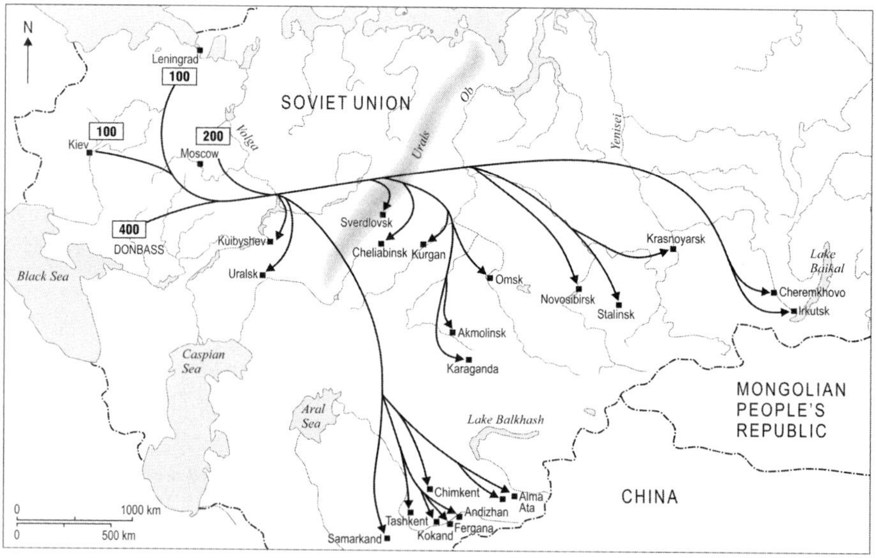

Figure 15.2. Wartime evacuation of industry in the Soviet Union. Source: Martin Gilbert, *The Routledge Atlas of Russian History* (London: Routledge, 2007), 121.

airplanes, and food.[40] Aluminum was an important mineral in war conditions because of its necessity in arms production. Another important metal, one of the so-called strategic minerals, was nickel. At the end of World War II, two more minerals—uranium and diamonds—also became important.

Occasionally, indigenous people became of importance during war. In the areas where snow made it impossible to use motorized vehicles, special reindeer-transport brigades run by indigenous peoples—the Saami, Nenets, and Komi—were organized at the front in such places as the Kola Peninsula. These brigades were used for many purposes, including transporting military goods, conveying the injured back from the front line, and contacting partisans. Indigenous peoples were also required to fish and hunt for military supply purposes.[41]

Other people had different fates. During World War II, many people were driven into internal exile within the Soviet Union, contributing to the northern expansion. The purposes of the forced migration were both to reduce security risks to the Soviet state and to Sovietize conquered peoples. According to Anne Applebaum, the expulsions constituted ethnic cleansing carried out under the pretense of war. The forced migrations were based on Stalin's grand plan for "settling the northern regions of the USSR." An estimated two million people were exiled from their home districts and transported to central Asia and Siberia. The largest group among the many different peoples and ethnic groups were the 1.2 million Volga Germans, who had been recruited to Russia as immigrants in the eighteenth century.[42] In addition to the internal "administrative transfers" (the term used by the official propaganda), hundreds of thousands of people from Soviet-occupied areas were transported to the Soviet Union. Furthermore, the displacement of hundreds of thousands of prisoners of war resulting from the defeat of the Axis powers also fueled the northern expansion. The total number of displaced persons was approximately 760,000, the largest group of which comprised people exiled from eastern Poland.[43]

The Soviet war economy was not dismantled after the war, continuing as actively as ever during the Cold War.[44] Enormous postwar infrastructure projects such as the Baikal-Amur Mainline railway; hydroelectric power stations in Siberia; military, air, and radar bases in Murmansk; industrial cities; nuclear and space research centers; and oil and other energy projects meant continued internal expansion to the north and east. It is no miracle that the population of Siberia increased four- to fivefold over the period 1926–56.[45]

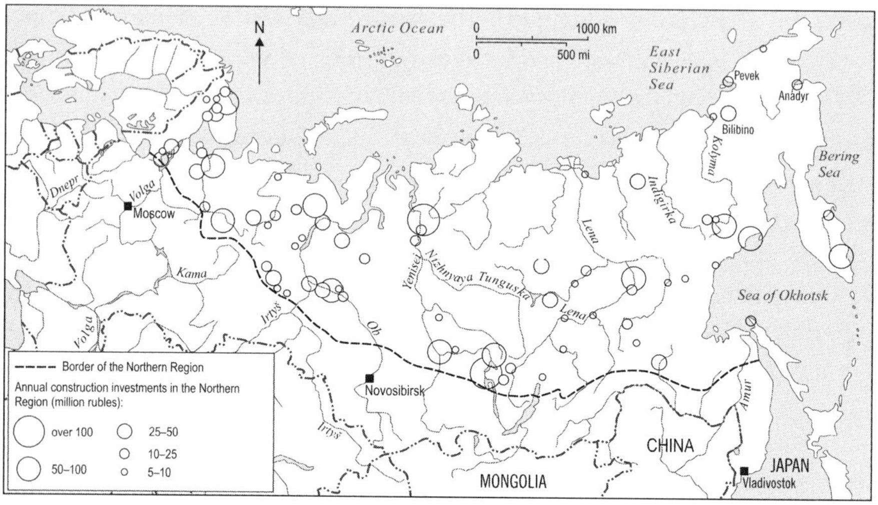

Figure 15.3. Main poles of development in the Soviet north in the 1970s presented in terms of annual investments. Source: M. E. Krinitshkaja, "Osobennosti razvitija industrialinoj bazi stroiteltsva v severnoj zone" in *Problemy severa: Problemy povyshenie efektivnosti stroitelstvo na svete* (M. Nauka: Moskva, 1974).

In all, a totalitarian and repressive version of the extractive economy developed in the Soviet Union during World War II. At least 20 million citizens perished, and "millions of soldiers engaged in pillaging and plunder, in scorched-earth policies and pitched battles, and they engaged in this conflict in the middle of the most productive agricultural lands of the Soviet Union and in some centers of some of the most vital industry."[46] But the Soviet Union defeated the Nazis and emerged from the war as one of the two postwar military superpowers. One of the most important reasons the Red Army was able to defeat the Wehrmacht was the expansion and north- and eastward shift of the industrial heart of the country. The transport connections of the northern and eastern peripheries greatly improved as a result. This expansion was based on forced labor and prison camps, but, at least after the war, it offered some real financial benefits for migrating workers, but in wartime Soviet Union, little thought was given to its incalculable human and ecological costs.[47]

NORTHERN FINLAND

Finland's weak position in world politics is illustrated in the Molotov-Ribbentrop Pact, signed on August 24, 1939, between Germany and the Soviet Union. In a secret clause of the agreement, Finland (along with eastern Poland, Estonia, Latvia, Lithuania, and part of Romania) was included in the Soviet sphere of influence.[48] The Winter War against the Soviet Union began with a Soviet offensive on November 30, 1939—three months after the German-Soviet invasion of Poland that launched World War II—and ended on March 13, 1940, with the signing of the Moscow Peace Treaty. During the Winter War, 23,000 Finnish soldiers were killed and more than 43,000 were wounded; however, the Soviet Union suffered much greater casualties and failed to conquer Finland. The Winter War proved that the Red Army had not mastered winter warfare. This might be the strongest root of the misconception of the Soviet Union as not being a first-class military power, and the sentiments in Germany that it no longer needed to view the Soviet Union as a serious military or economic threat.[49]

The focus of the Winter War was not in northern Finland but in the Karelian Isthmus, a neck of land 45 to 110 kilometers wide located between the Gulf of Finland and Lake Ladoga in northwestern Russia. In the Moscow Peace Treaty, Finland lost the Karelian Isthmus and the cities of Viipuri, Käkisalmi, and Sortavala, all in the eastern heartland of the country. More than 400,000 Karelians were evacuated to other parts of Finland. The treaty was seen in Finland as unjust and had an influence on later war-related decision making and activities in Finland.[50]

In a subsequent conflict, the Continuation War (June 25, 1941, to September 19, 1944), Finland not only recaptured territory lost in 1940 but also continued its offensive deeper into the Soviet Union. To this end, Finland allied actively with Nazi Germany, which had its own interests in enlisting Finland as its cobelligerent in an eastward advance. Germany also had a keen interest in the nickel deposits at Pechanga (Petsamo in Finnish) and the copper mine at Outokumpu, located in the province of eastern Finland known as North Karelia. Germany had only 30 percent self-sufficiency in copper during the war, and almost nonexistent self-sufficiency in nickel. Germany hoped that Pechanga would satisfy 70 percent of its wartime nickel needs. Finland, in turn, was able to acquire armaments and foodstuffs by selling its copper and nickel to Germany.[51] During the war, Finland was the second-largest supplier of nickel in the world.[52]

In preparation for the massive assault on the Soviet Union known as Operation Barbarossa, Finland and Germany agreed that German troops would be given responsibility for the northern part of Finland, which enabled the Finnish Army to concentrate its troops in the southern half of the country. On August 19, 1940, Finland agreed to grant access to its territory to the

Wehrmacht. Initially, this was for transporting troops and military equipment to and from northernmost Norway, but before long, minor bases sprouted up along the growing transport road. Germany's main goal for its presence in Finnish Lapland was the capture of the key Soviet port at Murmansk. This operation, known as Silberfuchs, was unsuccessful. Throughout the war, Finland kept its army outside the control of the German command structure despite numerous German attempts to knit them more tightly together. Consequently, Finland did not participate directly in the siege of Leningrad, which lasted nearly nine hundred days.[53]

German forces penetrated into two other Nordic countries, Denmark and Norway, on April 9, 1940. The most important reason for the invasion was to control the Baltic and North Atlantic naval and shipping routes. Germany also established priority access to marine food resources from Denmark, Norway, and other occupied countries.[54] Sweden was the only Nordic country to remain formally neutral during the war, but it did permit transit of German military goods and soldiers by an agreement reached on June 18, 1940. Germany's interest was securing the export of Swedish iron ore from the mining town of Kiruna in northern Sweden through Norway's rail terminus and the port of Narvik. More than 100,000 German military railway carriages and a total of 2.14 million German soldiers crossed Sweden before this traffic was officially suspended on August 20, 1943.[55] While half of Sweden's iron ore was shipped to Germany via Narvik and the North Sea, which was open for iron ore shipments all year round, the other half, excavated from Gällivare and other mines, was shipped to Germany via Swedish ports on the Baltic Sea when it was free of ice. This Baltic transport of iron ore to Germany did not come to an end until autumn 1944.[56]

World War II drastically altered social and environmental conditions in northern Finland. The Germans took responsibility for a five-hundred-kilometer stretch of the front in Finnish Lapland, stationing over 200,000 German soldiers in northern Finland at the peak of the military buildup. When added to the German and Soviet troops stationed in Norway and Soviet Lapland, this meant an increase of at least a half million inhabitants across transnational Lapland. World War II thus emerges as a highly exceptional, absurdly vigorous time in terms of population growth in the area. Such growth inevitably left ecological, genetic, and cultural traces. The Germans built nearly a hundred labor camps in the area, where they imprisoned some 30,000 Soviet soldiers. There were also some Finnish labor camps in the north. Because Lapland's infrastructure was poor, Russian prisoners of war and German convicts were used as a labor force to build and improve roads, bridges, prison camps, and defense sites. For example, a seventy-four-kilometer road from Karigasniemi to Kaamanen and a 193-kilometer road from Muonio to Kilpisjärvi, both in the northernmost parts of Lapland, were built or improved.[57] The Luftwaffe took over nine existing Finnish airports in Lapland and expanded and reequipped all of them significantly. In addition, the Luftwaffe built about ten small airstrips in the territories conquered from the Soviet Union.[58] Most German building sites, including the new airfields, were temporary and were wiped out during the subsequent Lapland War.

After German losses at Stalingrad in 1943 and the Normandy landings of June 6, 1944, Finland made several unsuccessful attempts to end the Continuation War. On June 9, 1944, the Red Army launched a massive attack on Finland, and by September 1944, the Soviet Union had forced Finland into a cease-fire agreement.[59] The Moscow armistice signed by Finland and the Soviet Union on September 19, 1944, ended the Continuation War, though the final peace treaty was not signed until 1947 in Paris. The conditions for peace were similar to those previously agreed in the Moscow Peace Treaty of 1940, with Finland being forced to cede parts of Finnish Karelia, part of Salla (in Finnish Lapland), and islands in the Gulf of Finland. Furthermore, the new armistice turned over Pechanga in its entirety to the Soviet Union and required that Finland drive German troops out of its territory. This demand was the cause of the third successive war in Finland, known as the Lapland War.[60]

The Lapland War resulted in a comparatively small number of Finnish military casualties: about one thousand men killed and three thousand wounded. Most of the civilian population of Lapland, totaling 168,000 people, was evacuated with the help of the German troops to Sweden and southern Finland, including a total of two thousand Saami people.[61] But the material losses were enormous. As they retreated, the German forces implemented scorched-earth tactics, devastating large areas of northern Finland. As a result, some 40 to 47 percent of the dwellings in the area were destroyed, and the provincial capital of Rovaniemi was burned to the ground, as were the villages of Savukoski and Enontekiö. Two-thirds of the buildings in the biggest villages were demolished, most of the bridges were blown up, all main roads were mined, and telephone lines were torn down. The Germans not only destroyed their former bases and camps, but also burned down every Finnish village within their reach, some

Figure 15.4. Long shadows on the main road of Rovaniemi town in northern Finland, which was burned to the ground by the German Army. Source: Finnish Defence Forces Photo Archive.

16,000 buildings in total. Over a thousand road bridges, some one hundred railroad bridges, and forty ferries were blown up; 170 kilometers of railroad, 9,500 kilometers of road, and almost three thousand culverts were destroyed; and most electricity poles were cut down. Tens of thousands of head of cattle and reindeer were slaughtered, and over 130,000 landmines and other explosives were planted in the terrain. In the years following the war, these mines would kill about two thousand people.[62]

As an outcome of the war, Finland lost Pechanga to the Soviet Union, and thereby its access to the Arctic Ocean and its nickel mines. These losses crushed the dreams, particularly prevalent among right-wing circles, of Finland as a new northern power.[63] Yet the postwar cession of territories to the Soviet Union, the loss of hydroelectric power from the Vuoksi River, and the recovering forest economy meant that the unharnessed rivers of northern Finland began to interest power companies. The impact of the war was also evident in both a 1951 report prepared by a national industrial committee and in a 1952 book on the industrialization of Lapland written by the then-prime minister, Urho Kekkonen. His Agrarian Union, an influential party in Finland at the time, supported these ideas, as did forest companies eager to exploit energy and timber resources in the north.[64] Discussions of the possibility of cultivating new arable land in Lapland were resumed because the food situation was poor after the war, and domestic politics dictated that promises made during the war to give to war veterans land of their own had to be fulfilled. Because of these projects, the population of Lapland increased slightly during the 1950s.[65]

In all, Denmark and Norway, the northern half of Finland, and to a lesser degree Sweden were colonized by Germany and became, at least partly, peripheries of the Nazi state. During the war, Germany received most of its marine food supplies from Norway and Denmark. Sweden allowed the Wehrmacht to use Swedish railways for transportation between Norway and Finland (June–July 1941) and sold iron ore, iron products, and ball bearings to Germany throughout the war. Finland tried to maintain its independence under pressure not only from Germany, but also from the Soviet Union. As a result of the three wars, the entire infrastructure of Finnish Lapland was destroyed, and Finland lost a tenth of its land area to the Soviet Union, including the strategic region of Pechanga. The country tried to compensate for these losses by putting its most prolific natural resources to use, not considering traditional trades and the conservation of resources. This northward expansion was an attempt by Finland's government to solve the severe economic problems brought on by the Winter War, the Continuation War, and the Lapland War, separately and together. As a consequence, the economic focus in Finnish Lapland shifted between one hundred and two hundred kilometers farther north from its prewar location, which is why it is accurate to talk about a government-run extractive economy in Finnish Lapland developing particularly after World War II.

CONCLUSION

In this essay, we tracked trends in the three separate national and subnational regions of the circumpolar north, each of which occupied a different position in the world system. The main players were Germany, the United States along with its allies Great Britain and Canada, and the Soviet Union, all exercising a powerful influence in the circumpolar world. The first important shift was Germany's penetration of Denmark, Norway, northern Finland, to a lesser extent Sweden, and finally the Soviet Union. The second was the expansion of the Soviet Union into Finland during the Winter War and then toward its own northern and eastern peripheries, particularly after the onslaught by Nazi

Germany. The third was the decision of the United States to take part in the war and consequently to colonize, more or less benignly, Canada, Greenland and Iceland to resist and counterattack the Axis powers.

These trends of increasing military and civilian presence in the Arctic continued during the Cold War until the collapse of the Soviet Union. Today we see a trend toward remilitarization, as the value of northern resources such as minerals and fossil fuels rises again, and global warming is evidently making the Arctic ice-free in summer. The remilitarization of the Arctic is now under way in several of the abovementioned nations: the United States, Canada, and Russia. In addition, Denmark and especially Norway have developed a considerable naval strength and military presence in the Arctic.[66]

One the longest shadows cast by World War II was this increasing militarization of the Arctic Ocean and the surrounding northern regions with the introduction of air and naval bases and weapons of mass destruction. Largely as a result of the war, militarization and the entwined extractive economies penetrated one of the largest remaining wildernesses in the world.

Notes

1 Terence Armstrong, George Rogers, and Graham Rowley, *The Circumpolar North: A Political and Economic Geography of the Arctic and Sub-Arctic* (London: Methuen, 1978); K. J. Rea, *The Political Economy of Northern Development* (Ottawa: Science Council of Canada, 1976); and Dolly Jørgensen and Sverker Sörlin, eds., *Northscapes: History, Technology and the Making of Northern Environments* (Vancouver: University of British Columbia Press, 2013).

2 In Europe: Sverker Sörlin, *Framtidslandet: Debatten om Norrland och naturresurserna under det industriella genombrottet* (Stockholm: Kungl. Skytteanska samfundets handlingar, 1988); Ilmo Massa, *Pohjoinen luonnonvalloitus: Suunnistus ympäristöhistoriaan Lapissa ja Suomessa* [The northern conquest of nature: the environmental history of Lapland and Finland] (Gaudeamus: Helsinki, 1994); Louwrens Hacquebord and Dag Avango, "Settlements in an Arctic Resource Frontier Region," *Arctic Anthropology* 46, no. 1–2 (January 2009): 1–2, 25–39. In the Soviet Union/Russia: Paul R. Josephson, *The Conquest of the Russian Arctic* (Cambridge, MA: Harvard University Press, 2014); and Paul Josephson et al., *An Environmental History of Russia* (Cambridge: Cambridge University Press, 2013). In the United States and Canada: Paul Charest, "Les ressources naturelles de la Côte-Nord ou la richesse des autres," *Recherches amérindiennes au Québec* 5, no. 2 (1975): 35–52; Paul Charest, "Les barrages hydro-électriques en territoire Montagnais et leur effets sur les communautés amérindiennes au Québec," *Recherches amérindiennes au Québec* 9, no. 4 (1980): 323–37; Cristian Morissonneau, *La terre promise: Le mythe du Nord québécois* (Montréal: Hurtubise, 1978); Patricia Marchak, *Green Gold: The Forest Industry in British Columbia* (Vancouver: University of British Columbia, 1983); Graeme Wynn, *Canada and Arctic North America: An Environmental History* (Santa Barbara, CA: ABC-CLIO, 2007); P. Whitney Lackenbauer and Matthew Farish, "The Cold War on Canadian Soil, Militarizing a Northern Environment," *Environmental History* 12, no. 4 (October 2007); Liza Piper, *The Industrial Transformation of Subarctic Canada* (Vancouver: University of British Columbia Press, 2009).

3 Joachim Radkau, *Nature and Power: A Global History of the Environment* (Cambridge: Cambridge University Press, 2008), 32–35.

4 Stephen G. Bunker, *Underdeveloping the Amazon: Extraction, Unequal Exchange, and the Failure of the Modern State* (Chicago: University of Chicago Press, 1988). Bunker uses the concept of extractive economy at the general level in describing the relationships between peripheral Amazonia's local economy and industrial "productive economy."

5 Simo Laakkonen and Richard Tucker, "War and Natural Resources in History: Introduction," in "World War II, the Cold War, and Natural Resources," ed. Simo Laakkonen and Richard Tucker, special issue, *Global Environment* 10 (2012): 14–15.

6 Ken Coates and William Morrison, *The Forgotten North: A History of Canada's Provincial Norths* (Toronto: James Lorimer, 1992), 63.

7 Steven G. Marks, *Road to Power: The Trans-Siberian Railroad and the Colonization of Asian Russia 1850–1917* (Ithaca, NY: Cornell University Press, 1991); Anthony Heywood, *Modernising Lenin's Russia: Economic Reconstruction, Foreign Trade and the Railways* (Cambridge: Cambridge University Press, 1999); Cynthia Ruder, *Making History for Stalin: The Story of the Belomor Canal* (Gainesville: University Press of Florida, 1998); I. I. Chuhin, *Kanaloarmeytsy: Istoriya stroitel´stva Belomorkanala v dokumentakh, tsifrakh, faktakh, svidetel´stvakh uchastnikov i ochevidtsev* (Petrozavodsk: Kareliya, 1990).

8 Samuil Venediktovich Slavin, *Osvoenie severa Sovetskogo soyuza* [Colonization of the north of the Soviet Union] (Moskva: Nauka, 1982), 155.

9 B. H. Liddell Hart, *History of the Second World War* (London: Pan Books, 1970), 253–56, 269–78.

10 Kari Levitt, *Silent Surrender: The American Economic Empire in Canada* (New York: Liveright, 1971).

11 Coates and Morrison, *Forgotten North*, 85.

12 Armstrong, Rogers, and Rowley, *Circumpolar North*, 87–88; Rea, *Political Economy of Northern Development*, 55; and Wynn, *Canada and Arctic North America*, 274–75, 317.

13 Morris Zaslow, *The Northward Expansion of Canada 1914–1967* (Toronto: McClelland and Stewart, 1988), 209; Lackenbauer and Farish, "Cold War on Canadian Soil," 5–6; Piper, *Industrial Transformation of Subarctic Canada*, 8.

14 Peter A. Coates, *The Trans-Alaskan Pipeline Controversy: Technology, Conservation and the Frontier* (London: University of Alaska Press, 1993), 62–80.

15 Coates and Morrison, *Forgotten North*, 63–64.

16 Rea, *Political Economy of Northern Development*, 3–13.

17 Coates and Morrison, *Forgotten North*, 63; Charles Emmerson, *The Future History of the Arctic* (London: Vintage Books, 2011), 199. The Canol pipeline was closed in 1945 because the oil was deemed too costly for peacetime purposes.

18 *Canadian Encyclopedia*, s.v. "Northwest Staging Route," accessed March 15, 2015, http://www.thecanadianencyclopedia.com/articles/northwest-staging-route.

19 Otis Hays Jr., *The Alaska-Siberia Connection: The World War II Air Route* (College Station: Texas A&M University Press, 1996).

20 Coates and Morrison, *Forgotten North*, 64–65.

21 Emmerson, *Future History of the Arctic*, 124, 137; Jørgen Taagholt, "Greenland's Future Development: A Historical and Political Perspective," *Polar Record* 21, no. 130 (January 1982): 23–24. In 2006, the United States ended over half a century of military presence in Iceland and ceded control of Keflavík Air Base to Iceland.

22 Wynn, *Canada and Arctic North America*, 274.

23 Piper, *Industrial Transformation of Subarctic Canada*, 174–79, 194–95.

24 Wynn, *Canada and Arctic North America*, 281–83.

25 Zaslow, *Northward Expansion of Canada 1914–1967*, 209; Lackenbauer and Farish, "Cold War on Canadian Soil," 5–6.

26 Zaslow, *Northward Expansion of Canada 1914–1967*, 237.

27 US President's Materials Policy Commission, *Resources for Freedom*, 5 vols. (Washington, DC: US Government Printing Office, 1952).

28 Marchak, *Green Gold*, 39; Wynn, *Canada and Arctic North America*, 315.

29 Zaslow, *Northward Expansion of Canada 1914–1967*, 253. For a more detailed description of provincial resource projects, see Ilmo Massa and Alla Bolotova, "The Opening of the Circumpolar World," *Finnish Journal of Environmental History* 3 (2012): 49–78.

30 Liddell Hart, *History of the Second World War*, 16–17, 54–61.

31 Mark Harrison, "The Second World War," in *The Economic Transformation of the Soviet Union 1913–1945*, ed. R. W. Davis, Mark Harrison, and S. G. Wheatcroft (Cambridge: Cambridge University Press, 1994/2008), 250; N. V. Butenina, *Lend-Liz: Sdelka veka* [Lend-Lease: trade of the century] (Moskva: GUVShE, 2004), 104–6, 292.

32 John R. Deane, *The Strange Alliance: The Story of Our Efforts at Wartime Co-operation with Russia* (New York: Viking Press, 1947).

33 A. H. Paperno, "O Lend-lize i tihookeanskoj transportnoj epopee" [Lend-lease and transports across the Pacific] *Otechestvennaja istorija* 2 (1997): 125–26; I. E. Negenblja, *Aljaska-Sibir: Trassa Muzhestva* [Alaska-Siberia: route of courage] (Jakutskij kraj: Jakutsk, 2000).

34 Nikolai Voznesenskiy, *Voennaya ekonomika SSSR v period velikoy otechestvennoy voyny* [The Soviet wartime economy in the Great Patriotic War] (Moskva: OGIZ, 1948), 42–43; *Narodonaselenie: Entsiklopedicheskiy slovar* (Moskva: BRE, 1994), 561; "Velikaya otechestvennaya voyna, 1941–1945," in *Entsiklopediya* [Encyclopedia of the Great Patriotic War] (Moskva: Sovetskaya Entsiklopediya, 1985), 739.

35 Harrison, "Second World War," 250–53.

36 Josephson et al., *Environmental History of Russia*, 104–5.

37 O. Azarov, "Po tundre, po zheleznoj doroge . . . " [On the tundra and railways . . .], in *Pokayanie: Martirolog*, vol. 2, ed. G. V. Nevskii (Syktyvkar: Komi knizhnoe izdatelstvo, 1999), 130–206; accessed May 29, 2012, http://pokayanie-komi.narod.ru/azarov_po_tundre.htm.

38 See A. I. Shirokov, *Dalstroj: Predystorija i pervoe desjatiletie* [Dalstroy: early history and the first decade] (Kordis: Magadan, 2000), 145–46; accessed May 29, 2012, http://www.kolyma.ru/magadan/index.php?newsid=49.

39 "Severnyj Morskoj Put," in *Istorija oteshvennogo sudostroenija*, vol. 4, *Sudostroenie v period pervyh pjatiletok i Velikoj otechestvennoj vojny, 1925–1945* [National history of shipbuilding, vol. 4, shipbuilding during the first five-year plans and the Great Patriotic War, 1925–1945], ed. I. D. Spasski (St. Petersburg: Sudostroenie, 1996), 9–10.

40 The Trans-Siberian Railway was completed in 1916 and opened up southern Siberia and the Russian Far East to economic development. See Marks, *Road to Power*, and Butenina, *Lend-Liz*, 104–6, 292.

41 A. A. Kiselev, *Voĭna v Zapolyarie* [War in the Arctic] (Murmansk, 1995).

42 Anne Applebaum, *Gulag: A History of the Soviet Camps* (Harmondsworth, UK: Penguin, 2004), 383–87.

43 Ibid. 383.

44 A. I. Kostanov, *Samaya vostochnaya doroga Rossii*, accessed May 29, 2012, http://vffs.narod.ru/sakh/tp/t03.htm, 15; Terence Armstrong, *Russian Settlement in the North* (Cambridge: Cambridge University Press, 1965), 135–41; Vadim Gritsenko and Vyacheslav Kalinin, *Istoriya "mertvoi dorogi"* [The history of the "dead road"] (Ekaterinburg: Basko, 2010).

45 Kerstin Eidlitz, *Revolution i norr: Om sovjetetnografi och minoritetspolitik* (Uppsala: Uppsala universitetet, 1979), 184; Armstrong, Rogers, and Rowley, *Circumpolar North*, 45–53.

46 Josephson et al., *Environmental History of Russia*, 112–13.

47 Armstrong, *Russian Settlement in the North*, 171.

48 Max Jakobson, *The Diplomacy of the Winter War: An Account of the Russo-Finnish War, 1939–1940* (Cambridge, MA: Harvard University Press, 1961); D. W. Spring, "The Soviet Decision for War against Finland, 30 November 1939," *Soviet Studies* 38, no. 2 (April 1986): 207–26.

49 Roger R. Reese, "Lessons of the Winter War: A Study in the Military Effectiveness of the Red Army, 1939–1940," *Journal of Military History* 72, no. 3 (July 2008): 852.

50 William R. Trotter, *A Frozen Hell: The Russo-Finnish Winter War of 1939–40* (Chapel Hill, NC: Algonquin Books, 1991); Ohto Manninen, *Talvisodan salatut taustat: Neuvostoliiton operatiiviset suunnitelmat 1939–1941 Suomen suunnalla* [The Soviet operation plans for the Finnish theatre of war in 1939–1941] (Porvoo: WSOY,1994).

51 Markku Kuisma, *Outokumpu 1910–1985* (Forssa: Outokumpu, 1985), 222–28, 234–35.

52 Raymond Dumett, "Africa's Strategic Minerals during the Second World War," *Journal of African History* 26, no. 4 (October 1985): table 3.

53 Olli Vehviläinen, *Finland in the Second World War: Between Germany and Russia* (New York: Palgrave, 2002).

54 Poul Holm, "World War II and the Great Acceleration of North Atlantic Fisheries," in "World War II, the Cold War, and Natural Resources," ed. Simo Laakkonen and Richard Tucker, special issue, *Global Environment* 10 (2012): 72–75.

55 Armstrong, Rogers, and Rowley, *Circumpolar North*, 212–13; Emmerson, *Future History of the Arctic*, 125.

56 Asbjørn Jaklin, *Nordfronten, Hitlers skjebneområde* (Oslo: Gyldendal, 2006), 110–12; Jan Linder and Lennart Lundberg, *Ofredens hav—Östersjön 1939–1992* (Stockholm: Infomanager Förlag Jan Linder, 2002), 148–51; Lennart Lundberg, *Krigsmalmens offer* (Danderyd: Åkeriförlaget, 1993), 220–24.

57 Martti Ursin, *Pohjois-Suomen tuhot ja jälleenrakennus saksalaissodan 1944–1945 jälkeen* [War damage and reconstruction in northern Finland after the Lapland War of 1944–1945] (Oulu: Pohjoinen, 1980), 398; Oula Seitsonen and Vesa-Pekka Herva, "Forgotten in the Wilderness: WWII German POW Camps in Finnish Lapland," in *Archaeologies of Internment*, ed. Adrian Myer and Gabriel Moshenska (New York: Springer, 2011), 171–90.

58 Jaakko Alakulppi, *Lapin ilmailuhistoria I* (Vaasa: Lapin maakuntamuseo, 1999), 431–32; Hannu Valtonen, *Pohjoinen ilmasota: Suomeen liittyviä sotatoimia syksystä 1944 kevääseen 1945* (Tikkakoski: Keski-Suomen ilmailumuseo, 1996), 126.

59 Vehviläinen, *Finland in the Second World War*, 65.

60 Sampo Ahto, *Aseveljet vastakkain: Lapin sota 1944–1945* (Helsinki: Kirjayhtymä, 1980).

61 Vehviläinen, *Finland in the Second World War*, 257.

62 Ursin, *Pohjois-Suomen tuhot ja jälleenrakennus saksalaissodan 1944–1945 jälkeen*, 395–412.

63 Jari Eloranta and Ilkka Nummela, "Finnish Nickel as a Strategic Metal, 1920–1944," *Scandinavian Journal of History* 32, no. 4 (2007): 322–45; see also Esko Vuorisjärvi, *Petsamon nikkeli kansainvälisessä politiikassa 1939–1944: Suomalainen todellisuus vastaan ulkomaiset myytit* [Pechenga Nickel in the International Politics 1939–1944: Finnish reality against foreign myths] (Helsinki: Otava, 1990). However, the key territory Finland lost during World War II was not in the north, but Karelia, in the southeastern part of the country.

64 Urho Kekkonen, *Onko maallamme malttia vaurastua?* [Does our country have the patience to prosper?] (Helsinki: Kustannustalo, 1952).

65 Uuno Varjo, "Development of Human Ecology in Lapland, Finland after World War II," *Geoforum* 2, no. 1 (1971): 17–71; Uuno Varjo, "Agriculture in North Lapland, Finland: Profitables and Trends since World War II," *Fennia* 132, no. 1 (1974): 36–66.

66 Emmerson, *Future History of the Arctic*, 136–37, 219; Laurence C. Smith, *The New North: The World in 2050* (London: Profile Books, 2011), 248–61.

Chapter 16
International Conservation after the Two World Wars

ANNA-KATHARINA WÖBSE

Today's global environmental regime has a long history dating to the beginning of the twentieth century. As some recent environmental history studies have shown, modern industrial war and the state of the natural environment are intrinsically linked, and so is the environmental movement.[1] War has shaped the evolution of environmentalism in the twentieth century. This essay explores the evolving international environmental discourse before, during, and after World War II by focusing on the relations between international organizations and some of their affiliated networks of experts and activists. Its main thesis is that while the beginning of World War II put an abrupt end to some promising liaisons and campaigns, it also triggered the making of a tight and influential network of environmental elites right after it ended in 1945. To understand the patterns of continuity and fracture, this essay explores the situation of the international elites and their constituencies immediately after the First World War. The "curious" green moment between 1945 and 1950 deserves special historical attention, as it illustrates how and why debates concerning the use and misuse of nature gained new dynamics because of the war experience.[2]

PRELUDE: THE BEGINNINGS OF ENVIRONMENTAL INTERNATIONALISM

The year 2013 marked one hundred years since the first Global Conference on Nature Protection convened in Bern. It was the first multinational conference with a global outlook and a broad and interdisciplinary scope. More recent accounts of the history of international environmentalism refer to the 1913 conference as one of the starting points in a process that would eventually lead to the building of global environmental regimes and institutions like the United Nations Environment Programme (UNEP).[3] The Swiss government had invited some twenty-two national delegations to debate the issue in Bern. There had been specific international meetings before—like the first international ornithological congress in 1884, the North American Conservation Conference under the auspices of Theodore Roosevelt in Washington in 1909, or an international conference for nature and so-called homeland protection in Paris in the same year. But the Bern meeting in 1913 aimed at creating a forum for nature protection worldwide.

The spiritus rector behind the gathering was the Swiss natural scientist Paul Sarasin. Watching the accelerating competition for natural resources such as whale oil and bird feathers as well as unrestricted industrialization, exploitation, and cultivation of vast tracts of land at home and abroad made him call for the establishment of a global clearinghouse for nature protection. He systematically referred to the violence men exercised against nature and used metaphors that described the disturbed human-nature relation.[4] Sarasin was neither the first nor the only critic who had detected the international dimension of the problem. But, being a gifted writer and orator, he provided a concise synopsis and narrative that met with the activities of expanding social movements dedicated to issues like protecting birds, safeguarding landscapes and scenery, and preserving African wildlife. In Bern, he managed to convince most of the delegates from about twenty-two countries to sign a preliminary declaration of intent to form a Commission for Global Nature Protection.[5] While Sarasin was on a boat to Spitsbergen in summer 1914 to prepare the first campaign of the commission (intending to turn the archipelago into an international nature reserve), Germany declared war on France. Any internationalistic ambitions were shattered.

The hostilities of the First World War stopped the process of establishing an international clearinghouse almost immediately. Europe saw the worst killing spree and resource depletion of its history. It experienced utter devastation on the western front. A new world order emerged in the aftermath. Although a victor's organization and the expression of a Eurocentric world, the League of Nations was received as a shiny flagship of a peaceful future. At least, that was what many protagonists of civil society thought when they approached the league to lobby for issues like the wise use of the seas, the protection of animals, the end of oil pollution, or a worldwide system of national parks. To

Figure 16.1. A league for nature? The League of Nations at its opening session in Geneva, November 1920. Source: UN Photo/Julien.

them, the league was a universal forum for any question the new community of nations and peoples was facing.

As soon as the war was over, Sarasin approached the League of Nations' secretariat to reestablish the Commission for Global Protection of Nature. In Geneva, he met allies like the vice director general, Japan's Inazo Nitobe, who enthusiastically supported the initiative to establish a commission that would be affiliated to the league. But the impact of the war had thwarted such plans. A deep gap was running through the community of experts, dividing them along the lines of former battlefields. The time was not yet ripe to place them at the same table. In 1920, when Sarasin asked the delegates of the Bern conference about his plan to revive the commission and seek the support of the league, his idea was met with reluctance—the answers reflected doubt, uncertainty, and some even sheer hostility. The Dutch activist Pieter van Tienhoven recommended patience; according to him, more time was needed for the wounds of the war to heal.[6] Inazo Nitobe, himself deeply disappointed that the plan of a central agency for nature protection was failing, showed some optimism: "As the general conditions of the world improve, I hope the time will come when the governments will take up a scheme for the global protection of nature."[7] Unsettled quarrels and hostilities hindered international cooperation.

Because of the war, the former elites of international preservation did not cooperate and thus did not manage to establish either a strong lobby group or an institution within the league's system. Thus the league lacked any agency it could ask for support, advice, and expertise. Sarasin's initiative had failed. Yet during the 1920s and the early 1930s, the league's technical organizations did actually discuss several international environmental issues like whaling, oil pollution, animal protection, and world natural heritage. The experts in Geneva prepared the ground for the several environmental conventions. But they could only be ratified after World War II. All of these problems concerning the human-nature relation were in one way or the other related to the war that just had ended and to the one that was yet to come. Whales provided the raw material for margarine, washing powder, and explosives—commodities that many devastated national economies longed for. And they were ready to fight for such exquisite resources. Oil pollution had been worsened by the modernization of navy ships. The pollution of the seas anticipated that the use of the global commons might be fiercely debated. Livestock was moved across Europe to meet the growing demand of a meat-hungry population. Animals too were used as test subjects for new chemicals designed for future wars. As new states were born and old ones had to rebuild or strengthen their domestic power, national parks were tools to define and to fence in national identity.[8]

The league was not successful at the time for various reasons: it lacked expertise and an efficient infrastructure, and it lacked a supporting and independent partner from civil society. But the league did host some initiatives that marked the beginning of a global environmental regime, like the whaling convention of 1931 and the first drafts for a convention on oil pollution in 1936. Both proved fundamental to the establishment of international rules regarding the use of the global commons, as their expertise and content provided the raw material for whaling and oil pollution regulations in the postwar years. At the beginning of the 1930s, when some of such debates resulted in tangible plans to create multilateral agreements, the next war was dawning. Germany and Japan left the League of Nations in 1933, and Italy followed in 1937. The final acts of the league before it ceased functioning was to expel the Soviet Union in December 1939 for its attack on Finland.

The growing diplomatic tensions in the world in the late 1930s pushed the early environmental debates back to the political periphery. The convention on whaling suffered from the fact that the league found itself stalemated and British diplomacy preferred to convene an "independent" conference in 1937.[9] When the league managed to draft a convention concerning oil pollution of the seas and started to organize a meeting to create the final version of the agreement, it found itself in the process of disarray. Sherman Strong Hayden, whose book on international preservation was published in 1942, reported on the situation of the league's debates on marine oil pollution. "In September 1936 the League was reported to be still brooding over the question, but worse things were in store for the League of Nations and no further word has been heard on oil pollution. . . . To the human victims of an inhuman day we may confidently add an untold number of maritime creatures."[10] Once more, war proved a deadlock for any environmental debate. As Hayden put it, "We cannot hope for much advance in wild life preservation when the preservation of any institution whatever seems almost in doubt."[11]

Meanwhile, French, Dutch, and Belgian activists had proposed a private initiative in 1927 to unify people and issues, proposing an International Federation of Nature Protection Societies to link the different groups interested in the protection of nature. It took some time to turn the initiative into a working office.[12] When the small network of preservationists interested in international cooperation eventually established the International Office for the Protection of Nature (IOPN), the league was a dying giant because of accelerating power plays and aggressive nationalism. Compared with the visions of Sarasin, the IOPN had modest ambitions.[13] The few activists lacked money and confidence. They set out to bring together data, knowledge, and relations by establishing sections for documentation, correlation, and education. Nevertheless, they ensured that there was something like an institutional forum for the movement.

One aspect is central for understanding the elitist history of international environmentalism. All the men organized in the IOPN were employed by the state or by a scientific institution. Although the IOPN mentioned sentimental and aesthetic reasons, scientific considerations were presented as the most important driving force behind the organization. They might diagnose that "the progress in agriculture, the development of industry, and the increase of the human population, have had for effect that in many regions of the world the fauna and flora are on the eve of total annihilation."[14] They would not, however, question the political or economic system responsible for such degradation. The IOPN was supportive of the states, inviting them to become members, and was subsidized financially by various governments. The IOPN saw itself as a consulting agency that governments would ask for advice.

It took years to conciliate international relations. At the end of the 1930s, not only had German and American key figures joined the office but also the financial situation had improved. But once again, any international cooperation stopped soon after Germany forced the rest of Europe into war, and the Second World War began. The office and its outstanding collection of books, journals, and law texts had to be moved from Belgium to Amsterdam. At the beginning of the war, the international community of preservationists tried to spare its issues from the hostilities. In a 1939 letter to all national sections, the French president of the International Committee for Bird Preservation, Jean Delacour, pleaded, "The preservation of nature is above all human quarrels."[15] His call was in vain. Soon Delacour himself was confronted with the outcome of such human quarrels. He emigrated to the United States when Germany invaded France in early 1940.[16]

WAR AND RECONSTRUCTION

The German military expansion brought death and destruction to both nature and people. Although the Nazi regime had tried to create a "green" image for itself, nature protection was a marginal issue that was neglected in times of war.[17] As far as the loss of the movement's people is concerned, we find here one of the yet unwritten dark chapters of international environmental history. Apparently many of the Belgian and French protagonists were members of the resistance movement, and many activists who had played decisive roles in the international conservation community were either arrested or killed. Among the victims were the Polish zoologist Michał Siedlecki, an expert on fishery and one of the key figures in IOPN, and his colleague Jerzy Smoleński, another eminent nature activist. Both were professors at the Jagiellonian-University of Krakow. They were among the 183 members of the University the Gestapo arrested in November 1939 and deported to the concentration camp Sachsenhausen near Berlin, where they soon died.[18] Another victim was Jean-Marie Derscheid, one of the key figures of the Belgian nature protection community, who had been working hard during the interwar period to bridge the gap between activists of the ex-belligerent nations and had managed to reintegrate the German players into the international community.[19] He was a member of the Belgian resistance and was arrested in 1941 and executed in 1944.[20] Among the French victims of German aggression were Dr.

Figure 16.2. Dr. Inazo Nitobe (1862–1933) was a Japanese nature lover, agricultural economist, educator, diplomat, and politician. He studied in Japan, the United States, and Germany. The Quaker and cosmopolitan served as under-secretary general of the League of Nations from 1919 to 1926. Source: National Diet Library, Japan.

Piton, killed in Dachau, and Vicomte de Masin, both members of the Société Nationale d'Acclimatation et de Protection de la Nature.[21] Another victim was Roger Heim, an eminent botanist and mycologist. Being a member of the resistance movement, he was denounced by a Gestapo agent in 1942 and deported to the Austrian concentration camp Mauthausen.[22] He survived and played an important role in building up the International Union for the Protection of Nature (IUPN). The Swiss zoologist Charles Bernard, another key figure in the postwar reconstruction process of international networks, also supported the French resistance. He served France as a secret liaison officer between the free and the occupied parts of France but was never arrested. The war radically changed biographies.[23] If one considers the state of things immediately after the Second World War, it is astonishing how quickly some activists were back on their feet and ready to revive international cooperation on environmental issues.

The general political situation in 1945 was similar to the one after the First World War. Nationalism had been the driving force behind the hostilities that had devastated Europe. The diplomatic systems had failed to prevent aggression, the concept of international understanding had been belittled, and the

League of Nations had been ridiculed. But the basic idea of institutionalizing a system of collective security and international understanding prevailed. The United Nations set out to maintain peace, to foster friendly relations among states, and to cooperate internationally in solving economic, social, humanitarian, and cultural problems. Immediately after the war, the first director general of UNESCO, the eminent British biologist Julian Huxley, pushed the preservation issue. He convinced the delegates of the member states that conservation should enter the organization's agenda by emphasizing the cultural and scientific aspects: "I persuaded the Conference that the enjoyment of nature was part of culture, and that the preservation of rare and interesting animals and plants was a scientific duty."[24] Apparently, UNESCO had learned the league's lesson. It made a strong effort to create and support an institution from civil society.

In November 1946, the head of the fledgling Scientific Unit of UNESCO, Joseph Needham, addressed a letter to one of the leading figures of the international nature preservation community, the British ornithologist Phyllis Barclay-Smith. He expressed his hope that soon "a good international organisation of wide scope in Nature Protection will . . . be formed." According to him, "UNESCO stands ready to give it all possible help, material as well as moral."[25] Only two weeks after its constitution had been ratified, UNESCO seemed ready to support any initiative from civil society that was willing to deal with the preservation and conservation of nature. It should soon play a decisive role in drafting the first environmental agenda of the United Nations by offering ambitious actors from nongovernmental organizations (NGOs) an international forum to stage their visions. It is doubtful whether the International Union for the Preservation of Nature, or IUPN (known as the International Union for Conservation of Nature, or IUCN, from 1956 onward)—which claimed to be the world's first global environmental organization, referring to the IOPN's establishment in 1927 and today's most influential and largest professional global conservation network—would have so quickly been established after the war without the fundamental support of UNESCO. In other words, UNESCO created allies it thought it would need when staging environmental issues. The activists were well aware that they might need institutional help, or, as the director of the International Union of Biological Sciences, M. J. Sirks, put it, "Principally: Nature Protection comprises not only biology, but also economy, sociology and even diplomatic problems." He continued, "Practically: Nature Protection needs for its international work much money."[26] Money was decisive.

The war tremendously influenced the history of environmentalism. But while individuals like Julian Huxley were optimistic to use the situation to establish new structures, large parts of the European constituency seemed exhausted, as the British activist Max Nicholson described. When some Swiss activists made a bid for the leadership in international conservation in 1946 and invited their colleagues to a "miniature impromptu international conference," the English activists could only "reserve their position. This was the more difficult because at that time the British economy was war-shattered, London itself was badly bomb-damaged, and even the initial plans for a British organization to undertake the conservation of nature remained to be fully formulated."[27] The difficult situation on the European continent surely fostered a growing influence of American protagonists. Compared with most of the European networkers, their American counterparts appeared to be determined and rather well off.

The clash of European and American mentalities and status right after the war was best portrayed in a personal account by Miriam Rothschild, published in *World Conservation,* in which she described her impressions while traveling from Britain to France, crossing war-torn Europe to attend the inaugurating conference of IUPN in 1948. She witnessed the clash of postwar European pessimism and fatigue with the American spirit of optimism and victory. Miriam Rothschild was the daughter of the Honorable Charles Rothschild, and she followed in the footsteps of her father, who had supported Sarasin's vision in 1913. The keen entomologist was one of the few women (besides Phyllis Barclay-Smith) involved in the international community. While the public movement for bird and animal protection had a strong female participation in the higher ranks of transnational networking, few women could raise their voices in international policy circles. In her account she illustrated how difficult it was to figure out where any such international project like the establishment of the IUPN would be going. Rothschild traveled from Britain via Calais, which "looked like a moonscape," to the picturesque site of Fontainebleau:

> The contrast between the concrete chaos of Calais and sunny, smiling Fontainebleau filled one with strange, conflicting emotions of doubt and optimism. I required a little encouragement since I had been

> originally excited at the thought of the rebirth of a plan for international conservation but had not found corresponding enthusiasm in England. . . . There was also a rather pathetic feeling among the British contingent that, despite the many years that had elapsed since the Bern conference . . . they had not had time to "get their national nature conservation straightened out" and could not yet contemplate a government-sponsored, grandiose international organization.

While listening to the speeches in Fontainebleau, Rothschild was deeply impressed by American delegates like George Brewer and William Vogt because they "not only thought internationally but were idealistic and optimistic. Talking to them one felt the future of the rhinoceros and the skylark depended on energetic visionaries, not reasonable men who discussed the basic question of our meagre resources." Rothschild concluded, "Surprisingly I left France in a sanguine mood. Julian Huxley . . . had put into words what the 150 delegates present all felt in their bones—the love and fascination of life other than our own, which must be protected. Maybe the necessary flow of energy would come from the New World? Despite the unmarked graves of

Figure 16.3. The construction of a new global perspective: the UN General Assembly Building in progress in New York in summer 1950. Source: UN Photo/MB.

Calais and the dark shadow of the Holocaust—yes, there had been something magical about the Conference at Fontainebleau."[28]

Her account reflects the moral situation of the European postwar environmentalists. They had to cope with devastation and despair. Some felt that their activism helped to distract them, for others activism was a way of coming to terms with their war experience, and for some the end of the war simply offered a unique chance to place their project of international nature protection on the global agenda. The fragility of wildlife and the vulnerability of nature in general resembled the human fragility most of the activists had just experienced. War had provided a tabula rasa. Where would this blend of traditional nature lovers, big game hunters, colonial game wardens, keen ornithologists, and young ecologists move to? They did not have much time to define a concise program—postwar planning forced them into action.

NATURE VERSUS RESOURCES? THE UNITED NATIONS CONFERENCES OF 1949

The famous biologist, ecologist, and evolutionary theorist Julian Huxley had actively supported the IOPN in Brussels. Now he was the one to open the door to global politics. He became UNESCO's first director general, with the political power to make virtual networks a reality. He helped revive the former IOPN as the new IUPN, and thus established a small but influential hybrid organization that was to provide UNESCO with environmental know-how and ecological expertise.[29] Many of the people involved in the young IUPN had colonial experience, like the Belgians Victor van Straelen and Jean-Paul Harroy, the American Harold Coolidge, and the Swiss Charles Bernard. Moreover, many of them had been in one way or the other involved with economical planning processes during the war. Some of them had been plunged into global operations.[30]

Besides the new planning paradigm, the war experience triggered a new "one-world" perspective,[31] which neatly correlated with the thriving ecological studies. "Interconnectedness" was one of the central terms used in these days. Looking back to the postwar years and the history of IUCN, Max Nicholson described the clash of traditional and modern activists. His contact with leading ecologists taught him that the "fast-unfolding knowledge of the biosphere and its living systems called for vigorous operational development of the field of nature conservation."[32] UNESCO helped design a political platform for discussing environmental issues. The experience of the infernal

Second World War had surely changed the tone of environmental debates. The atomic bomb, the pace of decolonization, and starving populations literally remodeled the worldview of the community and resonated in almost all papers and discussions. Disturbing apocalyptic forecasts and Malthusian worries like Fairfield Osborn's *Our Plundered Planet* and William Vogt's *Road to Survival* (both 1948) struck a new chord.[33] Nature protection was not only about rare species or thrilling views anymore, but it also had to deal with the basic question of how life on a rapidly changing planet could be saved.

The debate was pushed forward when the Economic and Social Council of the UN (ECOSOC) planned the United Nation's Scientific Conference on the Conservation and Utilization of Natural Resources (UNSCCUR) and asked UNESCO for organizational support.[34] The staff of Huxley's office challenged the biased approach of ECOSOC as tending toward a purely technical interpretation of natural resources. The ECOSOC conference had been kicked off by US president Harry Truman, who intended to put an end to armed conflicts caused by resource scarcity,[35] which implied open access to and security of supplies. To UNESCO, such a reduction of the resource debate seemed short-sighted and smelled of a postcolonial extension of the Western control of natural resources. Instead, the UNESCO staff emphasized the cultural, ethical, and moral aspect of the resource problem, and promoted the idea of a global common heritage.[36]

UNESCO insisted on organizing an embedded International Technical Conference on the Protection of Nature that would simultaneously take place at Lake Success near New York City in the summer of 1949. Although almost forgotten in today's environmental historiography, it was an amazing get-together. In collaboration with IUPN, UNESCO invited nearly three hundred institutions and organizations.[37] Eventually, 130 delegates gathered for ten days to discuss the environmental problems of the world. The majority of the participants were from the United States, which was the hosting nation but also dominated international programs and scientific institutions.[38] In the interwar period, American conservationists had actively supported the IOPN by paying many of its bills. Now they framed the agenda of IUPN. They had the expertise and also the money the Europeans lacked. The list of speakers was a who's who of international environmentalism, but American experts clearly dominated the conference. There was an atmosphere of hope and the will to change. The political and social tabula rasa caused by the war encouraged people to think in larger terms. The scope of the agenda contained the

Figure 16.4. Mrs. Indira Gandhi, prime minister of India, in June 1972, addressing the Conference on the Human Environment at Stockholm, which "considered the need for a common outlook and for common principles to inspire and guide the peoples of the world in the preservation and enhancement of the human environment." Source: UN Photo/ Yutaka Nagata.

worldwide education of the public regarding nature protection, international cooperation to promote ecological research, the balance of nature (threatened by the use of pesticides like dichlorodiphenyltrichloroethane, known as DDT), the danger of the uncontrolled introduction of exotic species, and the extinction of large mammals in Asia and Africa. In addition, a World Convention on Nature Conservation was discussed, as was the role of IUPN and the special problems arising from national parks bordering a frontier.[39]

For the next decades, UNESCO was to become the most important agency within the UN system as far as cooperation with environmental NGOs was concerned. Activists sensed that this offered a unique opportunity to gain influence and to place their issue on the international stage. The war experience definitely changed the self-perception of the experts. While they had fought the stigma of being nostalgic, sentimental nature lovers, they now set out to prove that issues like erosion, water supplies, nutrition, ecological equilibrium, environmental knowledge, and species diversity were necessary for future peace. The preservationists and ecologists dared to think big. The scope of the discussions reveals how ambitious they were. There was no homogeneous environmental community, and the protection of the rhinoceros and the creation of reserves went hand in hand with radical critique concerning pesticides, grand-scheme planning, and environmental education. For a short moment, the community showed what it was capable of, offering a new way of interpreting the human-nature relationship.

The two conferences mentioned above help explain why and how the seismographic talents of the postwar community got lost along the way.[40] They clearly belonged to a "green" vanguard, but they did not get the attention they deserved, at least in hindsight. While stemming from the same notion on the state of resource depletion, the conferences offered conflicting interpretations. The UNSCCUR set out to draft a status quo of the most pressing problems of resource depletion all over the world. Waste and scarcity on the one hand and wise use on the other were the themes running through all the sessions concerning mineral, fuel, energy, water, forest, land and wildlife, and fish resources.

Against the backdrop of the Cold War rivalry, the need to rebuild Europe and Japan, and the challenge of feeding the world, experts and delegates of UNSCCUR tended to concentrate on short-term solutions for the pressing problems of starvation and of how to guarantee the well-being of a growing world population. Their time frame did not exceed the subsequent decades. Strong emphasis was laid on the exploitation of the "last frontiers," such as the tropical rainforests or the "under-fished" oceans of the Southern Hemisphere. There was a certain tendency toward large-scale planning and universal instructions on agricultural management, while the preservationists and ecologists still struggled to frame the complex and yet unknown interaction of humans and the biosphere: while one pressed for the acceleration of the current course of action, the other insisted on the deceleration of certain technical interventions. In particular, when it came to the forced use of chemicals, large-scale planning projects, and the introduction of exotic species, the ecologists pleaded for conducting fundamental research before applying such strategies to the so-called underdeveloped world.[41] In retrospect, the two conferences exemplified the antagonism inherent in the UN and its promise of modernity. The UN triggered schemes and plans decisive for the "environmental crisis" of the 1970s. At the same time, it featured networks that would not only voice environmental concern, but also provide the empirical evidence and scientific knowledge underlying this concern.

With the Cold War intensifying, it soon became clear that the environmental avant-garde of the late 1940s did not overcome the deepening divide between conservation (applied power politics) and preservation (restricted to the protection of rare species and sites of natural beauty). As soon as the brief chance for "greening" the global agenda had passed and the Cold War featured a series of crises, environmental issues seemed to be pushed to the periphery.[42] In a way, the conservation elites volunteered to withdraw to their own reserves. Lacking funds and manpower, IUCN soon restricted itself to rather traditional issues. Standards and nomenclature for national parks, the protection of African wildlife, and the editing of data for the red lists became the centers of their campaigns. Instead of integrating new constituencies, the elites rather focused on saving exclusive wilderness. Even the initiative to "create" the new environmental organization World Wildlife Fund (WWF) in 1961, which was designed to get not only funds but also publicity, was a top-down project.[43] Perhaps their exclusiveness can be explained by the fact that the academic elite had been involved in wartime planning. Moreover, their experience of need, emergency, and misery may have supported their paternalistic approach. Nevertheless, they carried the torch; they provided ecological knowledge and offered a new language to cope with the phenomenon of interconnectedness. Their expertise was of high relevance for new and more integrative UN initiatives like the Man and the Biosphere Programme. It was decisive for setting the agenda of the UN's Stockholm conference in 1972.[44] At the end of the 1960s, their message met a sensitized audience.

The paradigms would eventually shift. The fledgling science of environmental ecology started to offer explanatory models for identifying and understanding systematic interrelations.[45] The new social movements mobilized new constituencies and blended environmental issues with debates on human rights, equality, and pacifism, and discovered new methods for campaigning that had nothing to do with the rather paternalistic concepts of the traditional elites. From the late 1960s onward, the terms "environment" or "environmental" suddenly offered a linguistic category to denote an issue cutting across all UN organizations in one way or the other. "Sustainable development" would provide the discursive base for merging the conflicting routes.

CONCLUSION

Was the postwar environmental awakening to a great extent a product of the war, as far as the international organizations and their affiliated environmental activists were concerned? Both wars, or at least the end of the wars, tremendously changed the history of the environmental avant-garde. They triggered the establishment of more or less powerful international agencies to which activists could turn. Both wars strengthened the position of experts and supported elites. Both wars encouraged activists to extend their scope and to demand new strategies for how to cope with global common goods.[46] So they definitely shaped environmentalism in the long run. At the same time, war had a devastating effect—any hostilities and diplomatic tension implied a serious setback for any agreement. The experience of war, violence, and destruction decisively changed many biographies. This is certainly true for the postwar generation of activists (and it would be worthwhile to study more systematically the biographies of some of the key figures). On one hand, war had trained them to work and think along new and complex planning models, which helped them understand the social and economic dimension of ecology. On the other hand, grand-scheme planning soon threatened some of the most beloved sites of the constituency. Last but not least, the Second World War triggered the Americanization and modernization of preservationism, and thus had a lasting impact on the evolution of environmentalism as we know it today.

Notes

1 Richard P. Tucker and Edmund Russell, eds., *Natural Enemy, Natural Ally: Toward an Environmental History of War* (Corvallis: Oregon State University Press, 2004); Edmund Russell, *War and Nature: Fighting Humans and Insects with Chemicals from World War I to* Silent Spring (New York: Cambridge University Press, 2001); J. R. McNeill and Corinna R. Unger, eds., *Environmental Histories of the Cold War* (Cambridge: Cambridge University Press, 2010); Charles E. Closmann, ed., *War and the Environment: Military Destruction in the Modern Age* (College Station: Texas A&M University Press, 2009).

2 On the "curious utopian moment," see Glenda Sluga, "UNESCO and the (One) World of Julian Huxley," *Journal of World History* 21, no. 3 (September 2010): 393–418, esp. 393.

3 Nico Schrijver, *Development without Destruction: The UN and Global Resource Management* (Bloomington: Indiana University Press, 2010); Martin Holdgate, *The Green Web: A Union for World Conservation* (London: Earthscan, 1999); Barbara J. Lausche, *Weaving a Web of Environmental Law* (Bonn: Erich Schmidt Verlag, 2008); Patricia Birnie, Alan Boyle, and Catherine Redgwell, *International Law and the Environment*, 3rd ed. (Oxford: Oxford University Press, 2009).

4 Paul Sarasin, "Weltnaturschutz—Global Protection of Nature" (paper presented at the 8th International Congress of Zoologists, Graz, August 16, 1910); reproduction translated from German, Basel 1910 (Bern: Schweizerische Landesbibliothek, 2005).

5 *Recueil des Procès-Verbaux de la Conférence Internationale pour la Protection de la Nature* (Bern: K. J. Wyss, 1914).

6 Copy of Tienhoven's letter of November 22, 1919, 40/17779/17779, League of Nations Archive, Geneva: "J'ai la conviction que dans quelques années les bords trop aiguisés seront polis et que les amis de la Nature pourront de noveau travailler main à main pour le but de protection nécessaire."

7 Nitobe to Sarasin, October 19, 1922, 13/24249/3514, League of Nations Archive.

8 Bernhard Gissibl, Sabine Höhler, and Patrick Kupper, "Towards a Global History of National Parks," in *Civilizing Nature: National Parks in Global Historical Perspective*, ed. Bernhard Gissibl, Sabine Höhler, and Patrick Kupper (Oxford: Berghahn Books, 2012).

9 Kurkpatrick Dorsey, "Compromising on Conservation: World War II and American Leadership in Whaling Diplomacy," in Tucker and Russell, *Natural Enemy, Natural Ally*, 252–69; Kurkpatrick Dorsey, *Whales and Nations: Environmental Diplomacy on the High Seas* (Seattle: University of Washington Press, 2013); Charlotte Epstein, *The Power of Words in International Relations: Birth of an Anti-Whaling Discourse* (Cambridge: Massachusetts Institute of Technology Press, 2008).

10 Sherman Strong Hayden, *The International Protection of Wild Life* (New York: Columbia University Press, 1942; reprint, 1970), 113.

11 Ibid., 175.

12 International Office for the Protection of Nature, Brochure 1936, Bundesarchiv Koblenz 245/201, German Federal Archives BAK.

13 The IOPN was a predecessor of the International Union for the Protection of Nature (IUPN), which in 1956 eventually turned into what is known today as the International Union for the Conservation of Nature (IUCN).

14 IOPN Brochure, Brussels 1936, Bundesarchiv Koblenz B 245/201, 167*ff*.

15 International Council for Bird Preservation, British Section and Polish Section, *Annual Report for 1939 and 1940* (London, 1940).

16 Jean Delacour, *The Living Air: The Memoirs of an Ornithologist* (London: Country Life Limited, 1966): Ernst Mayr, "In Memoriam: Jean (Theodore) Delacour," *Auk* 103 no. 3 (1986): 603–5.

17 See Simo Laakkonen, "Environmental Policies of the Third Reich," chapter 4 in this volume.

18 In France, Michel Siedlecki was honoured posthumously by his French colleagues with the Grande Medaille Isidore Geoffroy St. Hilaire by the Société d'Acclimatation. P. R., "La vie de la Société," *La Terre et la Vie—Revue d'Histoire Naturelle* 94, no. 2 (1947): 151.

19 Bundesarchiv Koblenz B 245/201.

20 Paul Brien (1971), Entry "Biography of Jean-Marie Derscheid," *Biographie nationale* 37 (1971): 211–35. Translated by Dan Reboussin, accessed July 10, 2012, http://ufdc.ufl.edu/IR00000442/00001/pdf.

21 P. R., "La vie de la Société," 147.

22 Roger Heim, *La Sombre Route* (Paris: José Corti, 1947).

23 On the legacy of the wars on French-German relations concerning the protection of nature, see Anna-Katharina Wöbse, "Les liaisons sinueuses: Les relations franco-allemandes en matière de protection de la nature dans la permière moitié du Xxe siècle," in *Une protection de l'environnement à la française?*, ed. Charles-Francois Mathis and Jean-François Mouhot (Paris: Champ Vallon, 2013), 108–9.

24 Julian Huxley, *Memories II* (Harmondsworth, UK: Penguin, 1978), 11.

25 Needham to Barclay-Smith, November 29, 1946, 502.7 A01 ICPB, UNESCO Archives.

26 Sirks to Needham, March 9, 1947, 502.7 A 01 "-66" IUCNNR, Part I, UNESCO Archives.

27 Max Nicholson, *The Environmental Revolution: A Guide for the New Masters of the Earth* (London: Hodder and Stoughton, 1970), 194–95.

28 Miriam Rothschild, "I Remember . . . ," *World Conservation* 3–4 (1998): 7.

29 Holdgate, *Green Web*, 18–19.

30 Peder Anker, *Imperial Ecology: Environmental Order in the British Empire, 1895–1945* (Cambridge, MA: Harvard University Press, 2001), 201*ff*.

31 Sluga, "UNESCO and the (One) World of Julian Huxley," 393–418.

32 Max Nicholson, "My Dilemmas with IUCN," *World Conservation* 3–4 (1998): 4–5.

33 Thomas Robertson, *The Malthusian Moment: Global Population Growth and the Birth of American Environmentalism* (New Brunswick, NJ: Rutgers University Press, 2012), 36–60.

34 United Nations, *Proceedings of the United Nations Scientific Conference on the Conservation and Utilization of Resources* (New York, 1950).

35 Björn-Ola Linnér, *The Return of Malthus: Environmentalism and Post-War Population-Resource Crises* (Isle of Harris, UK: White Horse Press, 2003).

36 Memorandum, Dr. Yeh Chu-Pei, ca. early 1947, 502.7 A 06 (73) "49," Part I Up to 30/IV/1948, UNESCO Archives.

37 Paris: International Technical Conference on the Protection of Nature, 1949, 502.7 A 06 (73) "49" 18, UNESCO Archives.

38 Michael Lewis, *Inventing Global Ecology: Tracking the Biodiversity Ideal in India, 1947–1997* (Athens: Ohio University Press, 2004).

39 Holdgate, *Green Web*, 41–43.

40 United Nations, *Proceedings of the United Nations Scientific Conference on the Conservation and Utilization of Resources*, vols. 1–8 (New York, 1950); UNESCO/IUPN: *International Technical Conference on the Protection of Nature, Proceedings and Papers* (Paris: Hayez, 1950); Anna-Katharina Wöbse, " 'The World after All Was One': The International Environmental Network of UNESCO and IUPN, 1945–1950," *Contemporary European History* 20, no. 3 (August 2011): 331–48; Linnér, *Return of Malthus.*

41 United Nations, *Proceedings of the United Nations Scientific Conference*, 177–89.

42 Tuomas Räsänen and Simo Laakkonen, "Cold War and the Environment: The Role of Finland in International Environmental Politics in the Baltic Sea region," *AMBIO* 36, no. 2–3 (April 2007): 223–30; J. R. McNeill and Corinna Unger, "The Big Picture," in McNeill and Unger, *Environmental Histories of the Cold War*, 1–18.

43 Alexis Schwarzenbach, *Der WWF—Eine Biografie* (Munich: Heyne, 2011).

44 Malcolm Hadley, "Nature to the Fore: The Early Years of UNESCO's Environmental Programme, 1945–1965," in *Sixty Years of Science at UNESCO, 1945–2005* (Paris: UNESCO, 2005), 201–23.

45 Ludwig Trepl, *Geschichte der Ökologie*, vol. 17, *Jahrhundert bis zur Gegenwart* (Frankfurt: Athenäum, 1987); Anker, *Imperial Ecology*; Lewis, *Inventing Global Ecology.*

46 Joachim Radkau, *Die Ära der Ökologie: Eine Weltgeschichte* (Munich: C. H. Beck, 2011), 101–18.

Part IV

CONCLUSIONS

Chapter 17
Hypotheses
World War II and Its Shadows

SIMO LAAKKONEN, RICHARD P. TUCKER,
AND TIMO VUORISALO

In August 1945, as the carnage of war ceased, wide regions of Europe, Asia, and the Pacific lay in ruins—an environmental as well as human catastrophe. While civil and regional wars continued in many places through the rest of the decade, the industrial giants, excluding the United States, lay crippled. In addition to a formal peace treaty, the end of World War II signified the change of borders in Asia and Europe, the expansion of Communist power in various parts of the world, and the birth of a new world order with two new superpowers. After the war, hunger, malnutrition, and poverty affected many nations. Ethnic cleansing took place in several regions. About fifty million people became refugees across the world—a number surpassed in absolute terms only seventy years later, in 2015.

Because of all of these changes, the Second World War cast short and long shadows not only on people but also on the environment. The more time that passes since this most overwhelming of wars, the more important it is to address the long shadows of the war. But because of the deeply contradictory nature of war-related socioecological changes, it is risky to draw global generalizations about this war's environmental impacts. Thus there is a need to construct a more inclusive socioecological perspective on the war. The essays in this book explore some of these shadows. The limits of this collection, and the subjects that these papers do not address in any detail, also suggest areas that invite further research. We need additional thematic and empirical case studies before well-grounded theoretical generalizations can be made.

Nevertheless, in the following we present some tentative hypotheses of the major environmental impacts of the largest military conflict in world history.

SHADOW I: PREPARING FOR THE WAR

As noted in the introduction and many of the chapters of this volume, knowledge of environmental history of the previous wars and military operations in the interwar years is vital for understanding the World War II period and its aftermath. Winston Churchill addressed the world wars and interwar period as "another Thirty Years' War."[1] Hence there is a need to construct a more inclusive socioecological perspective on the war, including its precursors and legacy. Yet until now there has been little study of the environmental history of military operations in the interwar years, though many of its components are well known.

The Versailles peace settlement left little resolved about the World War I conflagration.[2] The decade after the Versailles Treaty included continuing turmoil and fighting in several countries. In the early 1930s, the main industrial powers started to revive their arms industries. The environmental costs of that development have yet to be assessed. That massive economic shift, which moved the industrial world toward a permanent military-industrial complex, can be understood in terms of its intensification of natural resources for military purposes, the shift of industrial capacity toward long-term acceleration of resource exploitation, and a concomitant expansion of industrial wastes.[3]

Agricultural intensification was another environmentally significant dimension, as regimes and economic leaders feared that food resources for their populations would be inadequate.[4] Stalin's agricultural policy in the early 1930s was a disastrously intense variation on this effort, as the collectivization of farmland led to a crash of production and desperate malnutrition for masses of Russians.[5] Even in far reaches of the colonial world, that decade brought agricultural crises that nearly led to famine.[6] It is no wonder that erosion—"destruction of the world's thin living cover" by human mismanagement—was addressed as a global problem as early as the late 1930s.[7]

The depression decade also saw precursors of the global conflagration that followed. These wars were of two sorts; each brought environmental damage to new regions. One sort included the empire-building wars of conquest, as secondary powers attempted to compete more effectively with their powerful imperial rivals. Japan's conquest of China's industrial belt in Manchuria (1931) fits this category,[8] as did Italy's invasion of Abyssinia (Ethiopia) in

1935–36.[9] The other was ideological. Left/right conflict came to a head in Spain in the grisly civil war of 1936–39. Each of these regional conflicts left environmental scars, both rural and urban.[10]

Under the shadow of renewed total conflict, though, efforts at resource conservation were still possible until 1939. Simo Laakkonen in chapter 4 surveyed the contradictory poles of nature conservation and environmental protection under the Third Reich, before and during the war. Generally speaking, in Europe and North America, protection of soil, waters, and air from pollution had developed relatively rapidly, especially in the 1930s, on the municipal and national levels. Recycling programs also expanded in the same decade.[11] Municipal waterworks were nearly ubiquitous in the industrial countries, and wastewater treatment plants were widely discussed and constructed by pioneering cities.[12] Air pollution became an increasing problem in mill towns and cities. Anna-Katharina Wöbse in chapter 16 addressed international cooperation in terms of nature conservation and environmental protection prior to war.[13]

Had these developments continued, without the Second World War, the environmental awakening might well have accelerated earlier than it actually did (see Shadow IV, below). But nature conservation and environmental protection might have looked quite different from what they became after 1945. During the interwar period, governmental environmental administration was at the fledgling state, and most related movements were rather local and expert-oriented, socially exclusive, and politically neutral or moderate, and thus institutionally relatively weak. Therefore the outbreak of war became a major cause for the postwar reorientation and reorganizing of environmental policy making and movements. But it also redirected global environmental culture from a prewar multicultural and multilingual model toward an Anglo-American model.[14] The Second World War became one of the main forces that rewrote the environmental history of the latter part of the twentieth century.

SHADOW II: FROM DESTRUCTION TO RECONSTRUCTION

The most obvious and massive environmental dimension of the war was destruction—of cities, infrastructure, agricultural landscapes, and natural habitats. Whenever disasters, either natural or man-made, strike, the story of their environmental consequences is incomplete until the course of reconstruction is included. This dimension of urban-industrial environmental history during and immediately after war deserves greater attention.

Air war was the unprecedented key to the destruction of hundreds of urban and industrial centers. The first round was the German bombing of London and other British cities in the Battle of Britain of 1940. Then the British and American air forces retaliated massively against Germany, which virtually leveled more than a hundred German cities by war's end in mid-1945.[15] Ground warfare in France was more spasmodic, with less environmental damage to cities, farmlands, and forests.[16] Looking eastward toward Poland, the Baltic states, Belarus, Ukraine, and Russia, the environmental carnage of intensive and wide-ranging fighting still invites environmental historians, though Paul Josephson in chapter 5 describes what Russia and Belarus suffered.[17]

In East and Southeast Asia, Micah Muscolino and Richard Tucker in chapters 6 and 7, respectively, catalogue the appalling loss of human life, property, and natural environments.[18] In addition, William Tsutsui and Timo Vuorisalo in chapter 14 describe how Japan's struggle to wrest control over the resources of Southeast Asia and the Pacific Islands from the European colonial powers and the United States resulted in massive damage to the ecologically vulnerable oceanic islands and to maritime resources.[19] Carol MacLennan in chapter 8, on Pearl Harbor, illustrates the unique details of military operations in Hawai'i from the perspective of a century's time.[20]

World War II created a global dump. After the war, abandoned front lines and battlefields became dangerous landscapes, with a cryptic blend of contaminated soils, damaged trenches, pillboxes, personal belongings, decaying body parts and carcasses, weapons, shells, mines, unexploded ammunition, destroyed tanks, rusting barbed wire, and numerous pests. The Atlantic Wall and other major defense and fortification lines lost their military importance but still covered tens of thousands of miles and proved hard to deconstruct. Abandoned military bases, training areas, and proving grounds became in practice scrapyards with useless ships, vehicles, aircraft, collapsing hangars and barracks, spilled oil, leaking fuel storages, and tons of human waste.[21] Owing to the high number of sunken ships and planes, Allied sailors gave the name "Iron Bottom Sound" to a sound in the Pacific on the site of the Battle of Guadalcanal. Similar underwater grave- and scrapyards can be found in the Atlantic and other seas. Most wartime mines, production facilities, and industries had hastily improvised processes and dumping grounds.[22] A major part of the war-induced global landfill consisted of the above-mentioned villages, towns, and cities devastated by strategic bombings and battles. The resulting debris was gathered into giant mounds of rubble called

Schuttberge in German. But in many areas, ruins and debris were simply left on-site.[23] To make things worse, some dangerous military waste was concealed.[24] Consequently, a massive but unknown amount of war debris still lies scattered in watercourses, mires, forests, jungles, deserts, fields, gardens, suburban backyards, and city centers in Africa, America, Asia, Australia, and Europe. Some part of this dark heritage of the war scars landscape, people, and the rest of nature, while some part is beneficial or even valuable.[25] Therefore an increasing paradox today is whether the remains of the Second World War are problematic junk to be removed or historical remnants to be preserved.[26] In the aggregate, the environmental history of war rubble is one of the untold stories of the Second World War.

The suffering of civilian populations did not immediately abate in 1945, for the dislocation of populations and their sources of food and shelter would take years to resolve.[27] Reconstruction of the cities of Europe and Asia was an enormous task, requiring massive additional flows of natural resources.[28] Many of Japan's cities, which had been made largely of wood, were reduced to ashes by Allied wartime firebombing. They required the import of massive amounts of softwood, primarily from the northeastern shores of the Pacific in Canada and the United States, and provided a reliably expanding market for the already-injured forests of Southeast Asia. The vast Siberian forest was also opened to timber extraction on a large scale.[29] Even during the course of battle, people on the ground began recycling construction materials, first for survival and then for rebuilding designed environments.[30] All in all, the need for energy and construction materials and lack of control during and after the war led to excessive use of forests in different parts of the world. Also, wartime necessities for wood caused a turn toward industrial forest policy that prevailed after the war in most parts of the world, reducing overall biodiversity and disrupting the ecological integrity of forests.[31]

SHADOW III: GLOBAL EXPLOITATION OF NATURAL RESOURCES

What do the names Reykjavíkurflugvöllur, Sir Seewoosagur Ramgoolam, Svalbard, Seymour, and Frans Kaisiepo have in common? All are names of airports built during World War II on some of the world's most remote islands situated in different seas and oceans. The first one in the list was built by the Scottish Black Watch regiment in Reykjavík, Iceland; the second by the British on Mauritius; the third by the German Luftwaffe in Svalbard Islands; the fourth by the US Air Force in the Galápagos Islands, Ecuador;

and the last one by the Japanese Army in Biak, New Guinea, which now belongs to Indonesia.[32] In fact, the Second World War did not open the world with weapons but with spades, bulldozers, and engineers. Wartime urgencies made possible penetrating remote regions that had never experienced industrial development. Belligerent countries built roads, railroads, ports, airfields, storage facilities, industries, and towns all over the world during the Second World War.[33]

War entailed both destruction and construction. An army of soldiers, prisoners of war, and forced and slave laborers built new logistic infrastructure demanded by modern industrial warfare in harsh conditions in the Arctic, boreal, Mediterranean, and tropical ecosystems. In chapter 15, Ilmo Massa and Alla Bolotova describe the process for the circumpolar far north, and in chapter 7, Richard Tucker does the same for Asia.[34] The emergent military air, land, and maritime networks promptly opened whole new areas of the world for the spread of disease, alien species, large populations, urban-industrial culture, and natural wealth extraction. This war-created industrial network became the basis of accelerated extraction for the military demands of the Cold War and for the rising consumer demand of newly affluent economies, including the expansion of international mass tourism. In sum, the Second World War was a global infrastructure-building project with notable long-term environmental consequences.

Although World War II was not truly global in terms of the military struggle, it was nearly global in material and socioecological terms. The most geographically widespread dimension of the war was the contest for access to critical natural resources that highly industrialized warfare demanded.[35] The growth of armaments industry that required these resources had started already in the early 1930s, in the shadow of the Great Depression. The resulting quest for resources affected not only areas under military conflict but also Africa, Latin America, and the Arctic region. As Evan Mawdsley emphasizes in chapter 3, Anglo-American domination of the natural resources of the world was challenged by their competitors—not only Germany and Italy, but especially Japan in its penetration across the Pacific and into Southeast Asia.[36] No strategic resource was as fundamental for this war as petroleum. This decisively shaped the strategic goals of the major powers, and to a large extent the course of battle.[37] For Germany and the Soviet Union, the Ploesti oil fields in Romania and the Baku fields in Azerbaijan were the centers of conflict. Britain and the United States controlled the greatest petroleum reserves, in the Middle East and beyond into Southeast Asia and Latin America. Japan had little in the way of domestic oil sources; ominously for its ambitions, in the 1930s it imported 80 percent of its petroleum from the United States.[38] Consequently, oil became an effective weapon during war and increasingly a pollution problem after it.[39]

In the longer run, the arms industry expanded exponentially during the war, providing the base for long-term military-industrial societies: full mobilization of entire societies, entire civilian populations, for wartime production.[40] British industry, backed by the resources of its global empire, had begun full mobilization by 1939.[41] Then the United States, separated by oceans from direct fighting, rose to international economic domination between 1942 and 1945, beginning with providing a decisive military-industrial output to the beleaguered European Allies, especially Britain.[42] By 1943–44, Soviet industry, in a triumph of highly controlled social mobilization, succeeded in producing enough tanks, planes, and the entire range of war materiel to turn back the German invasion.[43] This set the Soviet economy on such intensive military footing that it continued into the Cold War decades with staggering production of pollution.

In addition to energy sources, the industrial armed forces required vast reserves of metals.[44] Nazi Germany was able to import iron ore from neutral Sweden and nickel from Finland.[45] Otherwise the list of strategic metals was constantly expanding, and many vital metals had to be shipped long distances from their sources to their end users. Allied scientists working on the Manhattan Project relied almost exclusively on uranium ore from the Belgian Congo. Aluminum had become critically important, especially for the decisive air forces of the combatants. Mining and processing of aluminum extracted heavy environmental costs, as Matthew Evenden reveals in chapter 11. Copper was also important: for the United States, even the world's greatest open-pit copper mine, in Montana, was insufficient to meet wartime industry's requirements.[46] Other copper reserves for meeting American and British needs were in Chile and southern Africa.[47] Europe's colonial empires tapped many of Africa's natural resources in an emergency expansion of the broad mining complex there, preparing the way for steady expansion of mining (including its human and environmental costs) in the postwar years.[48]

The world's forest resources also underwent heavy pressure for the war. The US government gave strategic support to American tropical timber firms to undertake surveys of the forest resources of most Latin American countries. For Latin America, Christopher Boyer in chapter 12 shows the consequences in Mexico, which had the most developed timber products industry in Latin America during the war years. Brazil sent tens of thousands of Brazilian "rubber soldiers" to tap the Amazonian forests during the war.[49] Nicolás Cuvi extends this to a description of how several South American governments cooperated closely with the United States in fostering surveys of forest and mineral wealth.[50] Japanese armies imposed intensive exploitation of forests in Southeast Asia, using forced labor from the Philippines to Burma. In Java's teak forests, "forest villagers who lived through this chaotic three and a half years today recall the rampant, and ubiquitous, cutting of trees. Everyone took part. Vast armies of forest laborers were put to work cutting and hauling timber under the Japanese forced labor system."[51] The attitude of conquerors toward nature proved to be as ruthless as it was toward people.

Across the rural landscapes of the world, the war saw a struggle to provide food foremost for soldiers, but also for their supportive civilian populations. Nazi Germany and imperial Japan tried to secure their food supply by means of exploiting resources of the occupied countries and sea areas.[52] Nevertheless, food deprivation intensified in most belligerent nations as the war continued. This resulted in complex environmental consequences of both intensified food production in urban and rural areas and the disruption of agroecosystems throughout the war zones and far beyond.[53] Several works give broader background to these studies.[54]

The race between capitalist agriculture and centrally controlled food production, both of which moved rapidly toward monocropping of food grains, was supported by use of chemical-intensive fertilizers and pesticides. In the United States, centralization of capital and technology meant that massive corporate agribusiness came to dominate the farming landscape.[55] In the Soviet Union, the collectivization of farming and the immediate postwar famine in 1946, and on to the Virgin Lands Campaign of the 1950s and 1960s, transformed vast grassland acreage into wheat in Kazakhstan and elsewhere.[56] In China and Vietnam, the wartime food crisis brought about by foreign imperialist powers explained to a high degree the success of Communists and consequent violent conflicts in the region.[57] Wartime experience of abundance of food in the United States and lack of food in the United Kingdom led to deep disagreements concerning whose interest policy of the newly established Food and Agriculture Organization (FAO) of the United Nations should serve.[58] Food policy remained a contested issue also after the war, when world hunger emerged as a new major international problem.

SHADOW IV: GLOBAL SYNCHRONIZATION OF ENVIRONMENTAL POLICY MAKING

The introduction to this volume argued that understanding the environmental significance of any modern war requires paying attention to the long shadows cast by urban-industrial development, industrial warfare, and previous wars. The Second World War continued in the industrial countries in the form of hostile coexistence, the Cold War, and, in many developing countries, in hot wars, representing almost half a century of bloody wars and armed conflicts. In many ways, the Cold War was a direct continuation of World War II, as a part of its long geopolitical shadow. In the Cold War, the struggle of the industrial powers for control of global resources intensified.[59] The rapidly developing military-industrial complex caused significant militarization of both physical and psychological landscapes all over the world.[60] Together, military and civilian industrial production and consumption created during and after the war a global landscape of synchronized socioenvironmental disasters. But World War II also had global repercussions for environmental ideas and activities, in two broad ways.

After World War I, the United States, the Soviet Union, and Japan practiced largely isolationist policies, while Europe was more or less divided into antagonistic democratic and authoritarian states. Thereby environmental discussions and activities were also in many ways fragmented in the industrial world and marginalized elsewhere. World War II unified world history and hence also environmental history. It involved almost all continents and compelled the world to focus on a single issue: the war. As a consequence, it synchronized global activities and generated surprisingly similar and simultaneous emergence and development of new environmental perceptions, organizations, and activities in the postwar period. World War II thus became a major turning point in environmental policy making over the twentieth century.

World War II was a stimulus for nature conservation movements on different continents (see chapter 16, by Anna-Katharina Wöbse), while in most fields (perhaps excluding materials recycling) the war was the lowest point for municipal and governmental environmental protection in the twentieth century.[61] In practice, the war and the resulting postwar problems delayed by almost two decades protection of waters but also of air and soil throughout the world. During postwar scarcity and reconstruction, however, environmental protection quickly revived, partly because of the severity of pollution associated with rapidly growing urban-industrial pollution, radioactive fallout, and chemicalization of agriculture. When the postwar era of economic growth started, both old and new environmental issues were introduced to the political agenda of industrialized nations. This common context and interconnected developments provided unprecedented impetus for subsequent environmental activities in a synchronizing manner.

SHADOW V: THE BIRTH OF THE TOXIC WORLD AND GLOBAL ENVIRONMENTAL CATASTROPHISM

World War II signified the birth of a more toxic world. Before the war, agriculture relied to a high extent on organic fertilizers and cultural methods of pest control. Widely used traditional pest-control methods included crop rotation, fallowing, and mixed cropping.[62] But the search for economic self-sufficiency before the war and the upheaval of international trade during the war years forced the combatants to find alternatives for traditional raw materials. During the war, the need for increased agricultural productivity in industrialized countries prompted a transition toward more intensive farming methods. The most important changes were increasing mechanization and emerging mass production of inorganic fertilizers and synthetic pesticides. Before 1940, there were only a few dozen chemical pesticides on sale.[63] Since 1950, the global use of pesticides has risen more than fiftyfold.[64]

World War II was the first "hi-tech" war in history. Modern warfare relied heavily on science and technology. Massive governmental funding and coordination of research and development promoted rapid development of new synthetic materials and products. The chemical industry expanded several times more rapidly than any other field of industry during the Second World War. New factories started to pour out plastics, synthetic pesticides such as dichlorodiphenyltrichloroethane (DDT), chlorofluorocarbons (CFCs), and other modern products. Emissions of pollutants into air, soil, and waters simultaneously increased.[65] The rapidly expanding chemical industries poured into the environment sulfur dioxide, nitrogen oxides, toxic heavy metals such as mercury and cadmium, and chlorine and fluorine compounds, including dioxins. The war-induced synthetic revolution helped such companies as BASF, DuPont, Shell, Dow, Union Carbide, and Monsanto expand their economic and political power in the postwar world. After the war, the world got to know the toxic power of these corporations. In the 1940s, the United States produced about one million tons of hazardous waste. Four decades later, the total amount had risen to over 250 million tons a year, which was about two-thirds of the worldwide total.[66] In terms of production and use of synthetic materials, plastics, and toxics, World War II was a global turning point.

World War II also meant the birth of global environmental catastrophism. First, the sheer scale of violence; the number of dead and wounded; mass murders of children, women, elderly, developmentally disabled people, and prisoners of war; the genocide of Jews and other ethnic groups; and news of concentration camps created a sense of unprecedented human catastrophism all over the world. Chapters 9 and 10, by Helene Laurent and Outi Ampuja, respectively, describe the range of physical and mental suffering from the years of conflict to the postwar period. Second, the wartime and postwar global social crisis—including rapid population growth, famine and undernourishment, poverty, and the largest refugee problem in world history—led to increased unrest, moving society toward the Malthusian fear of exhaustion of natural resources and ecological collapse.

Third, destruction caused by atomic warfare against Japan and the radioactive fallout problem caused by postwar atmospheric nuclear tests induced a completely new fear: the human-induced end of civilization. This fear was intensified by the new, unknown chemicals that contaminated the environment, and it created the second major environmental controversy after the radioactive fallout debate. If not to nuclear war or toxic substances, our species might well succumb to the "ecocatastrophe" caused by overpopulation.[67] In sum, World War II together with the emerging Cold War generated a new multidimensional cultural phenomenon—global environmental catastrophism—which has been a major element in environmental debates and policy making until today.

World War II not only created environmental catastrophism but also created the needed social prerequisites for hope of a better future. The outcome of World War II varied from country to country. Some old states disappeared as a result of the war, while new ones were created. The war brought democracy, work, and prosperity to some countries, whereas others were overrun by destruction, dictatorship, repression, and poverty. War gave rise to Communist dictatorships in Eastern Europe and Asia, violent neocolonialism in Southeast Asia and north Africa, and the Cold War, which became a threat to the continuity of human institutions and, beyond that, to the biosphere as a whole. But World War II ended Nazi Germany, imperial Japan, and fascist Italy and started an unprecedented era of economic growth, strong public policy, social welfare, and political democracy in the Western world, with its environmental concomitant of boundlessly expanding consumerism.

On the one hand, these deep structural changes caused environmental problems, but on the other hand, these changes also made first industrial and later industrializing societies structurally more receptive to new environmental ideas and reforms. Thus World War II explains to a high degree the emergence of both environmental catastrophism and hope for better, which was embodied in the institutional changes that were prerequisites of environmental awakening in the Cold War period. Shadows are, after all, created by light. Hence the shadows that the greatest of wars cast across the world in its aftermath were not all destructive; some of them provided the seeds of future improvements, which contemporary generations may now enjoy.

Notes

1 Winston Churchill, *The Second World War*, vol. 1 (London: Educational Book Company, 1954), ix.

2 David Reynolds, *The Long Shadow: The Great War and the Twentieth Century* (New York: Simon and Schuster, 2013). A global survey of the interwar years is Jeremy Black, *Avoiding Armageddon: From the Great War to the Fall of France, 1918–40* (London: Bloomsbury, 2012).

3 Eugene Staley, *Raw Materials in Peace and War* (New York: Council on Foreign Relations, 1937); Roger Chickering and Stig Förster, eds., *The Shadows of Total War: Europe, East Asia, and the United States, 1919–1939* (Cambridge: Cambridge University Press, 2003); J. Adam Tooze, *The Deluge: America and the Remaking of the Global Order, 1916–1931* (New York: Viking, 2014). Also Marcus Funck and Roger Chickering, eds., *Endangered Cities: Military Power and Urban Societies in the Era of the World Wars* (Boston: Brill, 2004); Williamson Murray and Allan R. Millett, eds., *Military Innovation in the Interwar Period* (Cambridge: Cambridge University Press, 1996); Joseph Maiolo, *Cry Havoc: How the Arms Race Drove the World to War, 1931–1941* (New York: Basic Books, 2010); Hugh Rockoff, *America's Economic Way of War* (Cambridge: Cambridge University Press, 2012); Corey Ross, "The Tin Frontier: Mining, Empire, and Environment in Southeast Asia, 1870s–1930s," *Environmental History* 19, no. 3 (July 2014): 454–79.

4 For background, see Giovanni Federico, *Feeding the World: An Economic History of World Agriculture, 1800–2000* (Princeton, NJ: Princeton University Press, 2005). For details, see Frank Trentmann and Flemming Just, eds., *Food and Conflict in Europe in the Age of the Two World Wars* (Basingstoke, UK: Palgrave Macmillan, 2006); Paul Brassley, Yves Segers, and Leen Van Molle, eds., *War, Agriculture, and Food: Rural Europe from the 1930s to the 1950s* (New York: Routledge, 2012); Sarah T. Phillips, *This Land, This Nation: Conservation, Rural America, and the New Deal* (Cambridge: Cambridge University Press, 2007).

5 Mark B. Tauger, "Natural Disaster and Human Actions in the Soviet Famine of 1931–1933," *Carl Beck Papers in Russian and East European Studies* 1506 (June 2001).

6 For example, see Gregory H. Maddox, "Njaa: Food Shortages and Famines in Tanzania between the Wars," *International Journal of African Historical Studies* 19, no. 1 (1986): 17–34.

7 A good example is the classic work of Graham Jacks and Robert Whyte, *The Rape of the Earth: A World Survey of Soil Erosion* (London: Faber and Faber, 1939), 21.

8 Patrick J. Caffrey, "Transforming the Forests of a Counterfeit Nation: Japan's 'Manchu Nation' in Northeast China," *Environmental History* 18, no. 2 (April 2013): 309–32.

9 David Nicolle, *The Italian Invasion of Abyssinia, 1935–36* (Oxford: Osprey, 1997).

10 Antony Beevor, *The Battle for Spain: The Spanish Civil War, 1936–1939* (London: Penguin, 1982).

11 Timothy Cooper, "Challenging the 'Refuse Revolution': War, Waste and the Rediscovery of Recycling, 1900–1950," *Historical Research* 81, no. 214 (November 2008): 710–31; Timothy Cooper and Sarah Bulmer, "Refuse and the 'Risk Society': The Political Ecology of Risk in Inter-War Britain," *Social History of Medicine* 26, no. 2 (May 2013): 246–66.

12 For interwar water pollution and protection in Finland, see Simo Laakkonen, Sari Laurila, and Marjatta Rahikainen, eds., *Harmaat aallot: Ympäristönsuojelun tulo Suomeen* [The gray waves: the introduction of environmental protection in Finland) (Vammala: Suomen Historiallinen Seura, 1999). For pioneering towns and cities in the Baltic Sea region, see "The Sea and the Cities: A Multidisciplinary Project on Environmental History," accessed October 22, 2015, http://www.helsinki.fi/envirohist/seaandcities/. The interwar period deserves more attention on all continents.

13 For a somewhat analogous development in Britain, see John Sheail, *Rural Conservation in Inter-War Britain* (Oxford: Clarendon Press, 1981). On conservation in the Soviet Union, see Douglas R. Weiner, *A Little Corner of Freedom: Russian Nature Protection from Stalin to Gorbachev* (Berkeley: University of California Press, 1999). On the wide Pacific region, where the Pacific Science Association had coordinated nature conservation efforts among many countries since the early 1920s, no broad survey has yet been published.

14 See Simo Laakkonen, "Polemosphere: The War, Society and the Environment," chapter 2 in this volume.

15 Keith Lowe, *Inferno: The Devastation of Hamburg, 1943* (London: Penguin, 2012); Randall Hansen, *Fire and Fury: The Allied Bombing of Germany, 1942–1945* (New York: NAL Caliber, 2009).

16 Chris Pearson, *Scarred Landscapes: War and Nature in Vichy France* (London: Palgrave Macmillan, 2008). The war around the Mediterranean region still lacks studies of environmental impacts. For the context of military operations, see Douglas Porch, *The Path to Victory: The Mediterranean Theater in World War II* (New York: Farrar, Straus and Giroux, 2004).

17 In addition to Paul Josephson's essay on Russia in this volume, see Nicolai N. Smirnov, "The Impact of Conventional War on Natural Areas of the USSR," *Environmental Conservation* 16, no. 4 (Winter 1989): 317–21. For background, see Timothy Snyder, *Bloodlands: Europe between Hitler and Stalin* (New York: Basic Books, 2010). For the impact of the massive tank battles, see the military geography study by Burl E. Self, "War in the Heartland: The Role of Geography in Operation Barbarossa, 1941–1942," in *Studies in Military Geography and Geology*, ed. Douglas R. Caldwell, Judy Ehlen, and Russell S. Harmon (Dordrecht: Kluwer, 2004), 215–28.

18 Micah Muscolino, "Refugees, Land Reclamation, and Militarized Landscapes in Wartime China: Huanglonghshan, Shaanxi, 1937–45," *Journal of Asian Studies* 69, no. 2 (May 2010): 453–78.

19 See also William Tsutsui, "Landscapes in the Dark Valley: Toward an Environmental History of Wartime Japan," in *Natural Enemy, Natural Ally: Toward an Environmental History of War*, ed. Richard P. Tucker and Edmund Russell (Corvallis: Oregon State University Press), 195–216. For the bombing of Japan's cities, see Kenneth P. Werrell, *Blankets of Fire: U.S. Bombers over Japan during World War II* (Washington, DC: Smithsonian Institution Press, 1996). The damage to lands and people in Southeast Asia is exemplified by B. M. Düng, "Japan's Role in the Vietnamese Starvation of 1944–45," *Modern Asian Studies* 29, no. 3 (1995): 573–618.

20 For equatorial Pacific islands, see Judith Bennett, *Natives and Exotics: World War II and Environment in the Southern Pacific* (Honolulu: University of Hawai'i Press, 2009). For impacts on wildlife in the ocean, see Kurkpatrick Dorsey, *Whales and Nations: Environmental Diplomacy on the High Seas* (Seattle: University of Washington Press, 2013), chap. 3; Kurkpatrick Dorsey, "Compromising on Conservation: World War II and American Leadership in Whaling Diplomacy," in Tucker and Russell, *Natural Enemy, Natural Ally*, 252–69. Yet the environmental history of seas and oceans has still been hardly explored.

21 For an interesting case in Alaska, see Jennifer Villeneuve, "Left to Waste: A Toxic Legacy Beneath the Arctic Snow," *Native Americans* 18, no. 2 (August 2001): 48–53.

22 See D. E. Wood, ed., *Proceedings Environmental Remediation '91: Cleaning Up the Environment for the 21st Century, Pasco, WA, 8–11 September 1991* (Washington, DC: US Government Printing Office, 1991).

23 See, for example, Stefan Robert Weissenborn, "Pacific Battlefield Tourism: A Dream Island Littered with Deadly Relics," *Spiegel Online*, February 8, 2012, http://www.spiegel.de/international/world/pacific-battlefield-tourism-a-dream-island-littered-with-deadly-relics-a-813883.html.

24 Peter Hongmei and Peter O'Meara Evans, "Social and Environmental Aspects of Abandoned Chemical Weapons in China," *Nonproliferation Review* 4, no. 3 (1997): 101–8.

25 Donovan Webster, *Aftermath: The Remnants of War* (New York: Pantheon, 1996); Gary R. W. Denton et al., "Impact of WWII Dumpsites on Saipan (CNMI): Heavy Metal Status of Soils and Sediments," *Environmental Science and Pollution Research* 23, no. 11 (June 2016): 11339–48.

26 Kirsten Emiko McAllister, "Captivating Debris: Unearthing a World War Two Internment Camp," *Cultural Values* 5, no. 1 (January 2001): 97–114; See Adrian Myers and Gabriel Moshenska, eds., *Archaeologies of Internment* (New York: Springer, 2011); Jung-Sun Hun, "Relics of Empire Underground: The Making of Dark Heritage in Contemporary Japan," *Asian Studies Review* 40, no. 2 (2016): 287–302; Suzie Thomas, Oula Seitsonen, and Vesa-Pekka Herva, "Nazi Memorabilia, Dark Heritage and Treasure Hunting as 'Alternative' Tourism: Understanding the Fascination with the Material Remains of World War II in Northern Finland," *Journal of Field Archaeology* 48, no. 2 (2016): http://www.tandfonline.com/doi/full/10.1080/00934690.2016.1168769.

27 Keith Lowe, *Savage Continent: Europe in the Aftermath of World War II* (New York: St. Martin's Press, 2012); Brett L. Walker, *Toxic Archipelago: A History of Industrial Disease in Japan* (Seattle: University of Washington Press, 2010); Jacob Darwin Hamblin, "The Vulnerability of Nations: Food Security in the Aftermath of World War II," in "World War II, the Cold War, and Natural Resources," ed. Simo Laakkonen and Richard Tucker, special issue, *Global Environment* 10 (2012): 42–65.

28 For example, Jeffrey M. Diefendorf, *In the Wake of War: The Reconstruction of German Cities after World War II* (New York: Oxford University Press, 1993); Donald Filtzer, *The Hazards of Urban Life in Late Stalinist Russia: Health, Hygiene, and Living Standards, 1943–1953* (Cambridge: Cambridge University Press, 2010).

29 Richard P. Tucker, "The World Wars and the Globalization of Timber Cutting," in Tucker and Russell, *Natural Enemy*, 110–41; Stephen Brain, *Song of the Forest: Russian Forestry and Stalinist Environmentalism, 1905–1953* (Pittsburgh: University of Pittsburgh Press, 2011).

30 Hugh Rockoff, *Keep on Scrapping: The Salvage Drives of World War II* (Cambridge, MA: National Board for Economic Research, 2007); Timothy Cooper, "War on Waste? The Politics of Waste and Recycling in Post-War Britain, 1950–1975," *Capitalism Nature Socialism* 20, no. 4 (December 2009): 53–72; Ruth Oldenziel and Heike Weber, eds., "Recycling and Re-Use in the Twentieth Century," special issue, *Contemporary European History* 22, no. 3 (August 2013): 347–526; Peter Thorsheim, *Waste into Weapons: Recycling in Britain during the Second World War* (Cambridge: Cambridge University Press, 2015).

31 Heikki Lindroos, "Maailmansodat ja Suomen metsät" [World wars and forests in Finland], in *Sodan ekologia: Nykyaikaisen sodankäynnin ympäristöhistoriaa*, ed. Simo Laakkonen and Timo Vuorisalo (Helsinki: Suomalaisen Kirjallisuuden Seura, 2007), 73–105; G. Tyler Miller Jr.,

Living in the Environment: Principles, Connections, and Solutions, 10th ed. (Belmont, CA: Wadsworth, 1998), 645.

32 Wikipedia is an excellent source for global airfield history. See *Wikipedia*, s.v. "World War II Airfields" and its subcategories (i.e., different theatres of war), accessed October 5, 2016, https://en.wikipedia.org/wiki/Category:World_War_II_airfields.

33 The United States alone has today about eight hundred military bases around the world, largely as a result of the Second World War and related wars. David Vine, *Base Nation: How U.S. Military Bases Abroad Harm America and the World* (New York: Metropolitan Books / Henry Holt, 2015).

34 See also Paul Josephson, *The Conquest of the Russian Arctic* (Cambridge, MA: Harvard University Press, 2014).

35 For the United States, see Staley, *Raw Materials in Peace and War*; Alfred E. Eckes Jr., *The United States and the Global Struggle for Minerals* (Austin: University of Texas Press, 1979); for Africa, see UNESCO, *Africa and the Second World War* (Paris, 1985).

36 For similar views, see Richard Overy, *Why the Allies Won* (New York: W. W. Norton, 1997); Jack Hayes, "World War I Environments, Military Actions, and Resource Management in East Asia," in *Environmental Histories of World War I*, ed. Richard P. Tucker et al. (unpublished manuscript), 2016.

37 W. G. Jensen, "The Importance of Energy in the First and Second World Wars," *Historical Journal* 11, no. 3 (September 1968): 538–54; Daniel Yergin, *The Prize* (New York: Touchstone, 1991); Brian C. Black, *Crude Reality: Petroleum in World History* (Lanham, MD: Rowman and Littlefield, 2012), chap. 5; David S. Painter, *Oil and the American Century: The Political Economy of U.S. Foreign Oil Policy, 1941–1954* (Baltimore: Johns Hopkins University Press, 1986); Toby Jones, *Desert Kingdom: How Oil and Water Forged Modern Saudi Arabia* (Cambridge, MA: Harvard University Press, 2010); Timothy Mitchell, *Carbon Democracy: Political Power in the Age of Oil* (London: Verso, 2011).

38 Robert Goralski and Russell W. Freeburg, *Oil & War: How the Deadly Struggle for Fuel in WWII Meant Victory or Defeat* (New York: Morrow, 1987); Leonard Caruana and Hugh Rockoff, "An Elephant in the Garden: The Allies, Spain, and Oil in World War II" (Working Paper No. 12228, National Bureau of Economic Research, New York, May 2006), http://www.nber.org/papers/w12228.

39 Sonia Zaide Pritchard, *Oil Pollution Control: A History* (London: Croom Helm, 1987); Simo Laakkonen and Antti Lehmuskoski, "Musta meri: Öljyonnettomuuksien ympäristöhistoriaa Suomessa vuoteen 1969" [Black Sea: environmental history of marine oil accidents in Finland prior to 1969], *Historiallinen Aikakauskirja* 4 (2005): 381–96; Anna-Katharina Wöbse, *Weltnaturschutz: Umweltdiplomatie in Völkerbund und Vereinten Nationen 1920–1950* (Frankfurt: Campus-Verlag, 2012).

40 David L. Gordon and Royden Dangerfield, *The Hidden Weapon: The Story of Economic Warfare* (New York: Harper and Brothers, 1947); Geoffrey T. Mills and Hugh Rockoff, eds., *The Sinews of War: Essays on the Economic History of World War II* (Ames: Iowa State University Press, 1993).

41 David Edgerton, *Britain's War Machine: Weapons, Resources, and Experts in the Second World War* (New York: Oxford University Press, 2011); David Edgerton, *Warfare State: Britain, 1920–1970* (Cambridge: Cambridge University Press, 2007); Alan S. Milward, *War, Economy and Society, 1939–1945* (Berkeley: University of California Press, 1977).

42 Arthur Herman, *Freedom's Forge: How American Business Produced Victory in World War II* (New York: Random House, 2012); Gerald D. Nash, *World War II and the West: Reshaping the Economy* (Lincoln: University of Nebraska Press, 1990); Charles K. Hyde, *Arsenal of Democracy: The American Automobile Industry in World War II* (Detroit: Wayne State University Press, 2013); Hugh Rockoff, *America's Economic Way of War: War and the U.S. Economy from the Spanish-American War to the Persian Gulf War* (Cambridge: Cambridge University Press, 2012).

43 For an excellent overall analysis, see Mark Harrison, ed., *The Economics of World War II: Six Great Powers in International Comparison* (Cambridge: Cambridge University Press, 1998).

44 Matthew Evenden, *Allied Power: Mobilizing Hydro-Electricity during Canada's Second World War* (Toronto: University of Toronto Press, 2015).

45 Martin Fritz, *German Steel and Swedish Iron Ore, 1939–1945* (Kungsbacka: Elanders Boktryckeri Aktiebolag, 1974).

46 James. W. Cortada, *United States-Spanish Relations, Wolfram and World War II* (Barcelona: Manuel Pareja, 1971); Ronald H. Limbaugh, *Tungsten in Peace and War, 1918–1946* (Reno: University of Nevada Press, 2010); Timothy J. LeCain, *Mass Destruction: The Men and Giant Mines That Wired America and Scarred the Planet* (New Brunswick, NJ: Rutgers University Press, 2009).

47 For example, Thomas M. Klubock, *Contested Communities: Class, Gender and Politics in Chile's El Teniente Copper Mines, 1904–51* (Durham, NC: Duke University Press, 1998).

48 Raymond Dumett, "Africa's Strategic Minerals during the Second World War," *Journal of African History* 26, no. 4 (October 1985): 381–408.

49 Xenia Vunovic Wilkinson, "Tapping the Amazon for Victory: Brazil's 'Battle for Rubber' of World War II" (PhD thesis, Georgetown University, Washington, DC, 2009). The estimations of the number of Brazilian rubber soldiers vary from 30,000 to 55,000; Seth Garfield, *In Search of the Amazon: Brazil, the United States, and the Nature of a Region* (Durham, NC: Duke University Press, 2013).

50 Nicolás Cuvi, "Big Science and the Enchantment of Growth in Latin America," *Global Environment* 10 (2012): 16–41.

51 Nancy Lee Peluso, *Rich Forests, Poor People: Resource Control and Resistance in Java* (Berkeley: University of California Press, 1992), 97; Jayamalar Kathirithamby-Wells, "The Environmental Impacts of Japan's Occupation of West Malaysia (1942–45) and Its Socio-Economic Implications," *Global Environment* 10 (2012): 92–115.

52 Poul Holm, "*World War II* and the 'Great Acceleration' of North Atlantic Fisheries," *Global Environment* 10 (2012): 66–91. For Japan, see William Tsutsui and Timo Vuorisalo, "Japanese Imperialism and Marine Resources," chapter 14 in this volume.

53 See Michah Muscolino, "Conceptualizing Wartime Flood and Famine in China," chapter 6, and Gregory Maddox, "Food Disruption and Agricultural Policy in Tanganyika," chapter 13, both in this volume; Char Miller, "In the Sweat of Our Brow: Citizenship in American Domestic Practice during WWII—Victory Gardens," *Journal of American Culture* 26, no. 3 (September 2006): 395–409; Rauno Lahtinen and Timo Vuorisalo, " 'It's War and Everyone Can Do as They Please!': An Environmental History of a Finnish City in Wartime," *Environmental History* 9, no. 4 (October 2004): 679–700; J. M. Tinley, *South African Food and Agriculture in World War II* (Stanford, CA: Stanford University Press, 1954).

54 See, for example, Lizzie Collingham, *The Taste of War: World War II and the Battle for Food* (New York: Penguin, 2012); Brian Short, Charles Watkins, and John Martin, eds., *The Front Line of Freedom: British Farming in the Second World War* (Exeter: British Agricultural History Society, 2006); Bernd Martin and Alan S. Milward, *Agriculture and Food Supply in the Second World War* (Ostfildern: Scripta Mercaturae Verlag, 1985); Katarzyna J. Cwiertka, ed., *Food and War in Mid-Twentieth-Century East Asia* (Burlington, VT: Ashgate, 2013).

55 Sarah T. Phillips, *The Price of Plenty: From Farm to Food Politics in Postwar America* (Oxford: Oxford University Press, 2015).

56 Zhores A. Medvedev, *Soviet Agriculture* (New York: Norton, 1987); Martin McCauley, *Khrushchev and the Development of Soviet Agriculture: The Debate on the Virgin Lands, 1953–1964* (London: Macmillan, 1976).

57 Geoffrey C. Gunn, *Rice Wars in Colonial Vietnam: The Great Famine and the Viet Minh Road to Power* (Lanham, MD: Rowman and Littlefield, 2014). For China, see Muscolino, "Conceptualizing Wartime Flood and Famine in China."

58 Jacob Darwin Hamblin, "The Vulnerability of Nations: Food Security in the Aftermath of World War II," *Global Environment* 10 (2012): 42–65.

59 Arthur Westing, *Global Resources and International Conflict: Environmental Factors in Strategic Policy and Action* (Oxford: Oxford University Press, 1986).

60 Two recent collections of essays provide initial perspectives on this subject: J. R. McNeill and Corinna R. Unger, eds., *Environmental Histories of the Cold War* (Cambridge: Cambridge University Press, 2010), and Simo Laakkonen and Richard Tucker, eds., "World War II, the Cold War, and Natural Resources," special issue, *Global Environment* 10 (2012): 8–115. In addition, Jacob Hamblin proposes five trends that World War II launched: "global struggle for natural resources, the natural transformations due to war mobilization, the environmental impacts of combat, the rise in environmental sciences, and the role of the natural world within political ideology." Jacob Darwin Hamblin, "Environmental Dimensions of World War II," in *A Companion to World War II*, ed. Thomas W. Zeiler (Oxford: Blackwell, 2013), 698–716. See quotation on p. 698.

61 See, for example, John Dower, *National Parks in England and Wales* (London: Her Majesty's Stationery Office, 1945); Arthur George Tansley, *Our Heritage of Wild Nature: A Plea for Organized Nature Conservation* (Cambridge: Cambridge University Press, 1945); John Sheail, "War and the Development of Nature Conservation in Britain," *Journal of Environmental Management* 44, no. 3 (June 1995): 267–83; Roderick P. Neumann, "The Postwar Conservation Boom in British Colonial Africa," *Environmental History* 7, no. 1 (2002): 22–47.

62 Andrew R. W. Jackson and Julie M. Jackson, *Environmental Science* (Harlow, UK: Longman, 1996), 346.

63 G. Tyler Miller Jr., *Living in the Environment: An Introduction to Environmental Science*, 6th ed. (Belmont, CA: Wadsworth, 1990), 550.

64 G. Tyler Miller Jr., *Living in the Environment: Principles, Connections, and Solutions* (Pacific Grove, CA: Brooks / Cole-Thomson Learning, 2005), 520.

65 For World War II and plastics, see Jenna R. Jambec et al., "Plastic Waste Inputs from Land into the Ocean," *Science* 347, no. 6223 (February 13, 2015): 768–71.

66 Clive Ponting, *A Green History of the World* (London: Penguin, 1991), 374.

67 The term "ecocatastrophe" was introduced and popularized by Paul R. Ehrlich and Anne H. Ehrlich in their book *Population, Resources, Environment: Issues in Human Ecology* (San Francisco: Freeman, 1970). For discussion among military strategists and scientists during the Cold War, see Jacob Hamblin, *Arming Mother Nature: The Birth of Catastrophic Environmentalism* (Oxford: Oxford University Press, 2013).

Contributors

OUTI AMPUJA, PhD, is an adjunct professor (docent) of Finnish and Scandinavian History at the University of Helsinki, Finland, and a member of the Finnish Academy of Technical Sciences. Ampuja's main fields of interest and expertise are the history of noise pollution problems and silence, soundscape studies, history of technology, city and traffic planning, and traffic noise control. At present, she works as a chief adviser at Finnish Transport Safety Agency.

ALLA BOLOTOVA, PhD, is a researcher at the Department of Anthropology, European University at St. Petersburg, St. Petersburg, Russia. Bolotova has also worked as a researcher at the Centre for Independent Social Research in St. Petersburg and at the Arctic Centre in Rovaniemi, Finland. Her research focuses on environmental history and anthropology, and on human-nature relations in northern industrial communities.

CHRISTOPHER R. BOYER is a professor of history and Latin American and Latino studies, and chair of the Department of History at the University of Illinois at Chicago. His most recent book, *Political Landscapes: Forests, Conservation, and Community in Mexico* (2015), investigates the social history of forest management between 1880 and 2000. He is coeditor of a University of Arizona Press book series on Latin American environmental history.

MATTHEW EVENDEN is a professor of geography and associate dean of research and graduate studies in the Faculty of Arts at the University of British Columbia. A specialist in the environmental history of rivers and water, Evenden explores in his work the conflicts, meanings, and contests that emerge over water perception, access, and use. His most recent book is *Allied Power: Mobilizing Hydro-Electricity during Canada's Second World War* (2015).

PAUL JOSEPHSON teaches history at Colby College and is a visiting professor at Tomsk State University. A specialist in environmental history and history of technology, he has written twelve books on various aspects of the Soviet and post-Soviet world, as well as others in environmental studies. Josephson is currently writing a history of nature transformation projects in Russia, from 1900 to present.

SIMO LAAKKONEN is a senior lecturer of landscape studies at the University of Turku, Finland. He is also an adjunct professor of social and economic history at the University of Helsinki. He was previously a temporary professor of environmental policy at the University of Eastern Finland. Recently he was a visiting fellow at Södertörn University in Stockholm and at the University of Kansas. He has explored urban environmental history and history of environmental policy, science, technology and media in the Baltic Sea Region.

HELENE LAURENT is a medical doctor with a degree from the University of Helsinki (1978). Laurent became a specialist in ear, nose, and throat diseases in 1988. She also has a master's degree in social sciences from the University of Helsinki (2006). Her main fields of interest include the social history of epidemics and public health during war. Laurent is currently working as epidemiologist.

CAROL MACLENNAN is an anthropologist and author of *Sovereign Sugar: Industry and Environment in Hawai'i* (2014). She teaches at Michigan Technological University in the Industrial Heritage and Environmental Policy Programs. She is currently working on an environmental history of complex contaminated industrial zones that include Pearl Harbor and historic copper mining districts in Michigan and New Mexico.

GREGORY MADDOX is a specialist in African and environmental history. He holds a BA from the University of Virginia and a PhD from Northwestern University. In addition to teaching at Texas Southern University for twenty-six years, he spent a year at the University of Dar es Salaam in Tanzania as a Fulbright senior lecturer. His most recent scholarly book, cowritten with Karl Ittmann and Dennis Cordell, is *The Demography of Empire* (2010).

ILMO MASSA, PhD, is an environmental social scientist working at the Department of Environmental Sciences, University of Helsinki. Massa has written the first doctoral theses related to environmental history in Finland (*Pohjoinen luonnonvalloitus* [The northern conquest of nature], 1994). His main fields of interest are the environmental history of the circumpolar north, ecological modernization, and theories of environmental social change.

EVAN MAWDSLEY was educated in the United States and Britain. Until 2010, he was a professor of international history at Glasgow University. His publications include *The Russian Civil War* (1987/2008); *Thunder in the East: The Nazi-Soviet War* (2005/2015); *World War II: A New History* (2009); and *December 1941* (2011). He was general editor of the *Cambridge History of the Second World War* (2015), and is currently writing a history of World War II at sea.

MICAH S. MUSCOLINO is a Jessica Rawson Fellow in modern Asian history and an associate professor of Chinese history at Merton College, University of Oxford. He has published *Fishing Wars and Environmental Change in Late Imperial and Modern China* (2009), as well as numerous journal articles. His most recent book is *The Ecology of War in China: Henan Province, the Yellow River, and Beyond, 1938–1950* (2015).

WILLIAM M. TSUTSUI is president and professor of history at Hendrix College in Conway, Arkansas. A specialist in the business, environmental, and cultural history of modern Japan, he is the author or editor of eight books, including *Manufacturing Ideology: Scientific Management in Twentieth-Century Japan* (1998) and *Godzilla on My Mind: Fifty Years of the King of Monsters* (2004).

RICHARD P. TUCKER is an adjunct professor of environmental history at the University of Michigan. His environmental history research centered first on the forest history of colonial India, and then on the ecological impact of American hegemony around the tropical world. Recently he has focused on the global environmental history of warfare and military operations.

TIMO VUORISALO has been an adjunct professor and senior lecturer of environmental science at the University of Turku, Finland, since 1992. In 1989–90, he was a postdoctoral fellow at Indiana University, Bloomington.

Vuorisalo's main research interests are evolutionary ecology, urban ecology, and environmental history.

ANNA-KATHARINA WÖBSE is an environmental historian, curator, and research fellow at the Justus Liebig University Giessen in Germany. She has published extensively on media and the environment as well as on human-animal relations. Her book *Weltnaturschutz: Umweltdiplomatie in Völkerbund und Vereinten Nationen, 1920–1950* (2012) explores the role of the League of Nations and the United Nations in the making of global environmental regimes.

Index